Lecture Notes in Physics

Volume 980

The Lecture Notes in Physics

The series Lecture Notes in Physics (LNP), founded in 1969, reports new developments in physics research and teaching-quickly and informally, but with a high quality and the explicit aim to summarize and communicate current knowledge in an accessible way. Books published in this series are conceived as bridging material between advanced graduate textbooks and the forefront of research and to serve three purposes:

- to be a compact and modern up-to-date source of reference on a well-defined topic;
- to serve as an accessible introduction to the field to postgraduate students and nonspecialist researchers from related areas;
- to be a source of advanced teaching material for specialized seminars, courses and schools.

Both monographs and multi-author volumes will be considered for publication. Edited volumes should however consist of a very limited number of contributions only. Proceedings will not be considered for LNP.

Volumes published in LNP are disseminated both in print and in electronic formats, the electronic archive being available at springerlink.com. The series content is indexed, abstracted and referenced by many abstracting and information services, bibliographic networks, subscription agencies, library networks, and consortia.

Proposals should be sent to a member of the Editorial Board, or directly to the responsible editor at Springer:

Dr Lisa Scalone
Springer Nature
Physics
Tiergartenstrasse 17
69121 Heidelberg, Germany
lisa.scalone@springernature.com

More information about this series at http://www.springer.com/series/5304

Harold Erbin

String Field Theory

A Modern Introduction

 Springer

Harold Erbin
Center for Theoretical Physics
Massachusetts Institute of Technology
Cambridge, MA, USA

CEA, LIST, Gif-sur-Yvette, France

ISSN 0075-8450   ISSN 1616-6361   (electronic)
Lecture Notes in Physics
ISBN 978-3-030-65320-0   ISBN 978-3-030-65321-7   (eBook)
https://doi.org/10.1007/978-3-030-65321-7

This Springer imprint is published by the registered company Springer Nature Switzerland AG.
The registered company address is: Gewerbestrasse 11, 6330 Cham, Switzerland

Preface

This book grew up from lectures delivered within the Elite Master Program "Theoretical and Mathematical Physics" from the Ludwig-Maximilians-Universität during the winter semesters 2017–2018 and 2018–2019.

The main focus of this book is the closed bosonic string field theory (SFT). While there are many resources available for the open bosonic SFT, a single review [2] has been written since the final construction of the bosonic closed SFT by Zwiebach [5]. For this reason, it makes sense to provide a modern and extensive study. Moreover, the usual approach to open SFT focuses on the cubic theory, which is so special that it is difficult to generalize the techniques to other SFTs. Finally, closed strings are arguably more fundamental than open strings because they are always present since they describe gravity, which further motivates my choice. However, the reader should not take this focus as denying the major achievements and the beauty of the open SFT; reading this book should provide most of the tools needed to feel comfortable also with this theory.

While part of the original goal of SFT is to provide a non-perturbative definition of string theory and to address important questions such as classifying consistent string backgrounds or understanding dualities, no progress on this front has been achieved so far. Hence, there is still much to understand, and the recent surge of developments provides a new chance to deepen our understanding of closed SFT. For example, several consistency properties of string theory have been proven rigorously using SFT. Moreover, the recent construction of the open–closed superstring field theory [3] together with earlier works [1, 4–6] shows that all types of string theories can be recast as a SFT. This is why, I believe, it is a good time to provide a complete book on SFT.

The goal of this book is to offer a self-contained description of SFT and all the tools necessary to build it. The emphasis is on describing the concepts behind SFT and to make the reader build intuitions on what it means. For this reason, there are relatively few applications.

The reader is assumed to have some knowledge of QFT and a basic knowledge of CFT and string theory (classical string, Nambu–Goto action, light-cone and old-covariant quantizations).

Organization

The text is organized on three levels: the main content (augmented with examples), computations and remarks. The latter two levels can be omitted in a first lecture. The examples, computations and remarks are clearly separated from the text (respectively, by a half-box on the left and bottom, by a vertical line on the left and by italics) to help the navigation.

Many computations have been set aside from the main text to avoid breaking the flow and to provide the reader with the opportunity to check by themself first. In some occasions, computations are postponed well below the corresponding formula to gather similar computations or to avoid breaking an argument. While the derivations contain more details than usual textbooks and may look pedantic to the expert, I think it is useful for students and newcomers to have complete references where to check each step. This is even more the case when there are many different conventions in the literature. The remarks are not directly relevant to the core of the text, but they make connections with other parts or topics. The goal is to broaden the perspectives of the main text.

General references can be found at the end of each chapter to avoid overloading the text. In-text references are reserved for specific points or explicit quotations (of a formula, a discussion, a proof, etc.) I did not try to be exhaustive in the citations and I have certainly missed important references: this should be imputed to my lack of familiarity with them and not to their value.

During the lectures, additional topics have been covered in tutorials, which can be found online (together with corrections):

https://www.physik.uni-muenchen.de/lehre/vorlesungen/wise_17_18/sft_ws_17_18/
exercises_sft/index.html.

My plan is to frequently update this book with new content. The last version of the draft can be accessed on my professional webpage, currently located at:

http://www.lpthe.jussieu.fr/~erbin/.

Acknowledgements

I have started to learn string field theory at HRI by attending lectures from Ashoke Sen. Since then, I have benefited from collaboration and many insightful discussions with him. Following his lectures has been much helpful in building an intuition that cannot be found in papers or reviews on the topic. Through this book, I hope being able to make some of these insights more accessible.

I am particularly grateful to Ivo Sachs who proposed me to teach this course and to Michael Haack for continuous support and help with the organization, and to both of them for many interesting discussions during the two years I have spent at LMU. Moreover, I have been very lucky to be assigned an excellent tutor for this course, Christoph Chiaffrino. After providing him with the topic and few references, Christoph has prepared all the tutorials and the corrections autonomously. His help brought a lot to the course.

I am particularly obliged to all the students who have taken this course at LMU for many interesting discussions and comments: Enrico Andriolo, Hrólfur Ásmundsson, Daniel Bockisch, Fabrizio Cordonnier, Julian Freigang, Wilfried Kaase, Andriana Makridou, Pouria Mazloumi, Daniel Panea and Martin Rojo.

I am also grateful to all the string theory community for many exchanges. For discussions related to the topics of this book, I would like to thank more particularly: Costas Bachas, Adel Bilal, Subhroneel Chakrabarti, Atish Dabholkar, Benoit Douçot, Ted Erler, Dileep Jatkar, Carlo Maccaferri, Juan Maldacena, Yuji Okawa, Sylvain Ribault, Raoul Santachiara, Martin Schnabl, Dimitri Skliros and Jakub Vošmera. I have received a lot of feedback during the different stages of writing this book, and I am obliged to all the colleagues who sent me feedback.

I am thankful to my colleagues at LMU for providing a warm and stimulating environment, with special thanks to Livia Ferro for many discussions around coffee. Moreover, the encouragements and advice from Oleg Andreev and Erik Plauschinn have been strong incentives for publishing this book.

The editorial process at Springer has been very smooth. I would also like to thank Christian Caron and Lisa Scalone for their help and efficiency during the publishing process. I am also indebted to Stefan Theisen for having supported the publication at Springer and for numerous comments and corrections on the draft.

Most part of this book was written at the Ludwig-Maximilians-Universität (LMU, Munich, Germany), where I was supported by a Carl Friedrich von Siemens Research Fellowship of the Alexander von Humboldt Foundation.

The final stage has been completed at the University of Turin (Italy).

My research is currently funded by the European Union's Horizon 2020 research and innovation programme under the Marie Skłodowska Curie grant agreement no. 891169.

Finally, writing this book would have been more difficult without the continuous and loving support from Corinne.

Boston, USA   Harold Erbin
November 2020

References

1. C. de Lacroix, H. Erbin, S.P. Kashyap, A. Sen, M. Verma, Closed superstring field theory and its applications. Int. J. Mod. Phys. A **32**(28–29), 1730021 (2017)
2. T. Erler, Four lectures on closed string field theory. Phys. Rep. S03701573203001 32 (2020)
3. S.F. Moosavian, A. Sen, M. Verma, Superstring field theory with open and closed strings. J. High Energy Phys. **2001**, 183 (2020). arXiv:1907.10632
4. A. Sen, BV master action for heterotic and type II string field theories. J. High Energy Phys. **2016**(2) (2016)

5. B. Zwiebach, Closed string field theory: quantum action and the BV master equation. Nucl. Phys. B **390**(1), 33–152 (1993)
6. B. Zwiebach, Oriented open-closed string theory revisited. Annal. Phys. **267**(2), 193–248 (1998)

Contents

About the Author

Harold Erbin obtained his PhD in theoretical physics from the Université Pierre et Marie Curie (Paris). Afterwards, he held postdoctoral positions at the Harish-Chandra Research Institute (Allahabad/Prayagraj), where he learned string field theory with Ashoke Sen, Ludwig-Maximilians-Universität (Munich) and Università di Torino.

He is currently a Marie Skłodowska-Curie fellow at MIT (Boston) and CEA-LIST (Paris). Besides string field theory, his contributions include works on black hole and supergravity, two-dimensional gravity, tensor models and applications of machine learning to theoretical physics.

Introduction

1

Abstract

In this chapter, we introduce the main motivations for studying string theory, and why it is important to design a string field theory. After describing the central features of string theory, we describe the most important concepts of the worldsheet formulation. Then, we explain the reasons leading to string field theory (SFT) and outline the ideas which will be discussed in the rest of the book.

1.1 Strings, a Distinguished Theory

The first and simplest reason for considering theories of fundamental p-branes (fundamental objects extended in p spatial dimensions) can be summarized by the following question: "Why would Nature just make use of point-particles?" There is no a priori reason forbidding the existence of fundamental extended objects and, according to Gell-Mann's totalitarian principle, "Everything not forbidden is compulsory." If a consistent theory cannot be built (after a reasonable amount of effort) or if it contradicts current theories (in their domains of validity) and experiments, then one can support the claim that only point-particles exist. On the other side, if such a theory can be built, it is of primary interest to understand it deeper and to see if it can solve the current problems in high-energy theoretical physics.

The simplest case after the point-particle is the string, so it makes sense to start with it. It happens that a consistent theory of strings can be constructed, and that

© Springer Nature Switzerland AG 2021
H. Erbin, *String Field Theory*, Lecture Notes in Physics 980,
https://doi.org/10.1007/978-3-030-65321-7_1

the latter (in its supersymmetric version) contains all the necessary ingredients for a fully consistent high-energy model:[1]

- quantum gravity (quantization of general relativity plus higher-derivative corrections);
- grand unification (of matter, interactions and gravity);
- no divergences, UV finiteness (finite and renormalizable theory);
- fixed number of dimensions ($26 = 25 + 1$ for the bosonic string, $10 = 9 + 1$ for the supersymmetric version);
- existence of all possible branes;
- no dimensionless parameters and one dimensionful parameter (the string length ℓ_s).

It can be expected that a theory of fundamental strings (1-branes) occupies a distinguished place among fundamental p-branes for the following reasons.

Interaction Non-locality   In a QFT of point-particles, UV divergences arise because interactions (defined as the place where the number and/or nature of the objects change) are arbitrarily localized at a spacetime point. In Feynman graphs, such divergences can be seen when the momentum of a loop becomes infinite (two vertices collide): this happens when trying to concentrate an infinite amount of energy at a single point. However, these divergences are expected to be reduced or absent in a field theory of extended objects: whereas the interaction between particles is perfectly local in spacetime and agreed upon by all observers (Fig. 1.1), the spatial extension of branes makes the interactions non-local. This means that two different observers will neither agree on the place of the interactions (Fig. 1.2), nor on the part of the diagram which describes one or two branes.

The string lies at the boundary between too much local and too much non-local: in any given frame, the interaction is local in space, but not in spacetime. The reason is that a string is one-dimensional and splits or joins along a point. For $p > 1$, the brane needs to break/join along an extended spatial section, which looks non-local.

Another consequence of the non-locality is a drastic reduction of the possible interactions. If an interaction is Lorentz invariant, Lorentz covariant objects can be attached at the vertex (such as momentum or gamma matrices): this gives Lorentz invariants after contracting with indices carried by the field. But, this is impossible if the interaction itself is non-local (and thus not invariant): inserting a covariant object would break Lorentz invariance.

---

[1]There are also indications that a theory of membranes (2-branes) in $10 + 1$ dimensions, called M-theory, should exist. No direct and satisfactory description of the latter has been found and we will thus focus on string theory in this book.

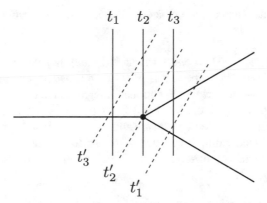

Fig. 1.1 Locality of a particle interaction: two different observers always agree on the interaction point and which parts of the worldline are 1- and 2-particle states

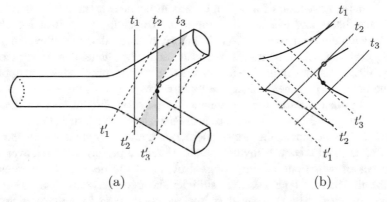

Fig. 1.2 Non-locality of string interaction: two different observers see the interaction happening at different places (denoted by the filled and empty circles) and they do not agree on which parts of the worldsheet are 1- and 2-string states (the litigation is denoted by the grey zone). (**a**) Observers at rest and boosted. (**b**) Observers close to the speed of light moving in opposite directions. The interactions are widely separated in each case

Brane Degrees of Freedom The higher the number of spatial dimensions of a p-brane, the more possibilities it has to fluctuate. As a consequence, it is expected that new divergences appear as p increases due to the proliferations of the brane degrees of freedom. From the worldvolume perspective, this is understood from the fact that the worldvolume theory describes a field theory in $(p + 1)$ dimensions, and UV divergences become worse as the number of dimensions increases. The limiting case happens for the string ($p = 1$) since two-dimensional field theories are well-behaved in this respect (for example, any monomial interaction for a scalar field is power-counting renormalizable). This can be explained by the low-dimensionality of the momentum integration and by the enhancement of symmetries in two dimensions. Hence, strings should display nice properties and are thus of special interest.

Worldvolume Theory The point-particle (0-brane) and the string (1-brane) are also remarkable in another aspect: it is possible to construct a simple worldvolume field theory (and the associated functional integral) in terms of a worldvolume metric. All components of the latter are fixed by gauge symmetries (diffeomorphisms for the particle, diffeomorphisms and Weyl invariance for the string). This ensures the reparametrization invariance of the worldvolume without having to use a complicated action. Oppositely, the worldvolume metric cannot be completely gauge fixed for $p > 1$.

Summary As a conclusion, strings achieve an optimal balance between spacetime and worldsheet divergences, as well as having a simple description with reparametrization invariance.

Since the construction of a field theory is difficult, it is natural to start with a worldsheet theory and to study it in the first-quantization formalism, which will provide a guideline for writing the field theory. In particular, this allows to access the physical states in a simple way and to find other general properties of the theory. When it comes to the interactions and scattering amplitudes, this approach may be hopeless in general since the topology of the worldvolume needs to be specified by hand (describing the interaction process). In this respect, the case of the string is again exceptional: because Riemann surfaces have been classified and are well-understood, the arbitrariness is minimal. Combined with the tools of conformal field theory, many computations can be performed. Moreover, since the modes of vibrations of the strings provide all the necessary ingredients to describe the Standard model, it is sufficient to consider only one string field (for one type of strings), instead of the plethora found in point-particle field theory (one field for each particle). Similarly, non-perturbative information (such as branes and dualities) could be found only due to the specific properties of strings.

Coming back to the question which opened this section, higher-dimensional branes of all the allowed dimensions naturally appear in string theory as bound states. Hence, even if the worldvolume formulation of branes with $p > 1$ looks pathological,[2] string theory hints towards another definition of these objects.

---

[2]Entering in the details would take us too far away from the main topic of this book. Some of the problems found when dealing with ($p > 2$)-branes are: how to define a Wick rotation for 3-manifolds, the presence of Lorentz anomalies in target spacetime, problems with the spectrum, lack of renormalizability, impossibility to gauge-fix the worldvolume metric [1–7, 13, 16–18, 44, 58, 61–63, 71, 78].

1.2   String Theory

1.2.1   Properties

The goal of this section is to give a general idea of string theory by introducing some concepts and terminology. The reader not familiar with the points described in this section is advised to follow in parallel some standard worldsheet string theory textbooks.

Worldsheet CFT

A string is characterized by its worldsheet field theory (Chap. 2).[3] The worldsheet is parametrized by coordinates $\sigma^a = (\tau, \sigma)$. The simplest description is obtained by endowing the worldsheet with a metric $g_{ab}(\sigma^a)$ ($a = 0, 1$) and by adding a set of D scalar fields $X^\mu(\sigma^a)$ living on the worldsheet ($\mu = 0, \ldots, D - 1$). The latter represents the position of the string in the D-dimensional spacetime. From the classical equations of motion, the metric g_{ab} is proportional to the metric induced on the worldsheet from its embedding in spacetime. More generally, one ensures that the worldsheet metric is non-dynamical by imposing that the action is invariant under (worldsheet) diffeomorphisms and under Weyl transformations (local rescalings of the metric). The consistency of these conditions at the quantum level imposes that $D = 26$, and this number is called the *critical dimension*. Gauge fixing the symmetries, and thus the metric, leads to the conformal invariance of the resulting worldsheet field theory: a conformal field theory (CFT) is a field theory (possibly on a curved background) in which only angles and not distances can be measured (Chaps. 5–7). This simplifies greatly the analysis since the two-dimensional conformal algebra (called the Virasoro algebra) is infinite-dimensional.

CFTs more general than D free scalar fields can be considered: fields taking non-compact values are interpreted as non-compact dimensions while compact or Grassmann-odd fields are interpreted as compact dimensions or internal structure, like the spin.

While the light-cone quantization allows to find quickly the states of the theory, the simplest covariant method is the BRST quantization (Chap. 8). It introduces ghosts (and super-ghosts) associated to the gauge fixing of diffeomorphisms (and local supersymmetry). These (super)ghosts form a CFT which is universal (independent of the matter CFT).

The trajectory of the string is denoted by $x_c(\tau, \sigma)$. It begins and ends respectively at the geometric shapes parametrized by $x_c(\tau_i, \sigma) = x_i(\sigma)$ and by $x_c(\tau_f, \sigma) = x_f(\sigma)$. Note that the coordinate system on the worldsheet itself is arbitrary. The spatial section of a string can be topologically closed (circle) or open (line) (Fig. 1.3), leading to cylindrical or rectangular worldsheets as illustrated in Figs. 1.4

---

[3]We focus mainly on the bosonic string theory, leaving aside the superstring, except when differences are important.

(a)   (b)

Fig. 1.3 (a) Open and (b) closed strings

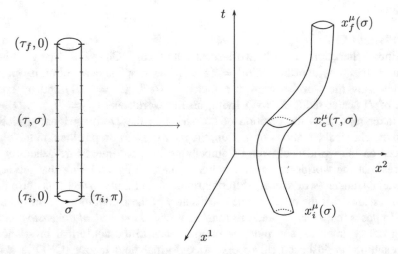

Fig. 1.4 Trajectory $x_c^\mu(\tau, \sigma)$ of a closed string in spacetime (worldsheet). It begins and ends at the circles parametrized by $x_i(\sigma)$ and $x_f(\sigma)$. The worldsheet is topologically a cylinder and is parametrized by $(\tau, \sigma) \in [\tau_i, \tau_f] \times [0, 2\pi)$

and 1.5. To each topology is associated different boundary conditions and types of strings:

- closed: periodic and anti-periodic boundary conditions;
- open: Dirichlet and Neumann boundary conditions.

While a closed string theory is consistent by itself, an open string theory is not and requires closed strings.

Spectrum

In order to gain some intuition for the states described by a closed string, one can write the Fourier expansion of the fields X^μ (in the gauge $g_{ab} = \eta_{ab}$ and after imposing the equations of motion)

$$X^\mu(\tau, \sigma) \sim x^\mu + p^\mu \tau + \frac{i}{\sqrt{2}} \sum_{n \in \mathbb{Z}^*} \frac{1}{n} \left(\alpha_n^\mu e^{-in(\tau-\sigma)} + \bar{\alpha}_n^\mu e^{-in(\tau+\sigma)} \right), \qquad (1.1)$$

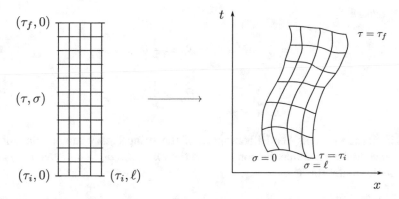

Fig. 1.5 Trajectory $x_c^\mu(\tau, \sigma)$ of an open string in spacetime (worldsheet). It begins and ends at the lines parametrized by $x_i(\sigma)$ and $x_f(\sigma)$. The worldsheet is topologically a rectangle and is parametrized by $(\tau, \sigma) \in [\tau_i, \tau_f] \times [0, \ell]$

where x^μ is the centre-of-mass position of the string and p^μ its momentum.[4] Canonical quantization leads to the usual commutator:

$$[x^\mu, p^\nu] = i\eta^{\mu\nu}. \tag{1.2}$$

With respect to a point-particle for which only the first two terms are present, there are an infinite number of oscillators α_n^μ and $\bar\alpha_n^\mu$ which satisfy canonical commutation relations for creation $n < 0$ and annihilation operators $n > 0$

$$[\alpha_m^\mu, \alpha_n^\nu] = m\,\eta^{\mu\nu}\delta_{m+n,0}. \tag{1.3}$$

The non-zero modes are the Fourier modes of the excitations of the embedded string. The case of the open string is simply obtained by setting $\bar\alpha_n = \alpha_n$ and $p \to 2p$. The Hamiltonian for the closed and open strings read respectively

$$H_{\text{closed}} = -\frac{m^2}{2} + N + \bar N - 2, \tag{1.4a}$$

$$H_{\text{open}} = -m^2 + N - 1, \tag{1.4b}$$

where $m^2 = -p^\mu p_\mu$ is the mass of the state (in Planck units), N and $\bar N$ (level operators) count the numbers N_n and $\bar N_n$ of oscillators α_n and $\bar\alpha_n$ weighted by their

---

[4]In the introduction, we set $\alpha' = 1$.

mode index n:

$$N = \sum_{n \in \mathbb{N}} n N_n , \qquad N_n = \frac{1}{n} \alpha_{-n} \cdot \alpha_n ,$$

$$\bar{N} = \sum_{n \in \mathbb{N}} n \bar{N}_n , \qquad \bar{N}_n = \frac{1}{n} \bar{\alpha}_{-n} \cdot \bar{\alpha}_n . \tag{1.5}$$

With these elements, the Hilbert space of the string theory can be constructed. Invariance under reparametrization leads to the *on-shell condition*, which says that the Hamiltonian vanishes:

$$H |\psi\rangle = 0 \tag{1.6}$$

for any physical state $|\psi\rangle$. Another constraint for the closed string is the *level-matching condition*

$$(N - \bar{N}) |\psi\rangle = 0 . \tag{1.7}$$

It can be understood as fixing an origin on the string.

The ground state $|k\rangle$ with momentum k is defined to be the eigenstate of the momentum operator which does not contain any oscillator excitation:

$$p^\mu |k\rangle = k^\mu |k\rangle , \qquad \forall n > 0 : \quad \alpha_n^\mu |k\rangle = 0 . \tag{1.8}$$

A general state can be built by applying successively creation operators

$$|\psi\rangle = \prod_{n>0} \prod_{\mu=0}^{D-1} (\alpha_{-n}^\mu)^{N_{n,\mu}} |k\rangle , \tag{1.9}$$

where $N_{n,\mu} \in \mathbb{N}$ counts excitation level of the oscillator α_{-n}^μ. In the rest of this section, we describe the first two levels of states.

The ground state is a tachyon (faster-than-light particle) because the Hamiltonian constraint shows that it has a negative mass (in the units where $\alpha' = 1$):

$$\text{closed}: \quad m^2 = -4 , \qquad\qquad \text{open}: \quad m^2 = -1 . \tag{1.10}$$

The first excited state of the open string is found by applying α_{-1} on the vacuum $|k\rangle$:

$$\alpha_{-1}^\mu |k\rangle . \tag{1.11}$$

This state is massless:

$$m^2 = 0 \qquad (1.12)$$

and since it transforms as a Lorentz vector (spin 1), it is identified with a U(1) gauge boson. Writing a superposition of such states

$$|A\rangle = \int d^D k \, A_\mu(k) \, \alpha_{-1}^\mu \, |k\rangle \,, \qquad (1.13)$$

the coefficient $A_\mu(k)$ of the Fourier expansion is interpreted as the spacetime field for the gauge boson. Reparametrization invariance is equivalent to the equation of motion

$$k^2 A_\mu = 0 \,. \qquad (1.14)$$

One can prove that the field obeys the Lorentz gauge condition

$$k^\mu A_\mu = 0 \,, \qquad (1.15)$$

which results from gauge fixing the U(1) gauge invariance

$$A_\mu \longrightarrow A_\mu + k_\mu \lambda \,. \qquad (1.16)$$

It can also be checked that the low-energy action reproduces the Maxwell action.

The first level of the closed string is obtained by applying both α_{-1} and $\bar\alpha_{-1}$ (this is the only way to match $N = \bar N$ at this level)

$$\alpha_{-1}^\mu \bar\alpha_{-1}^\nu \, |k\rangle \qquad (1.17)$$

and the corresponding states are massless

$$m^2 = 0 \,. \qquad (1.18)$$

These states can be decomposed into irreducible representations of the Lorentz group

$$\left(\alpha_{-1}^\mu \bar\alpha_{-1}^\nu + \alpha_{-1}^\nu \bar\alpha_{-1}^\mu - \frac{1}{D} \eta^{\mu\nu} \alpha_{-1} \cdot \bar\alpha_{-1} \right) |p\rangle \,,$$

$$\left(\alpha_{-1}^\mu \bar\alpha_{-1}^\nu - \alpha_{-1}^\nu \bar\alpha_{-1}^\mu \right) |p\rangle \,, \qquad \frac{1}{D} \eta_{\mu\nu} \alpha_{-1}^\mu \bar\alpha_{-1}^\nu |p\rangle$$

$$(1.19)$$

which are respectively associated to the spacetime fields $G_{\mu\nu}$ (metric, spin 2), $B_{\mu\nu}$ (Kalb–Ramond 2-form) and Φ (dilaton, spin 0). The appearance of a massless spin

2 particle (with low-energy action being the Einstein–Hilbert action) is a key result and originally raised interest for string theory.

Remark 1.1 (Reparametrization Constraints) Reparametrization invariance leads to other constraints than $H = 0$. They imply in particular that the massless fields have the correct gauge invariance and hence the correct degrees of freedom.

Note that, after taking into account these constraints, the remaining modes correspond to excitations of the string in the directions transverse to it.

Hence, each vibrational mode of the string corresponds to a spacetime field for a point-particle (and linear superpositions of modes can describe several fields). This is how string theory achieves unification since a single type of string (of each topology) is sufficient for describing all the possible types of fields encountered in the standard model and in gravity. They correspond to the lowest excitation modes, the higher massive modes being too heavy to be observed at low energy.

Bosonic string theory includes tachyons and is thus unstable. While the instability of the open string tachyon is well understood and indicates that open strings are unstable and condense to closed strings, the status of the closed string tachyon is more worrisome (literally interpreted, it indicates a decay of spacetime itself). In order to solve this problem, one can introduce supersymmetry: in this case, the spectrum does not include the tachyon because it cannot be paired with a supersymmetric partner.

Moreover, as its name indicates, the bosonic string possesses only bosons in its spectrum (perturbatively), which is an important obstacle to reproduce the standard model. By introducing spacetime fermions, supersymmetry also solves this problem. The last direct advantage of the superstring is that it reduces the number of dimensions from 26 to 10, which makes the compactification easier.

1.2.2 Classification of Superstring Theories

In this section, we describe the different superstring theories (Chap. 17). In order to proceed, we need to introduce some new elements.

The worldsheet field theory of the closed string is made of two sectors, called the left- and right-moving sectors (the α_n and $\bar{\alpha}_n$ modes). While they are treated symmetrically in the simplest models, they are in fact independent (up to the zero-mode) and the corresponding CFT can be chosen to be distinct.

The second ingredient already evoked earlier is supersymmetry. This symmetry associates a fermion to each boson (and conversely) through the action of a supercharge Q

$$|\text{boson}\rangle = Q \, |\text{fermion}\rangle . \tag{1.20}$$

More generally, one can consider N supercharges which build up a family of several bosonic and fermionic partners. Since each supercharge increases the spin by $1/2$ (in $D = 4$), there is an upper limit for the number of supersymmetries—for interacting theories with a finite number of fields[5]—in order to keep the spin of a family in the range where consistent actions exist:

- $N_{max} = 4$ without gravity ($-1 \le$ spin ≤ 1);
- $N_{max} = 8$ with gravity ($-2 \le$ spin ≤ 2).

This counting serves as a basis to determine the maximal number of supersymmetries in other dimensions (by relating them through dimensional reductions).

Let us turn our attention to the case of the two-dimensional worldsheet theory. The number of supersymmetries of the closed left- and right-moving sectors can be chosen independently, and the number of charges is written as (N_L, N_R) (the index is omitted when statements are made at the level of the CFT). The critical dimension (absence of quantum anomaly for the Weyl invariance) depends on the number of supersymmetry

$$D(N = 0) = 26, \qquad D(N = 1) = 10. \tag{1.21}$$

Type II superstrings have $(N_L, N_R) = (1, 1)$ and come in two flavours called IIA and IIB according to the chirality of the spacetime gravitini chiralities. A theory is called *heterotic* if $N_L > N_R$; we will mostly be interested in the case $N_L = 1$ and $N_R = 0$.[6] In such theories, there cannot be open strings since both sectors must be equal in the latter. Since the critical dimensions of the two sectors do not match, one needs to get rid of the additional dimensions of the right-moving sector; this leads to the next topic—gauge groups.

Gauge groups associated with spacetime gauge bosons appear in two different places. In heterotic models, the compactification of the unbalanced dimensions of the left sector leads to the appearance of a gauge symmetry. The possibilities are scarce due to consistency conditions which ensure a correct gluing with the right-sector. Another possibility is to add degrees of freedom—known as Chan–Paton indices—at the ends of open strings: one end transforms in the fundamental representation of a group G, while the other end transforms in the anti-fundamental. The modes of the open string then reside in the adjoint representation, and the massless spin-1 particles become the gauge bosons of the non-Abelian gauge symmetry.

Finally, one can consider oriented or unoriented strings. An oriented string possesses an internal direction, i.e. there is a distinction between going from the left to the right (for an open string) or circling in clockwise or anti-clockwise direction (for a closed string). Such an orientation can be attributed globally to the spacetime history of all strings (interacting or not). The unoriented string is obtained by quotienting the theory by the $\mathbb{Z}_2$ worldsheet parity symmetry which exchanges the left- and right-moving sectors. Applying this to the type IIB gives the type I theory.

---

[5]These conditions exclude the cases of free theories and higher-spin theories.

[6]The case $N_L < N_R$ is identical up to exchange of the left- and right-moving sectors.

Table 1.1 List of the consistent tachyon-free (super)string theories. The bosonic theory is added for comparison. There are additional heterotic theories without spacetime supersymmetry, but they contain a tachyon and are thus omitted

	Worldsheet susy	D	Spacetime susy	Gauge group	Open string	Oriented	Tachyon
Bosonic	$(0, 0)$	26	0	Any[a]	Yes	Yes/no	Yes
Type I	$(1, 1)$	10	$(1, 0)$	SO(32)	Yes	No	No
Type IIA	$(1, 1)$	10	$(1, 1)$	U(1)	(Yes)[b]	Yes	No
Type IIB	$(1, 1)$	10	$(2, 0)$	None	(Yes)[b]	Yes	No
Heterotic SO(32)	$(1, 0)$	10	$(1, 0)$	SO(32)	No	Yes	No
Heterotic E_8	$(1, 0)$	10	$(1, 0)$	$E_8 \times E_8$	No	Yes	No
Heterotic SO(16)	$(1, 0)$	10	$(0, 0)$	SO(16) × SO(16)	No	Yes	No

[a]UV divergences beyond the tachyon (interpreted as closed string dilaton tadpoles) cancel only for the unoriented open plus closed strings with gauge group $SO(2^{13}) = SO(8192)$

[b]The parenthesis indicates that type II theories do not have open strings in the vacuum: they require a D-brane background. This is expected since there is no gauge multiplet in $d = 10$ $(1, 1)$ or $(2, 0)$ supergravities (the D-brane breaks half of the supersymmetry)

The tachyon-free superstring theories together with the bosonic string are summarized in Table 1.1.

1.2.3 Interactions

Worldsheet and Riemann Surfaces

After having described the spectrum and the general characteristics of string theory comes the question of interactions. The worldsheets obtained in this way are Riemann surfaces, i.e. one-dimensional complex manifolds. They are classified by the numbers of handles (or holes) g (called the genus) and external tubes n. In the presence of open strings, surfaces have boundaries: in addition to the handles and tubes, they are classified by the numbers of disks b and of strips m.[7] A particularly important number associated to each surface is the Euler characteristics

$$\chi = 2 - 2g - b, \tag{1.22}$$

which is a topological invariant. It is remarkable that there is a single topology at every loop order when one considers only closed strings, and just a few more in the presence of open strings. The analysis is greatly simplified in contrast to QFT, for which the number of Feynman graphs increases very rapidly with the number of loops and external particles.

---

[7]We ignore unoriented strings in this discussion. The associated worldsheets can have cross-caps which make the surfaces non-orientables.

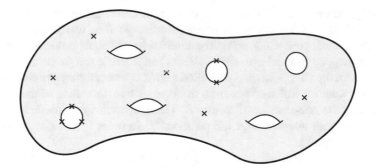

Fig. 1.6 General Riemann surfaces with boundaries and punctures

Due to the topological equivalence between surfaces, a conformal map can be used in order to work with simpler surfaces. In particular, the external tubes and strips are collapsed to points called punctures (or marked points) on the corresponding surfaces or boundaries. A general amplitude then looks like a sphere from which holes and disks have been removed and to which marked points have been pierced (Fig. 1.6).

Amplitudes

In order to compute an amplitude for the scattering of n strings (Chaps. 3 and 4), one must sum over all the inequivalent worldsheets through a path integral weighted by the CFT action chosen to describe the theory.[8] At fixed n, the sum runs over the genus g, such that each term is described by a Riemann surface $\Sigma_{g,n}$ of genus g with n punctures.

The interactions between strings follow from the graph topologies: since the latter are not encoded into the action, the dependence in the coupling constant must be added by hand. For closed strings, there is a unique cubic vertex with coupling g_s. A direct inspection shows that the correct factor is g_s^{n-2+2g}:

- for $n = 3$ there is one factor g_s, and every additional external string leads to the addition of one vertex with factor g_s, since this process can be obtained from the $n - 1$ process by splitting one of the external string in two by inserting a vertex;
- each loop comes with two vertices, so g-loops provide a factor g_s^{2g}.

Remark 1.2 (Status of g_s as a Parameter) It was stated earlier that string theory has no dimensionless parameter, but g_s looks to be one. In reality it is determined by the expectation value of the dilaton $g_s = e^{\langle \Phi \rangle}$. Hence the coupling constant is not a parameter defining the theory but is rather determined by the dynamics of the theory.

---

[8]For simplicity we focus on closed string amplitudes in this section.

Finally, the external states must be specified: this amounts to prescribe boundary conditions for the path integral or to insert the corresponding wave functions. Under the conformal mapping which brings the external legs to punctures located at z_i, the states are mapped to local operators $V_i(k_i, z_i)$ inserted at the points z_i. The latter are built from the CFT fields and are called vertex operators: they are characterized by a momentum k^μ which comes from the Fourier transformation of the X^μ fields representing the non-compact dimensions. These operators are inserted inside the path integral with integrals over the positions z_i in order to describe all possible conformal mappings.

Ultimately, the amplitude (amputated Green function) is computed as

$$A_n(k_1, \ldots, k_n) = \sum_{g \geq 0} g_s^{n-2+2g} A_{g,n}, \tag{1.23}$$

where

$$A_{g,n} = \int \prod_{i=1}^{n} d^2 z_i \int dg_{ab} d\Psi \, e^{-S_{cft}[g_{ab}, \Psi]} \prod_{i=1}^{n} V_i(k_i, z_i) \tag{1.24}$$

is the g-loop n-point amplitude (for simplicity we omit the dependence on the states beyond the momentum). Ψ denotes collectively the CFT fields and g_{ab} is the metric on the surface.

The integration over the metrics and over the puncture locations contain a huge redundancy due to the invariance under reparametrizations, which means that one integrates over many equivalent surfaces. To avoid this, Faddeev–Popov ghosts must be introduced and the integral is restricted to only finitely many (real) parameters t_λ. They form the *moduli space* $\mathcal{M}_{g,n}$ of the Riemann surfaces $\Sigma_{g,n}$ whose dimension is

$$\dim_{\mathbb{R}} \mathcal{M}_{g,n} = 6g - 6 + 2n. \tag{1.25}$$

The computation of the amplitude $A_{g,n}$ can be summarized as

$$A_{g,n} = \int_{\mathcal{M}_{g,n}} \prod_{\lambda=1}^{6g-6+2n} dt_\lambda \, F(t). \tag{1.26}$$

The function $F(t)$ is a correlation function in the worldsheet CFT defined on the Riemann surface $\Sigma_{g,n}$

$$F(t) = \left\langle \prod_{i=1}^{n} V_i \times \text{ghosts} \times \text{super-ghosts} \right\rangle_{\Sigma_{g,n}}. \tag{1.27}$$

Note that the (super)ghost part is independent of the choice of the matter CFT.

Divergences and Feynman Graphs

Formally the moduli parameters are equivalent to Schwinger (proper-time) parameters s_i in usual QFT: these are introduced in order to rewrite propagators as

$$\frac{1}{k^2 + m^2} = \int_0^\infty ds\, e^{-s(k^2 + m^2)}, \tag{1.28}$$

such that the integration over the momentum k becomes a Gaussian times a polynomial. This form of the propagator is useful to display the three types of divergences which can be encountered:

1. IR: regions $s_i \to \infty$ (for $k^2 + m^2 \le 0$). These divergences are artificial for $k^2 + m^2 < 0$ and means that the parametrization is not appropriate. Divergences for $k^2 + m^2 = 0$ are genuine and translates the fact that quantum effects shift the vacuum and the masses. Taking these effects into account necessitates a field theory framework in which renormalization can be used.
2. UV: regions $s_i \to 0$ (after integrating over k). Such divergences are absent in string theories because these regions are excluded from the moduli space $\mathcal{M}_{g,n}$ (see Fig. 1.7 for the example of the torus).[9]
3. Spurious: regions with finite s_i where the amplitude diverges. This happens typically only in the presence of super-ghosts and it translates a breakdown of

Fig. 1.7 Moduli space of the torus: $\mathrm{Re}\,\tau \in [-1/2, 1/2]$, $\mathrm{Im}\,\tau > 0$ and $|\tau| > 1$

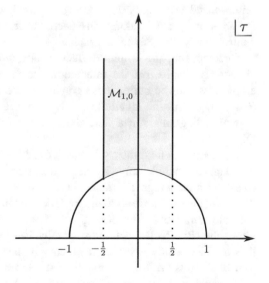

---

[9]There is a caveat to this statement: UV divergences reappear in string field theory in Lorentzian signature due to the way the theory is formulated. The solution requires a generalization of the Wick rotation.

Moreover, this does not hold for open strings whose moduli spaces contain those regions: in this case, the divergences are reinterpreted in terms of closed strings propagating.

the gauge fixing condition.[10] Since these spurious singularities of the amplitudes are not physical, one needs to ensure that they can be removed, which is indeed possible to achieve.

Hence, only IR divergences present a real challenge to string theory. Dealing with these divergences requires renormalizing the amplitudes, but this is not possible in the standard formulation of worldsheet string theory since the states are on-shell.[11]

1.3   String Field Theory

1.3.1   From the Worldsheet to Field Theory

The first step is to solve the IR divergences problem is to go *off-shell* (Chaps. 11 and 13). This is made possible by introducing *local coordinates* around the punctures of the Riemann surface (Chap. 12).

The IR divergences originate from Riemann surfaces close to *degeneration*, that is, surfaces with long tubes. The latter can be of *separating* and *non-separating* types, depending on whether the Riemann surface splits in two pieces if the tube is cut (Fig. 1.8). By exploring the form of the amplitudes in this limit (Chap. 14), the expression naturally separates into several pieces, to be interpreted as two amplitudes (of lower n and g) connected by a propagator. The latter can be reinterpreted as a standard $(k^2 + m^2)^{-1}$ term, hence solving the divergence problem for $k^2 + m^2 < 0$. Taking this decomposition seriously leads to identify each contribution with a Feynman graph.

Decomposing the amplitude recursively, the next step consists in finding the elementary graphs, i.e. the interaction vertices from which all other graphs (and amplitudes) can be built. These graphs are the building blocks of the field theory (Chap. 15), with the kinetic term given by the inverse of the propagator. Having Feynman diagrams and a field theory allows to use all the standard tools from QFT.

However, this field theory is gauge fixed because on-shell amplitudes are gauge invariant and include only physical states. For this reason, one needs to find how to re-establish the gauge invariance. Due to the complicated structure of string theory, the full-fledged *Batalin–Vilkovisky* (BV) formalism must be used (Chap. 15): it basically amounts to introduce ghosts before the gauge fixing. The final stage is to obtain the 1PI effective action from which the physics is more easily extracted. But, it is useful to study first the free theory (Chaps. 9 and 10) to gain some insights. The book ends with a discussion of the momentum-space representation and of background independence (Chaps. 16 and 18).

---

[10]Such spurious singularities are also found in supergravity.

[11]The on-shell condition is a consequence of the BRST and conformal invariance. While the first will be given up, the second will be maintained to facilitate the computations.

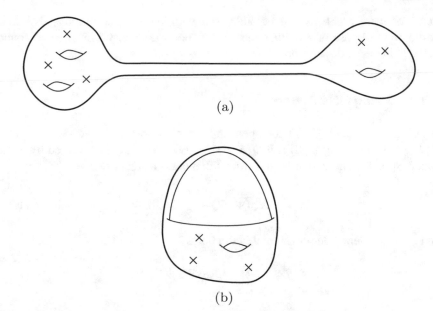

Fig. 1.8 Degeneration of Riemann surfaces. (**a**) Separating. (**b**) Non-separating

   The procedure we will follow is a kind of reverse engineering: we know what is the final result and we want to study backwards how it is obtained:

   on-shell amplitude → off-shell amplitude → Feynman graphs

   → gauge fixed field theory → BV field theory

In standard QFT, one follows the opposite process.

Remark 1.3 There are some prescriptions (using for example analytic continuation, the optical theorem, some tricks...) to address the problems mentioned above, but there is no general and universally valid procedure. A field theory is much more satisfactory because it provides a unique and complete framework.

   We can now summarize the disadvantages of the worldsheet approach over the spacetime field one:

- no natural description of (relativistic) multi-particle states;
- on-shell states:
  - lack of renormalization,
  - presence of infrared divergences,
  - scattering amplitudes only for protected states;
- interactions added by hand;
- hard to check consistency (unitarity, causality...);
- absence of non-perturbative processes.

Some of these problems can be addressed with various prescriptions, but it is desirable to dispose of a unified and systematic procedure, which is to be found in the field theory description.

1.3.2   String Field Action

A string field theory (SFT) for open and closed strings is based on two fields $\Phi[X(\sigma)]$ (open string field) and $\Psi[X(\sigma)]$ (closed string field) governed by some action $S[\Phi, \Psi]$. This action is built from a diagonal kinetic term

$$S_0 = \frac{1}{2} K_\Psi(\Psi, \Psi) + \frac{1}{2} K_\Phi(\Phi, \Phi) \tag{1.29}$$

and from an interaction polynomial in the fields

$$S_{\text{int}} = \sum_{m,n} \mathcal{V}_{m,n}(\Phi^m, \Psi^n), \tag{1.30}$$

where $\mathcal{V}_{m,n}$ is an appropriate product mapping m closed and n open string states to a number (the power is with respect to the tensor product). In particular, it contains the coupling constant. Contrary to the worldsheet approach where the cubic interaction looks sufficient, higher-order elementary interactions with $m, n \in \mathbb{N}$ are typically needed. A second specific feature is that the products also admit a loop (or genus g) expansion: a fundamental n-point interaction is introduced at every loop order g. These terms are interpreted as (finite) counter-terms needed to restore the gauge invariance of the measure. These two facts come from the decomposition of the moduli spaces in pieces (Sect. 1.2.3).

Writing an action for a field $\Psi[X(\sigma)]$ for which reparametrization invariance holds is highly complicated. The most powerful method is to introduce a functional dependence in ghost fields $\Psi[X(\sigma), c(\sigma)]$ and to extend the BRST formalism to the string field, leading ultimately to the BV formalism. While the latter formalism is the most complete and ensures that the theory is consistent at the quantum level, it is difficult to characterize the interactions explicitly. Several constructions which exploit different properties of the theory have been proposed:

- direct computation by reverse engineering of worldsheet amplitudes;
- specific parametrization of the Riemann surfaces (hyperbolic, minimal area);
- analogy with Chern–Simons and Wess–Zumino–Witten (WZW) theories;
- exploitation of the L_∞ and A_∞ algebra structures.

It can be shown that these constructions are all equivalent. For the superstring, the simplest strategy is to dress the bosonic interactions with data from the super-ghost sector, which motivates the study of the bosonic SFT by itself. The main difficulty in working with SFT is that only the first few interactions have been constructed explicitly. Finally, the advantage of the first formulation is that it provides a general formulation of SFT at the quantum level, from which the general structure can be studied.

1.3.3 Expression with Spacetime Fields

To obtain a more intuitive picture and to make contact with the spacetime fields, the field is expanded in terms of 1-particle states in the momentum representation

$$|\Psi\rangle = \sum_n \int \frac{d^D k}{(2\pi)^D} \, \psi_\alpha(k) \, |k, \alpha\rangle \,, \tag{1.31}$$

where α denotes collectively the discrete labels of the CFT eigenstates. The coefficients $\psi_\alpha(k)$ of the CFT states $|k, \alpha\rangle$ are spacetime fields, the first ones being the same as those found in the first-quantized picture (Sect. 1.2.1)

$$\psi_\alpha = \{T, G_{\mu\nu}, B_{\mu\nu}, \Phi, \ldots\}. \tag{1.32}$$

Then, inserting this expansion in the action gives an expression like $S[T, G_{\mu\nu}, \ldots]$. The exact expression of this action is out of reach and only the lowest terms are explicitly computable for a given CFT background. Nonetheless, examining the string field action indicates what is the generic form of the action in terms of the spacetime fields. One can then study the properties of such a general QFT: since it is more general than the SFT (expanded) action, any result derived for it will also be valid for SFT. This approach is very fruitful for studying properties related to consistency of QFT (unitarity, soft theorems...) and this can provide helpful phenomenological models.

In conclusion, SFT can be seen as a regular QFT with the following properties:

- infinite number of fields;
- non-local interaction (proportional to $e^{-k^2 \#}$);
- the amplitudes agree with the worldsheet amplitudes (when the latter can be defined);
- genuine (IR) divergences agree but can be handled with the usual QFT tools.

1.3.4 Applications

The first aspect is the possibility to use standard QFT techniques (such as renormal-ization) to study—and to make sense of—string amplitudes. In this sense, SFT can be viewed as providing recipes for computing quantities in the worldsheet theory which are otherwise not defined. This program has been pushed quite far in the last years.

Another reason to use SFT is gauge invariance: it is always easier to describe a system when its gauge invariance is manifest. We have explained that string theory contains Yang–Mills and graviton fields with the corresponding (spacetime) gauge invariances (non-Abelian gauge symmetry and diffeomorphisms). In fact, these symmetries are enhanced to an enormous gauge invariance when taking into account the higher-spin fields. This invariance is hidden in the standard formulation and cannot be exploited fully. On the other hand, the full gauge symmetry is manifest in string field theory.

Finally, the worldvolume description of p-brane is difficult because there is no analogue of the Polyakov action. If one could find a first-principle description of SFT which does not rely on CFT and first-quantization, then one may hope to generalize it to build a brane field theory.

We can summarize the general motivations for studying SFT:

- field theory (second-quantization);
- more rigorous and constructive formulation;
- make gauge invariance explicit (L_∞ algebras et al.);
- use standard QFT techniques (renormalization, analyticity...)
  $\to$ remove IR divergences, prove consistency (Cutkosky rules, unitarity, soft theorems, background independence...);
- worldvolume theory ill-defined for ($p > 1$)-branes.

Beyond these general ideas, SFT has been developed in order to address different questions:

- worldsheet scattering amplitudes;
- effective actions;
- map of the consistent backgrounds (classical solutions, marginal deformations, RR fluxes...);
- collective, non-perturbative, thermal, dynamical effects;
- symmetry breaking effects;
- dynamics of compactification;
- proof of dualities;
- proof of the AdS/CFT correspondence.

The last series of points is still out of reach within the current formulation of SFT. However, the last two decades have seen many important developments:

- construction of the open, closed and open-closed superstring field theories:
  - 1PI and BV actions and general properties [26, 67, 68, 79, 80, 82, 84, 86, 88, 91, 92, 96],
  - dressing of bosonic products using the WZW construction and homotopy algebra [8, 9, 20–23, 29–33, 35, 38, 45, 46, 50–56, 70],
  - light-cone super-SFT [39–42],
  - supermoduli space [69, 97];

- hyperbolic and minimal area constructions [12, 36, 37, 64–66, 72];
- open string analytic solutions [27, 28];
- level-truncation solutions [47–49];
- field theory properties [10, 15, 57, 73, 87, 89, 90];
- spacetime effective actions [19, 59, 60, 98];
- defining worldsheet scattering amplitudes [74–76, 81, 82, 85, 93–95];
- marginal and RR deformations [11, 95, 98].

Recent reviews are [14, 24, 25].

1.4 Suggested Readings

For references about different aspects in this chapter:

- Differences between the worldvolume and spacetime formalisms—and of the associated first- and second-quantization—for the particle and string [43, chap. 1, 100, chap. 11].
- General properties of relativistic strings [34, 100].
- Divergences in string theory [14, 83, 99, sec. 7.2].
- Motivations for building a string field theory [77, sec. 4].

References

1. I. Bars, First massive level and anomalies in the supermembrane. Nucl. Phys. B **308**(2), 462–476 (1988). https://doi.org/10.1016/0550-3213(88)90573-1
2. I. Bars, Is there a unique consistent unified theory based on extended objects? (1988), pp. 693–698. https://inspirehep.net/literature/264587
3. I. Bars, Issues of topology and the spectrum of membranes, pp. 209–212, Jan 1990. https://inspirehep.net/literature/285041
4. I. Bars, Membrane symmetries and anomalies. Nucl. Phys. B **343**(2), 398–417 (1990). https://doi.org/10.1016/0550-3213(90)90476-T
5. I. Bars, C.N. Pope, Anomalies in super P-branes. Class. Quantum Gravity **5**(9), 1157 (1988). https://doi.org/10.1088/0264-9381/5/9/002
6. I. Bars, C.N. Pope, E. Sezgin, Massless spectrum and critical dimension of the supermembrane. Phys. Lett. B **198**(4), 455–460 (1987). https://doi.org/10.1016/0370-2693(87)90899-9
7. I. Bars, C.N. Pope, E. Sezgin, Central extensions of area preserving membrane algebras. Phys. Lett. B **210**(1), 85–91 (1988). https://doi.org/10.1016/0370-2693(88)90354-1
8. N. Berkovits, Super-Poincare invariant superstring field theory. Nucl. Phys. B **450**(1–2), 90–102 (1995). https://doi.org/10.1016/0550-3213(95)00259-U. arXiv: hep-th/9503099
9. N. Berkovits, Y. Okawa, B. Zwiebach, WZW-like action for heterotic string field theory. J. High Energy Phys. **2004**(11), 038–038 (2004). https://doi.org/10.1088/1126-6708/2004/11/038. arXiv: hep-th/0409018
10. S. Chakrabarti, S.P. Kashyap, B. Sahoo, A. Sen, M. Verma, Subleading soft theorem for multiple soft gravitons. J. High Energy Phys. **2017**(12) (2017). https://doi.org/10.1007/JHEP12(2017)150. arXiv: 1707.06803

11. M. Cho, S. Collier, X. Yin, Strings in Ramond-Ramond backgrounds from the Neveu-Schwarz-Ramond formalism, Oct 2018. arXiv: 1811.00032
12. K. Costello, B. Zwiebach, Hyperbolic string vertices, Aug 2019. arXiv: 1909.00033
13. J.A. de Azcárraga, J.M. Izquierdo, P.K. Townsend, Classical anomalies of supersymmetric extended objects. Phys. Lett. B **267**(3), 366–373 (1991). https://doi.org/10.1016/0370-2693(91)90947-O
14. C. de Lacroix, H. Erbin, S.P. Kashyap, A. Sen, M. Verma, Closed superstring field theory and its applications. Int. J. Mod. Phys. A **32**(28n29), 1730021 (2017). https://doi.org/10.1142/S0217751X17300216. arXiv: 1703.06410
15. C. de Lacroix, H. Erbin, A. Sen, Analyticity and crossing symmetry of superstring loop amplitudes. J. High Energy Phys. **2019**(5), 139 (2019). https://doi.org/10.1007/JHEP05(2019)139. arXiv: 1810.07197
16. B. de Wit, J. Hoppe, H. Nicolai, On the quantum mechanics of supermembranes. Nucl. Phys. B **305**(4), 545–581 (1988). https://doi.org/10.1016/0550-3213(88)90116-2
17. B. de Wit, M. Lüscher, H. Nicolai, The supermembrane is unstable. Nucl. Phys. B **320**(1), 135–159 (1989). https://doi.org/10.1016/0550-3213(89)90214-9
18. M.J. Duff, T. Inami, C.N. Pope, E. Sezgin, K.S. Stelle, Semiclassical quantization of the supermembrane. Nucl. Phys. B **297**(3), 515–538 (1988). https://doi.org/10.1016/0550-3213(88)90316-1
19. H. Erbin, C. Maccaferri, J. Vošmera, Localization of effective actions in heterotic string field theory. J. High Energy Phys. **2020**(2), 59 (2020). https://doi.org/10.1007/JHEP02(2020)059. arXiv: 1912.05463
20. T. Erler, Relating Berkovits and A_∞ superstring field theories; large Hilbert space perspective. J. High Energy Phys. **1602**, 121 (2015). https://doi.org/10.1007/JHEP02(2016)121. arXiv: 1510.00364
21. T. Erler, Relating Berkovits and A_∞ superstring field theories; small Hilbert space perspective. J. High Energy Phys. **1510**, 157 (2015). https://doi.org/10.1007/JHEP10(2015)157. arXiv: 1505.02069
22. T. Erler, Superstring field theory and the Wess-Zumino-Witten action. J. High Energy Phys. **2017**(10), 57 (2017). https://doi.org/10.1007/JHEP10(2017)057. arXiv: 1706.02629
23. T. Erler, Supersymmetry in open superstring field theory. J. High Energy Phys. **2017**(5), 113 (2017). https://doi.org/10.1007/JHEP05(2017)113. arXiv: 1610.03251
24. T. Erler, Four lectures on analytic solutions in open string field theory, Dec 2019. arXiv: 1912.00521
25. T. Erler, Four lectures on closed string field theory. Phys. Rep. **851**, S0370157320300132 (2020). https://doi.org/10.1016/j.physrep.2020.01.003. arXiv: 1905.06785
26. T. Erler, S. Konopka, Vertical integration from the Large Hilbert space. J. High Energy Phys. **2017**(12), 112 (2017). https://doi.org/10.1007/JHEP12(2017)112. arXiv: 1710.07232
27. T. Erler, C. Maccaferri, String field theory solution for any open string background. J. High Energy Phys. **2014**(10), 29 (2014). https://doi.org/10.1007/JHEP10(2014)029. arXiv: 1406.3021
28. T. Erler, C. Maccaferri, String field theory solution for any open string background. II. J. High Energy Phys. **2020**(1), 21 (2020). https://doi.org/10.1007/JHEP01(2020)021. arXiv: 1909.11675
29. T. Erler, S. Konopka, I. Sachs, NS-NS sector of closed superstring field theory. J. High Energy Phys. **2014**(8), 158 (2014). https://doi.org/10.1007/JHEP08(2014)158. arXiv: 1403.0940
30. T. Erler, S. Konopka, I. Sachs, Resolving Witten's superstring field theory. J. High Energy Phys. **2014**(4), 150 (2014). https://doi.org/10.1007/JHEP04(2014)150. arXiv: 1312.2948
31. T. Erler, S. Konopka, I. Sachs, Ramond equations of motion in superstring field theory. J. High Energy Phys. **2015**(11), 199 (2015). https://doi.org/10.1007/JHEP11(2015)199. arXiv: 1506.05774
32. T. Erler, Y. Okawa, T. Takezaki, A_∞ structure from the Berkovits formulation of open superstring field theory, May 2015. arXiv: 1505.01659

33. T. Erler, Y. Okawa, T. Takezaki, Complete action for open superstring field theory with cyclic A_∞ structure. J. High Energy Phys. **2016**(8), 12 (2016). https://doi.org/10.1007/JHEP08(2016)012. arXiv: 1602.02582
34. P. Goddard, J. Goldstone, C. Rebbi, C.B. Thorn, Quantum dynamics of a massless relativistic string. Nucl. Phys. B **56**(1), 109–135 (1973). https://doi.org/10.1016/0550-3213(73)90223-X
35. K. Goto, H. Kunitomo, Construction of action for heterotic string field theory including the Ramond sector, June 2016. arXiv: 1606.07194
36. M. Headrick, B. Zwiebach, Convex programs for minimal-area problems, June 2018. arXiv: 1806.00449
37. M. Headrick, B. Zwiebach, Minimal-area metrics on the Swiss cross and punctured torus. Commun. Math. Phys. **24**, 1–57 (2020). https://doi.org/10.1007/s00220-020-03734-z. arXiv: 1806.00450
38. Y. Iimori, T. Noumi, Y. Okawa, S. Torii, From the Berkovits formulation to the Witten formulation in open superstring field theory. J. High Energy Phys. **2014**(3), 44 (2014). https://doi.org/10.1007/JHEP03(2014)044. arXiv: 1312.1677
39. N. Ishibashi, Multiloop amplitudes of light-cone gauge string field theory for type II superstrings, Oct 2018
40. N. Ishibashi, K. Murakami, Multiloop amplitudes of light-cone Gauge Bosonic string field theory in noncritical dimensions. J. High Energy Phys. **2013**(9), 53 (2013). https://doi.org/10.1007/JHEP09(2013)053. arXiv: 1307.6001
41. N. Ishibashi, K. Murakami, Worldsheet theory of light-cone Gauge noncritical strings on higher genus Riemann surfaces. J. High Energy Phys. **2016**(6), 87 (2016). https://doi.org/10.1007/JHEP06(2016)087. arXiv: 1603.08337
42. N. Ishibashi, K. Murakami, Multiloop amplitudes of light-cone Gauge superstring field theory: odd spin structure contributions. J. High Energy Phys **2018**(3), 63 (2018). https://doi.org/10.1007/JHEP03(2018)063 arXiv: 1712.09049
43. M. Kaku, *Introduction to Superstrings and M-Theory* (Springer, Berlin, 1999)
44. K. Kikkawa, M. Yamasaki, Can the membrane be a unification model? Prog. Theor. Phys. **76**(6), 1379–1389 (1986). https://doi.org/10.1143/PTP.76.1379
45. S. Konopka, I. Sachs, Open superstring field theory on the restricted Hilbert space. J. High Energy Phys. **2016**(4), 1–12 (2016). https://doi.org/10.1007/JHEP04(2016)164. arXiv: 1602.02583
46. M. Kroyter, Y. Okawa, M. Schnabl, S. Torii, B. Zwiebach, Open superstring field theory I: Gauge fixing, ghost structure, and propagator. J. High Energy Phys. **2012**(3), 30 (2012). https://doi.org/10.1007/JHEP03(2012)030. arXiv: 1201.1761
47. M. Kudrna, C. Maccaferri, BCFT moduli space in level truncation. J. High Energy Phys. **2016**(4), 57 (2016). https://doi.org/10.1007/JHEP04(2016)057. arXiv: 1601.04046
48. M. Kudrna, M. Schnabl, Universal solutions in open string field theory, Dec 2018. arXiv: 1812.03221
49. M. Kudrna, T. Masuda, Y. Okawa, M. Schnabl, K. Yoshida, Gauge-invariant observables and marginal deformations in open string field theory. J. High Energy Phys. **2013**(1), 103 (2013)
50. H. Kunitomo, First-order equations of motion for heterotic string field theory. Prog. Theor. Exp. Phys. **2014**(9), 93B07–0 (2014). https://doi.org/10.1093/ptep/ptu125. arXiv: 1407.0801
51. H. Kunitomo, The Ramond sector of heterotic string field theory. Prog. Theor. Exp. Phys. **2014**(4), 43B01–0 (2014). https://doi.org/10.1093/ptep/ptu032. arXiv: 1312.7197
52. H. Kunitomo, Symmetries and Feynman rules for Ramond sector in heterotic string field theory. Prog. Theor. Exp. Phys. **2015**(9), 093B02 (2015). https://doi.org/10.1093/ptep/ptv117. arXiv: 1506.08926
53. H. Kunitomo, Space-time supersymmetry in WZW-like open superstring field theory. Prog. Theor. Exp. Phys. **2017**(4), 043B04 (2017). https://doi.org/10.1093/ptep/ptx028. arXiv: 1612.08508
54. H. Kunitomo, Y. Okawa, Complete action for open superstring field theory. Prog. Theor. Exp. Phys. **2016**(2), 023B01 (2016). https://doi.org/10.1093/ptep/ptv189. arXiv: 1508.00366

55. H. Kunitomo, T. Sugimoto, Heterotic string field theory with cyclic L-infinity structure. Prog. Theor. Exp. Phys. **2019**(6), 063B02 (2019). https://doi.org/10.1093/ptep/ptz051. arXiv: 1902.02991

56. H. Kunitomo, T. Sugimoto, Type II superstring field theory with cyclic L-infinity structure, March 2020. arXiv: 1911.04103.

57. A. Laddha, A. Sen, Sub-subleading soft graviton theorem in generic theories of quantum gravity. J. High Energy Phys. **2017**(10), 065 (2017). https://doi.org/10.1007/JHEP10(2017)065. arXiv: 1706.00759

58. H. Luckock, I. Moss, The quantum geometry of random surfaces and spinning membranes. Class. Quantum Gravity **6**(12), 1993–2027 (1989). https://doi.org/10.1088/0264-9381/6/12/025

59. C. Maccaferri, A. Merlano, Localization of effective actions in open superstring field theory. J. High Energy Phys. **2018**(3), 112 (2018). https://doi.org/10.1007/JHEP03(2018)112. arXiv: 1801.07607

60. C. Maccaferri, A. Merlano, Localization of effective actions in open superstring field theory: small Hilbert space. J. High Energy Phys. **1906**, 101 (2019). https://doi.org/10.1007/JHEP06(2019)101. arXiv: 1905.04958

61. U. Marquard, M. Scholl, Conditions of the embedding space of P-branes from their constraint algebras. Phys. Lett. B **209**(4), 434–440 (1988). https://doi.org/10.1016/0370-2693(88)91169-0

62. U. Marquard, M. Scholl, Lorentz algebra and critical dimension for the bosonic membrane. Phys. Lett. B **227**(2), 227–233 (1989). https://doi.org/10.1016/S0370-2693(89)80027-9

63. U. Marquard, R. Kaiser, M. Scholl, Lorentz algebra and critical dimension for the supermembrane. Phys. Lett. B **227**(2), 234–238 (1989). https://doi.org/10.1016/S0370-2693(89)80028-0

64. S.F. Moosavian, R. Pius, Hyperbolic geometry of superstring perturbation theory, March 2017. arXiv: 1703.10563

65. S.F. Moosavian, R. Pius, Hyperbolic geometry and closed bosonic string field theory I: the string vertices via hyperbolic Riemann surfaces. J. High Energy Phys. **1908**, 157 (2019). arXiv: 1706.07366

66. S.F. Moosavian, R. Pius, Hyperbolic geometry and closed bosonic string field theory II: the rules for evaluating the quantum BV master action. J. High Energy Phys. **1908**, 177 (2019). arXiv: 1708.04977

67. S.F. Moosavian, Y. Zhou, On the existence of heterotic-string and type-II-superstring field theory vertices, Nov 2019. arXiv: 1911.04343

68. S.F. Moosavian, A. Sen, M. Verma, Superstring field theory with open and closed strings. J. High Energy Phys. **2001**, 183 (2020). arXiv:1907.10632

69. K. Ohmori, Y. Okawa, Open superstring field theory based on the supermoduli space. J. High Energy Phys. **2018**(4), 35 (2018). https://doi.org/10.1007/JHEP04(2018)035. arXiv: 1703.08214

70. Y. Okawa, B. Zwiebach, Heterotic string field theory. J. High Energy Phys. **2004**(07), 042–042 (2004). https://doi.org/10.1088/1126-6708/2004/07/042. arXiv: hep-th/0406212

71. F. Paccanoni, P. Pasti, M. Tonin, Some remarks on the consistency of quantum supermembranes. Mod. Phys. Lett. A **4**(09), 807–814 (1989). https://doi.org/10.1142/S0217732389000940

72. R. Pius, Quantum closed superstring field theory and hyperbolic geometry I: construction of string vertices, Aug 2018. arXiv: 1808.09441

73. R. Pius, A. Sen, Cutkosky rules for superstring field theory. J. High Energy Phys. **2016**(10), 24 (2016) https://doi.org/10.1007/JHEP10(2016)024. arXiv: 1604.01783

74. R. Pius, A. Rudra, A. Sen, Mass renormalization in string theory: general states. J. High Energy Phys. **2014**(7), 62 (2014). https://doi.org/10.1007/JHEP07(2014)062. arXiv: 1401.7014

75. R. Pius, A. Rudra, A. Sen, Mass renormalization in string theory: special states. J. High Energy Phys. **2014**(7), 58 (2014). https://doi.org/10.1007/JHEP07(2014)058. arXiv: 1311.1257

76. R. Pius, A. Rudra, A. Sen, String perturbation theory around dynamically shifted vacuum. J. High Energy Phys. **2014**(10), 70 (2014). https://doi.org/10.1007/JHEP10(2014)070. arXiv: 1404.6254

77. J. Polchinski, What is string theory? Nov 1994. arXiv: hep-th/9411028

78. J. Polchinski, *String Theory: Volume 1, An Introduction to the Bosonic String* (Cambridge University Press, Cambridge, 2005)

79. A. Sen, Gauge invariant 1PI effective action for superstring field theory. J. High Energy Phys. **1506**, 022 (2015). https://doi.org/10.1007/JHEP06(2015)022. arXiv: 1411.7478

80. A. Sen, Gauge invariant 1PI effective superstring field theory: inclusion of the Ramond sector. J. High Energy Phys. **2015**(8), 25 (2015). https://doi.org/10.1007/JHEP08(2015)025. arXiv: 1501.00988

81. A. Sen, Off-shell amplitudes in superstring theory. Fortschr. Phys. **63**(3–4), 149–188 (2015). https://doi.org/10.1002/prop.201500002. arXiv: 1408.0571

82. A. Sen, Supersymmetry restoration in superstring perturbation theory. J. High Energy Phys. **2015**(12), 1–19 (2015). https://doi.org/10.1007/JHEP12(2015)075. arXiv: 1508.02481

83. A. Sen, Ultraviolet and infrared divergences in superstring theory, Nov 2015. arXiv: 1512.00026

84. A. Sen, BV master action for heterotic and type II string field theories. J. High Energy Phys. **2016**(2), 87 (2016). https://doi.org/10.1007/JHEP02(2016)087. arXiv: 1508.05387

85. A. Sen, One loop mass renormalization of unstable particles in superstring theory. J. High Energy Phys. **2016**(11), 50 (2016). https://doi.org/10.1007/JHEP11(2016)050. arXiv: 1607.06500

86. A. Sen, Reality of superstring field theory action. J. High Energy Phys. **2016**(11), 014 (2016). https://doi.org/10.1007/JHEP11(2016)014. arXiv: 1606.03455

87. A. Sen, Unitarity of superstring field theory. J. High Energy Phys. **2016**(12), 115 (2016). https://doi.org/10.1007/JHEP12(2016)115. arXiv: 1607.08244

88. A. Sen, Equivalence of two contour prescriptions in superstring perturbation theory. J. High Energy Phys. **2017**(04), 25 (2017). https://doi.org/10.1007/JHEP04(2017)025. arXiv: 1610.00443

89. A. Sen, Soft theorems in superstring theory. J. High Energy Phys. **2017**(06), 113 (2017). https://doi.org/10.1007/JHEP06(2017)113. arXiv: 1702.03934

90. A. Sen, Subleading soft graviton theorem for loop amplitudes. J. High Energy Phys. **2017**(11), 1–8 (2017). https://doi.org/10.1007/JHEP11(2017)123. arXiv: 1703.00024

91. A. Sen, Wilsonian effective action of superstring theory. J. High Energy Phys. **2017**(1), 108 (2017). https://doi.org/10.1007/JHEP01(2017)108. arXiv: 1609.00459

92. A. Sen, Background independence of closed superstring field theory. J. High Energy Phys. **2018**(2), 155 (2018). https://doi.org/10.1007/JHEP02(2018)155. arXiv: 1711.08468

93. A. Sen, String field theory as world-sheet UV regulator. J. High Energy Phys. **2019**(10), 119 (2019). https://doi.org/10.1007/JHEP10(2019)119. arXiv: 1902.00263

94. A. Sen, D-instanton perturbation theory, May 2020. arXiv: 2002.04043

95. A. Sen, Fixing an ambiguity in two dimensional string theory using string field theory. J. High Energy Phys. **2020**(3), 5 (2020). https://doi.org/10.1007/JHEP03(2020)005. arXiv: 1908.02782

96. A. Sen, E. Witten, Filling the gaps with PCO's. J. High Energy Phys. **1509**, 004 (2015). https://doi.org/10.1007/JHEP09(2015)004. arXiv: 1504.00609

97. T. Takezaki, Open superstring field theory including the Ramond sector based on the supermoduli space (2019). arXiv: 1901.02176

98. J. Vošmera, Generalized ADHM equations from marginal deformations in open superstring field theory. J. High Energy Phys. **2019**(12), 118 (2019). https://doi.org/10.1007/JHEP12(2019)118. arXiv: 1910.00538

99. E. Witten, Superstring perturbation theory revisited, Sept 2012. arXiv: 1209.5461

100. B. Zwiebach, *A First Course in String Theory*, 2nd edn. (Cambridge University Press, Cambridge, 2009)

Part I

Worldsheet Theory

Worldsheet Path Integral: Vacuum Amplitudes

2

Abstract

In this chapter, we develop the path integral quantization for a generic closed string theory in worldsheet Euclidean signature. We focus on the vacuum amplitudes, leaving scattering amplitudes for the next chapter. This allows to focus on the definition and gauge fixing of the path integral measure.

The exposition differs from most traditional textbooks in three ways: (1) we consider a general matter CFT, (2) we consider the most general treatment (for any genus) and (3) we do not use complex coordinates but always a covariant parametrization.

The derivation is technical and the reader is encouraged to not stop at this chapter in case of difficulties and to proceed forward: most concepts will be reintroduced from a different point of view later in other chapters of the book.

2.1 Worldsheet Action and Symmetries

The string worldsheet is a Riemann surface $\mathcal{W} = \Sigma_g$ of genus g: the genus counts the number of holes or handles. Coordinates on the worldsheet are denoted by $\sigma^a = (\tau, \sigma)$. When there is no risk of confusion, σ denotes collectively both coordinates. Since closed strings are considered, the Riemann surface has locally the topology of a cylinder, with the spatial section being circles S^1 with radius taken to be 1, such that

$$\sigma \in [0, 2\pi), \qquad \sigma \sim \sigma + 2\pi. \tag{2.1}$$

The string is embedded in the D-dimensional spacetime $\mathcal{M}$ with metric $G_{\mu\nu}$ through maps $X^\mu(\sigma^a) : \mathcal{W} \to \mathcal{M}$ with $\mu = 0, \ldots, D-1$.

© Springer Nature Switzerland AG 2021
H. Erbin, *String Field Theory*, Lecture Notes in Physics 980,
https://doi.org/10.1007/978-3-030-65321-7_2

The Nambu–Goto action is the starting point of the worldsheet description:

$$S_{\mathrm{NG}}[X^\mu] = \frac{1}{2\pi\alpha'} \int d^2\sigma \sqrt{\det G_{\mu\nu}(X) \frac{\partial X^\mu}{\partial\sigma^a} \frac{\partial X^\nu}{\partial\sigma^b}}, \tag{2.2}$$

where α' is the Regge slope (related to the string tension and string length). However, quantizing this action is difficult because it is highly non-linear. To solve this problem, a Lagrange multiplier is introduced to remove the squareroot. This auxiliary field corresponds to an intrinsic worldsheet metric $g_{ab}(\sigma)$. The worldsheet dynamics is described by the Polyakov action:

$$S_{\mathrm{P}}[g, X^\mu] = \frac{1}{4\pi\alpha'} \int d^2\sigma \sqrt{g}\, g^{ab} G_{\mu\nu}(X) \frac{\partial X^\mu}{\partial\sigma^a} \frac{\partial X^\nu}{\partial\sigma^b}, \tag{2.3}$$

which is classically equivalent to the Nambu–Goto action (2.2). In this form, it is clear that the scalar fields $X^\mu(\sigma)$ ($\mu = 0, \ldots D - 1$) characterize the string theory under consideration in two ways. First, by specifying some properties of the spacetime in which the string propagates (for example, the number of dimensions is determined by the number of fields X^μ), second, by describing the internal degrees of freedom (vibration modes).[1]

But, nothing prevents to consider a more general matter content in order to describe a different spacetime or different degrees of freedom. In Polyakov's formalism, the worldsheet geometry is endowed with a metric $g_{ab}(\sigma)$ together with a set of matter fields living on it. The scalar fields X^μ can be described by a general sigma model which encodes the embedding of the string in the D non-compact spacetime dimensions, and other fields can be added, for example to describe compactified dimensions or (spacetime) spin. Different sets of fields (and actions) correspond to different string theories. However, to describe precisely the different possibilities, we first have to understand the constraints on the worldsheet theories and to introduce conformal field theories (Part I). In this chapter (and in most of the book), the precise matter content is not important and we will denote the fields collectively as $\Psi(\sigma)$.

Before discussing the symmetries, let us introduce a topological invariant which will be needed throughout the text: the *Euler characteristics*. It is computed by integrating the Riemann curvature R of the metric g_{ab} over the surface Σ_g:

$$\chi_g := \chi(\Sigma_g) := 2 - 2g = \frac{1}{4\pi} \int_{\Sigma_g} d^2\sigma \sqrt{g}\, R, \tag{2.4}$$

where g is the genus of the surface. Oriented Riemann surfaces without boundaries are completely classified (topologically or as complex manifolds) by their Euler characteristics χ_g, or equivalently by their genus g.

---

[1] Obviously, the vibrational modes are also constrained by the spacetime geometry.

In order to describe a proper string theory, the worldsheet metric $g_{ab}(\sigma)$ should not be dynamical. This means that the worldsheet has no intrinsic dynamics and that no supplementary degrees of freedom are introduced when parametrizing the worldsheet with a metric. A solution to remove these degrees of freedom is to introduce gauge symmetries with as many gauge parameters as there are of degrees of freedom. The simplest symmetry is invariance under diffeomorphisms: indeed, the worldsheet theory is effectively a QFT coupled to gravity and it makes sense to require this invariance. Physically, this corresponds to the fact that the worldsheet spatial coordinate σ used along the string and worldsheet time are arbitrary. However, diffeomorphisms alone are not sufficient to completely fix the metric. Another natural candidate is *Weyl invariance* (local rescalings of the metric).

A diffeomorphism $f \in \text{Diff}(\Sigma_g)$ acts on the fields as

$$\sigma'^a = f^a(\sigma^b), \qquad g'(\sigma') = f^* g(\sigma), \qquad \Psi'(\sigma') = f^* \Psi(\sigma), \tag{2.5}$$

where the star denotes the pullback by f: this corresponds simply to the standard coordinate transformation where each tensor index of the field receives a factor $\partial \sigma^a / \partial \sigma'^b$. In particular, the metric and scalar fields transform explicitly as

$$g'_{ab}(\sigma') = \frac{\partial \sigma^c}{\partial \sigma'^a} \frac{\partial \sigma^d}{\partial \sigma'^b} g_{cd}(\sigma), \qquad X'^\mu(\sigma') = X^\mu(\sigma). \tag{2.6}$$

The index μ is inert since it is a target spacetime index: from the worldsheet point of view, it just labels a collection of worldsheet scalar fields. Infinitesimal variations are generated by vector fields on Σ_g:

$$\delta_\xi \sigma^a = \xi^a, \qquad \delta_\xi \Psi = \mathcal{L}_\xi \Psi, \qquad \delta_\xi g_{ab} = \mathcal{L}_\xi g_{ab}, \tag{2.7}$$

where $\mathcal{L}_\xi$ is the Lie derivative[2] with respect to the vector field $\xi \in \mathfrak{diff}(\Sigma_g) \simeq T\Sigma_g$. The Lie derivative of the metric is

$$\mathcal{L}_\xi g_{ab} = \xi^c \partial_c g_{ab} + g_{ac} \partial_b \xi^c + g_{bc} \partial_a \xi^c = \nabla_a \xi_b + \nabla_b \xi_a. \tag{2.8}$$

The Lie algebra generates only transformations in the connected component $\text{Diff}_0(\Sigma_g)$ of the diffeomorphism group which contains the identity.

Transformations not contained in $\text{Diff}_0(\Sigma_g)$ are called large diffeomorphisms: this includes reflections, for example. The quotient of the two groups is called the modular group Γ_g (also mapping class group or MCG):

$$\Gamma_g := \pi_0\big(\text{Diff}(\Sigma_g)\big) = \frac{\text{Diff}(\Sigma_g)}{\text{Diff}_0(\Sigma_g)}. \tag{2.9}$$

---

[2]For our purpose here, it is sufficient to accept the definition of the Lie derivative as corresponding to the infinitesimal variation.

It depends only on the genus g of the Riemann surface, but not on the metric. It is an infinite discrete group for genus $g \geq 1$ surfaces; in particular, $\Gamma_1 = \mathrm{SL}(2, \mathbb{Z})$.

A Weyl transformation $e^{2\omega} \in \mathrm{Weyl}(\Sigma_g)$ corresponds to a local rescaling of the metric and leaves the other fields unaffected[3]

$$g'_{ab}(\sigma) = e^{2\omega(\sigma)} g_{ab}(\sigma), \qquad \Psi'(\sigma) = \Psi(\sigma). \tag{2.10}$$

The exponential parametrization is generally more useful, but one should remember that it is $e^{2\omega}$ and not ω which is an element of the group. The infinitesimal variation reads

$$\delta_\omega g_{ab} = 2\omega\, g_{ab}, \qquad \delta_\omega \Psi = 0 \tag{2.11}$$

where $\omega \in \mathfrak{weyl}(\Sigma) \simeq \mathcal{F}(\Sigma_g)$ is a function on the manifold. Two metrics related in this way are said to be conformally equivalent. The *conformal structure* of the Riemann surface is defined by

$$\mathrm{Conf}(\Sigma_g) := \frac{\mathrm{Met}(\Sigma_g)}{\mathrm{Weyl}(\Sigma_g)}, \tag{2.12}$$

where $\mathrm{Met}(\Sigma_g)$ denotes the space of all metrics on Σ_g. Each element is a class of conformally equivalent metrics.

Diffeomorphisms have two parameters ξ^a (vector field) and Weyl invariance has one, ω (function). Hence, this is sufficient to locally fix the three components of the metric (symmetric matrix) and the total gauge group of the theory is the semi-direct product

$$G := \mathrm{Diff}(\Sigma_g) \ltimes \mathrm{Weyl}(\Sigma_g). \tag{2.13}$$

Similarly, the component connected of the identity is written as

$$G_0 := \mathrm{Diff}_0(\Sigma_g) \ltimes \mathrm{Weyl}(\Sigma_g). \tag{2.14}$$

The semi-direct product arises because the Weyl parameter is not inert under diffeomorphisms. Indeed, the combination of two transformations is

$$g' = f^*\left(e^{2\omega} g\right) = e^{2f^*\omega} f^* g, \tag{2.15}$$

such that the diffeomorphism acts also on the conformal factor.

---

[3] For simplicity, we consider only fields which do not transform under Weyl transformations, which excludes fermions.

The combination of transformations (2.15) can be chosen to fix the metric in a convenient gauge. For example, the *conformal gauge* reads

$$g_{ab}(\sigma) = e^{2\phi(\sigma)} \hat{g}_{ab}(\sigma), \tag{2.16}$$

where $\hat{g}_{ab}$ is some (fixed) *background metric* and $\phi(\sigma)$ is the conformal factor, also called the *Liouville field*. Fixing only diffeomorphisms amounts to keep ϕ arbitrary: the latter can then be fixed with a Weyl transformation. For instance, one can adopt the conformally flat gauge

$$\hat{g}_{ab} = \delta_{ab}, \qquad \phi \text{ arbitrary} \tag{2.17}$$

with a diffeomorphism, and then reach the flat gauge

$$\hat{g}_{ab} = \delta_{ab}, \qquad \phi = 0 \tag{2.18}$$

with a Weyl transformation. Another common choice is the uniformization gauge where $\hat{g}$ is taken to be the metric of constant curvature on the sphere ($g = 0$), on the plane ($g = 1$) or on the hyperbolic space ($g > 1$). All these gauges are covariant (both in spacetime and worldsheet).

Remark 2.1 (Active and Passive Transformations) Usually, symmetries are described by *active* transformations, which means that the field is seen to be changed by the transformation. On the other hand, gauge fixing is seen as a *passive* transformation, where the field is expressed in terms of other fields (i.e. a different parametrization). These are mathematically equivalent since both cases correspond to inverse elements, and one can choose the most convenient representation. We will use indifferently the same name for the parameters to avoid introducing minus signs and inverse.

Remark 2.2 (Topology and Gauge Choices) While it is always possible to adopt locally the flat gauge (2.18), it may not be possible to extend it globally. Indeed, Riemann surfaces are curved (with the sign of the curvature given by the sign of $1 - g$), so it can be described by a flat metric only locally.

The final step is to write an action $S_m[g, \Psi]$ for the matter fields. According to the previous discussion, it must have the following properties:

- local in the fields;
- renormalizable;
- non-linear sigma models for the scalar fields;
- periodicity conditions;
- invariant under diffeomorphisms (2.5);
- invariant under Weyl transformations (2.10).

The latter two conditions are summarized by

$$S_m[f^*g, f^*\Psi] = S_m[g, \Psi], \qquad S_m[e^{2\omega}g, \Psi] = S_m[g, \Psi]. \tag{2.19}$$

The invariance under diffeomorphisms is straightforward to enforce by using only covariant objects. Since the scalar fields represent embedding of the string in spacetime, the non-linear sigma model condition means that spacetime is identified with the target space of the sigma model, of which D dimensions are non-compact, and the spacetime metric appears in the matter action as in (2.3). The isometries of the target manifold metric become global symmetries of S_m: while they are not needed in this chapter, they will have their importances in other chapters. Finally, to make the action consistent with the topology of the worldsheet, the fields must satisfy appropriate boundary conditions. For example, the scalar fields X^μ must be periodic for the closed string:

$$X^\mu(\tau, \sigma) \sim X^\mu(\tau, \sigma + 2\pi). \tag{2.20}$$

Remark 2.3 (2d Gravity) The setup in two-dimensional gravity is exactly similar, except that the system is, in general, not invariant under Weyl transformations. As a consequence, one component of the metric (usually taken to be the Liouville mode) remains unconstrained: in the conformal gauge, (2.16) only $\hat{g}$ is fixed.

The symmetries (2.19) of the action have an important consequence: they imply that the matter action is conformally invariant on flat space $g_{ab} = \delta_{ab}$. A two-dimensional conformal field theory (CFT) is characterized by a *central charge c_m*: roughly, it is a measure of the quantum degrees of freedom. The central charge is additive for decoupled sectors. In particular, the scalar fields X^μ contribute as D, and it is useful to define the perpendicular CFT with central charge $c_\perp$ as the matter which does not describe the non-compact dimensions:

$$c_m = D + c_\perp. \tag{2.21}$$

This will be discussed in length in Part I. For this chapter and most of the book, it is sufficient to know that the matter is a CFT of central charge c_m and includes D scalar fields X^μ:

$$\text{matter CFT parameters: } D, c_m. \tag{2.22}$$

The energy–momentum is defined by

$$T_{m,ab} := -\frac{4\pi}{\sqrt{g}} \frac{\delta S_m}{\delta g^{ab}}. \tag{2.23}$$

The variation of the action under the transformations (2.7) vanishes on-shell if the energy–momentum tensor is conserved

$$\nabla^a T_{m,ab} = 0 \qquad \text{(on-shell)}. \tag{2.24}$$

On the other hand, the variation under (2.11) vanishes off-shell (i.e. without using the equations of motion) if the energy–momentum tensor is traceless:

$$g^{ab} T_{m,ab} = 0 \qquad \text{(off-shell)}. \tag{2.25}$$

The conserved charges associated to the energy–momentum tensor generate world-sheet translations

$$P^a := \int d\sigma \, T_m^{0a}. \tag{2.26}$$

The first component is identified with the worldsheet Hamiltonian $P^0 = H$ and generates time translations; the second component generates spatial translations.

Remark 2.4 (Tracelessness of the Energy–Momentum Tensor) In fact, the trace can also be proportional to the curvature

$$g^{ab} T_{m,ab} \propto R. \tag{2.27}$$

Then, the equations of motion are invariant since the integral of R is topological. The theory is invariant even if the action is not. Importantly, this happens for fields at the quantum level (Weyl anomaly), for the Weyl ghost field (Sect. 2.4) and for the Liouville theory (two-dimensional gravity coupled to conformal matter).

2.2   Path Integral

The quantization of the system is achieved by considering the path integral, which yields the genus-g vacuum amplitude (or partition function):

$$Z_g := \int \frac{d_g g_{ab}}{\Omega_{\text{gauge}}[g]} \, Z_m[g], \qquad Z_m[g] := \int d_g \Psi \, e^{-S_m[g,\Psi]} \tag{2.28}$$

at fixed genus g (not to be confused with the metric). The integration over g_{ab} is performed over all metrics of the genus-g Riemann surface Σ_g: $g_{ab} \in \text{Met}(\Sigma_g)$. The factor $\Omega_{\text{gauge}}[g]$ is a normalization inserted in order to make the integral finite: it depends on the metric (but only through the moduli parameters, as we will show later) [11, p. 931], which explains why it is included after the integral sign. Its value will be determined in the next section by requiring the cancellation of the infinities due to the integration over the gauge parameters. This partition function corresponds to the g-loop vacuum amplitude: interactions and their associated scattering amplitudes are discussed in Sect. 3.1.

In order to perform the gauge fixing and to manipulate the path integral (2.28), it is necessary to define the integration measure over the fields. Because the space is infinite-dimensional, this is a difficult task. One possibility is to define the measure implicitly through Gaussian integration over the field tangent space (see also Appendix C.1). A Gaussian integral involves a quadratic form, that is, an inner product (or equivalently a metric) on the field space. The explanation is that a metric also defines a volume form, and thus a measure. To reduce the freedom in the definition of the inner product, it is useful to introduce three natural assumptions:

1. *ultralocality:* the measure is invariant under reparametrizations and defined point-wise, which implies that it can depend on the fields but not on their derivatives;
2. *invariant* measure: the measure for the matter transforms trivially under any symmetry of the matter theory by contracting indices with appropriate tensors;
3. *free-field* measure: for fields other than the worldsheet metric and matter (like ghosts, Killing vectors, etc.), the measure is the one of a free field.

This means that the inner product is obtained by contracting the worldsheet indices of the fields with a tensor built only from the worldsheet metric, by contracting other indices (like spacetime) with some invariant tensor (like the spacetime metric), and finally by integrating over the worldsheet.

We need to distinguish the matter fields from those appearing in the gauge fixing procedure. The matter fields live in the representation of some group under which the inner product is invariant: this means that it is not possible to define each field measure independently if the exponential of inner products does not factorize. As an example, on a curved background: $dX \neq \prod_\mu dX^\mu$. However, we will not need to write explicitly the partition function for performing the gauge fixing: it is sufficient to know that the matter is a CFT. In the gauge fixing procedure, different types of fields (including the metric) appear which do not carry indices (beyond the worldsheet indices). Below, we focus on defining a measure for each of those single fields (and use free-field measures according to the third condition).

Considering the finite elements $\delta\Phi_1$ and $\delta\Phi_2$ of tangent space at the point Φ of the state of fields, the inner product $(\cdot, \cdot)_g$ and its associated norm $|\cdot|_g$ read

$$(\delta\Phi_1, \delta\Phi_2)_g := \int d^2\sigma \sqrt{g}\, \gamma_g(\delta\Phi_1, \delta\Phi_2), \qquad |\delta\Phi|_g^2 := (\delta\Phi, \delta\Phi)_g, \qquad (2.29)$$

where γ_g is a metric on the $\delta\Phi$ space. It is taken to be flat for all fields except the metric itself, that is, independent of Φ. The dependence in the metric ensures that the inner product is diffeomorphism invariant, which in turn will lead to a metric-dependent but diffeomorphism invariant measure. The functional measure is then normalized by a Gaussian integral:

$$\int d_g \delta\Phi \, e^{-\frac{1}{2}(\delta\Phi,\delta\Phi)_g} = \frac{1}{\sqrt{\det \gamma_g}}. \tag{2.30}$$

This, in turn, induces a measure on the field space itself:

$$\int d\Phi \sqrt{\det \gamma_g}. \tag{2.31}$$

The determinant can be absorbed in the measure, such that

$$\int d_g \delta\Phi \, e^{-\frac{1}{2}(\delta\Phi,\delta\Phi)_g} = 1. \tag{2.32}$$

In fact, this normalization and the definition of the inner product are ambiguous, but the ultralocality condition allows to fix uniquely the final result (Sect. 2.3.4). Moreover, such a free-field measure is invariant under field translations

$$\Phi(\sigma) \longrightarrow \Phi'(\sigma) = \Phi(\sigma) + \varepsilon(\sigma). \tag{2.33}$$

The most natural inner products for single scalar, vector and symmetric tensor fields are

$$(\delta f, \delta f)_g := \int d^2\sigma \sqrt{g} \, \delta f^2 \tag{2.34a}$$

$$(\delta V^a, \delta V^a)_g := \int d^2\sigma \sqrt{g} \, g_{ab} \delta V^a \delta V^b, \tag{2.34b}$$

$$(\delta T_{ab}, \delta T_{ab})_g := \int d^2\sigma \sqrt{g} \, G^{abcd} \delta T_{ab} \delta T_{cd}, \tag{2.34c}$$

where the (DeWitt) metric for the symmetric tensor is

$$G^{abcd} := G_\perp^{abcd} + u \, g^{ab} g^{cd}, \qquad G_\perp^{abcd} := g^{ac} g^{bd} + g^{ad} g^{bc} - g^{ab} g^{cd}, \tag{2.35}$$

with u a constant. The first term $G_\perp$ is the projector on the traceless component of the tensor. Indeed, consider a traceless tensor $g^{ab} T_{ab} = 0$ and a pure trace tensor Λg_{ab}, then we have

$$G^{abcd} T_{cd} = G_\perp^{abcd} T_{cd} = 2T_{ab}, \qquad G^{abcd} (\Lambda g_{cd}) = 2u \, (\Lambda g_{ab}). \tag{2.36}$$

While all measures are invariant under diffeomorphisms, only the vector measure is invariant under Weyl transformations. This implies the existence of a quantum anomaly (the *Weyl* or *conformal anomaly*): the classical symmetry is broken by quantum effects because the path integral measure cannot respect all the classical symmetries. Hence, one can expect difficulties for imposing it at the quantum level and ensuring that the Liouville mode in (2.16) remains without dynamics.

The metric variation (symmetric tensor) is decomposed in its trace and traceless parts

$$\delta g_{ab} = g_{ab}\,\delta\Lambda + \delta g_{ab}^{\perp}, \qquad \delta\Lambda = \frac{1}{2}\,g^{ab}\delta g_{ab}, \qquad g^{ab}\delta g_{ab}^{\perp} = 0. \tag{2.37}$$

In this decomposition, both terms are decoupled in the inner product

$$|\delta g_{ab}|_g^2 = 4u|\delta\Lambda|_g^2 + |\delta g_{\mu\nu}^{\perp}|_g^2, \tag{2.38}$$

where the norm of $\delta\Lambda$ is the one of a scalar field (2.34a). The norm for $\delta g_{ab}^{\perp}$ is equivalent to (2.34c) with $u = 0$ (since it is traceless). Requiring positivity of the inner product for a non-traceless tensor imposes the following constraint on u:

$$u > 0. \tag{2.39}$$

One can absorb the coefficient with u in $\delta\Lambda$, which will just contribute as an overall factor: its precise value has no physical meaning. The simple choice $u = 1/4$ sets the coefficient of $|\delta\Lambda|_g^2$ to 1 in (2.38) (another common choice is $u = 1/2$). Ultimately, this implies that the measure factorizes as

$$d_g\,g_{ab} = d_g\Lambda\,d_g\,g_{ab}^{\perp}. \tag{2.40}$$

Computation: Equation (2.38)

$$
\begin{aligned}
G^{abcd}\,\delta g_{ab}\delta g_{cd} &= \left(G_{\perp}^{abcd} + u\,g^{ab}g^{cd}\right)\left(g_{ab}\,\delta\Lambda + \delta g_{ab}^{\perp}\right)\left(g_{cd}\,\delta\Lambda + \delta g_{cd}^{\perp}\right) \\
&= \left(2u\,g^{cd}\,\delta\Lambda + G_{\perp}^{abcd}\delta g_{ab}^{\perp}\right)\left(g_{cd}\,\delta\Lambda + \delta g_{cd}^{\perp}\right) \\
&= 4u\,(\delta\Lambda)^2 + G_{\perp}^{abcd}\delta g_{ab}^{\perp}\delta g_{cd}^{\perp} \\
&= 4u\,\delta\Lambda^2 + 2g^{ac}\,g^{bd}\delta g_{ab}^{\perp}\delta g_{cd}^{\perp}.
\end{aligned}
$$

Remark 2.5 Another common parametrization is

$$G^{abcd} = g^{ac}g^{bd} + c\,g^{ab}g^{cd}. \tag{2.41}$$

It corresponds to (2.35) up to a factor $1/2$ and setting $u = 1 + 2c$.

Remark 2.6 (Matter and Curved Background Measures) As explained previously, matter fields carry a representation and the inner product must yield an invariant combination. In particular, spacetime indices must be contracted with the spacetime metric $G_{\mu\nu}(X)$ (which is the non-linear sigma model metric appearing in front of the kinetic term) for a general curved background. For example, the inner product for the scalar fields X^μ is

$$(\delta X^\mu, \delta X^\mu)_g = \int \mathrm{d}^2\sigma \sqrt{g}\, G_{\mu\nu}(X)\delta X^\mu \delta X^\nu. \tag{2.42}$$

It is not possible to normalize anymore the measure to set $\det G(X) = 1$ like in (2.32) since it depends on the fields. On the other hand, this factor is not important for the manipulations performed in this chapter. Any ambiguity in the measure will again correspond to a renormalization of the cosmological constant [11, p. 923]. Moreover, as explained above, it is not necessary to write explicitly the matter partition function as long as it describes a CFT.

2.3 Faddeev–Popov Gauge Fixing

The naive integration over the space $\mathrm{Met}(\Sigma_g)$ of all metrics of Σ_g (note that the genus is fixed) leads to a divergence of the functional integral since equivalent configurations

$$(f^*g, f^*\Psi) \sim (g, \Psi), \qquad (\mathrm{e}^{2\omega}g, \Psi) \sim (g, \Psi) \tag{2.43}$$

give the same contribution to the integral. This infinite redundancy causes the integral to diverge, and since the multiple counting is generated by the gauge group, the infinite contribution corresponds to the volume of the latter. The Faddeev–Popov procedure is a means to extract this volume by separating the integration over the gauge and physical degrees of freedom

$$\mathrm{d(fields)} = \mathrm{Jacobian} \times \mathrm{d(gauge)} \times \mathrm{d(physical)}. \tag{2.44}$$

The space of fields (g, Ψ) is divided into equivalence classes and one integrates over only one representative of each class (gauge slice), see Fig. 2.1. This change of variables introduces a Jacobian which can be represented by a partition function with ghost fields (fields with a wrong statistics). This program encounters some complications since G is a semi-direct product and is non-connected.

Example 2.1: Gauge Redundancy

A finite-dimensional integral which mimics the problem is

$$Z = \int_{\mathbb{R}^2} \mathrm{d}x\, \mathrm{d}y\, \mathrm{e}^{-(x-y)^2}. \tag{2.45}$$

Fig. 2.1 The space of metrics decomposed in gauge orbits. Two metrics related by a gauge transformation lie on the same orbit. Choosing a gauge slice amounts to pick one metric in each orbit, and the projection gives the space of metric classes

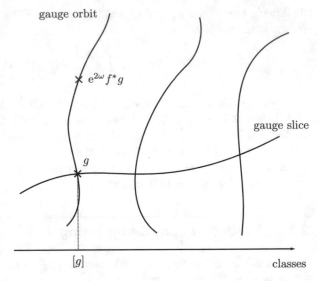

One can perform the change of variables

$$r = x - y, \qquad y = a \tag{2.46}$$

such that

$$Z = \int_{\mathbb{R}} da \int_0^\infty e^{-r^2} = \frac{\sqrt{\pi}}{2} \, \mathrm{Vol}(\mathbb{R}), \tag{2.47}$$

and $\mathrm{Vol}(\mathbb{R})$ is to be interpreted as the volume of the gauge group (translation by a real number a). ◄

Remark 2.7 Mathematically, the Faddeev–Popov procedure consists in identifying the orbits (class of equivalent metrics) under the gauge group G and to write the integral in terms of G-invariant objects (orbits instead of individual metrics). This can be done by decomposing the tangent space into variations generated by G and its complement. Then, one can define a foliation of the field space which equips it with a fibre bundle structure: the base is the push-forward of the complement and the fibre corresponds to the gauge orbits. The integral is then defined by selecting a section of this bundle.

2.3.1 Metrics on Riemann Surfaces

According to the above procedure, each metric $g_{ab} \in \mathrm{Met}(\Sigma_g)$ has to be expressed in terms of gauge parameters (ξ and ω) and of a metric $\hat{g}_{ab}$ which contains the remaining gauge-independent degrees of freedom. As there are as many gauge parameters as metric components (Sect. 2.1), one could expect that there are no remaining physical parameters and then that $\hat{g}$ is totally fixed. But, this is not the case and the metric $\hat{g}$ depends on a finite number of parameters t_i (moduli). The reason for this is topological: while locally it is always possible to completely fix the metric, topological obstructions may prevent doing it globally. This means that not all conformal classes in (2.12) can be (globally) related by a diffeomorphism.

The quotient of the space of metrics by gauge transformations is called the *moduli space*

$$\mathcal{M}_g := \frac{\text{Met}(\Sigma_g)}{G}. \tag{2.48}$$

Accordingly, its coordinates t_i with $i = 1, \ldots, \dim_{\mathbb{R}} \mathcal{M}_g$ are called moduli parameters. The *Teichmüller space* $\mathcal{T}_g$ is obtained by taking the quotient of $\text{Met}(\Sigma_g)$ with the component connected to the identity

$$\mathcal{T}_g := \frac{\text{Met}(\Sigma_g)}{G_0}. \tag{2.49}$$

The space $\mathcal{T}_g$ is the covering space of $\mathcal{M}_g$:

$$\mathcal{M}_g = \frac{\mathcal{T}_g}{\Gamma_g}, \tag{2.50}$$

where Γ_g is the modular group defined in (2.9). Both spaces can be endowed with a complex structure and are finite-dimensional [26]:

$$M_g := \dim_{\mathbb{R}} \mathcal{M}_g = \dim_{\mathbb{R}} \mathcal{T}_g = \begin{cases} 0 & g = 0, \\ 2 & g = 1, \\ 6g - 6 & g \geq 2, \end{cases} \tag{2.51}$$

In particular, their volumes are related by

$$\int_{\mathcal{M}_g} \mathrm{d}^{M_g} t = \frac{1}{\Omega_{\Gamma_g}} \int_{\mathcal{T}_g} \mathrm{d}^{M_g} t, \tag{2.52}$$

where Ω_{Γ_g} is the volume of Γ_g.

We will need to extract volumes of different groups, so it is useful to explain how they are defined. A natural measure on a connected group G is the Haar measure $\mathrm{d}g$, which is the unique left-invariant measure on G. Integrating the measure gives the volume of the group

$$\Omega_G := \int_G \mathrm{d}g = \int_G \mathrm{d}(hg), \tag{2.53}$$

for any $h \in G$. Given the Lie algebra $\mathfrak{g}$ of the group, a general element of the algebra is a linear combination of the generators T_i with coefficients α^i

$$\alpha = \alpha^i T_i. \tag{2.54}$$

Group elements can be parametrized in terms of α through the exponential map. Moreover, since a Lie group is a manifold, it is locally isomorphic to $\mathbb{R}^n$: this motivates the use of a flat metric for the Lie algebra, such that

$$\Omega_G = \int d\alpha := \int \prod_i d\alpha^i. \tag{2.55}$$

Finally, it is possible to perform a change of coordinates from the Lie parameters to coordinates x on the group: the resulting Jacobian is the Haar measure for the coordinates x.

Remark 2.8 While $\mathcal{T}_g$ is a manifold, this is not the case of $\mathcal{M}_g$ for $g \geq 2$, which is an orbifold: the quotient by the modular group introduces singularities [27].

Remark 2.9 (Moduli Space and Fundamental Domain) Given a group acting on a space, a fundamental domain for a group is a subspace such that the full space is generated by acting with the group on the fundamental domain. Hence, one can view the moduli space $\mathcal{M}_g$ as a fundamental domain (sometimes denoted by $\mathcal{F}_g$) for the group Γ_g and the space $\mathcal{T}_g$.

In the conformal gauge (2.16), the metric g_{ab} can be parametrized by

$$g_{ab} = \hat{g}_{ab}^{(f,\phi)}(t) := e^{2f^*\phi} f^* \hat{g}_{ab}(t) = f^*\big(e^{2\phi} \hat{g}_{ab}(t)\big), \tag{2.56}$$

where $\phi := \omega$ and t denotes the dependence in the moduli parameters. To avoid surcharging the notations, we will continue to write g when there is no ambiguity. In coordinates, this is equivalent to

$$g_{ab}(\sigma) = \hat{g}_{ab}^{(f,\phi)}(\sigma; t) := e^{2\phi(\sigma)} \hat{g}'_{ab}(\sigma; t), \qquad \hat{g}'_{ab}(\sigma; t) = \frac{\partial \sigma'^c}{\partial \sigma^a} \frac{\partial \sigma'^d}{\partial \sigma^b} \hat{g}_{cd}(\sigma'; t). \tag{2.57}$$

Remark 2.10 Strictly speaking, the matter fields also transform and one should write $\Psi = \Psi^{(f)} := f^*\hat{\Psi}$ and include them in the change of integration measures of the following sections. But, this does not bring any particular benefits since these changes are trivial because the matter is decoupled from the metric.

Remark 2.11 Although the metric cannot be completely gauge fixed, having just a finite-dimensional integral is much simpler than a functional integral. In higher dimensions, the gauge fixing does not reduce that much the degrees of freedom and a functional integral over $\hat{g}$ remains (in similarity with Yang–Mills theories).

The corresponding infinitesimal transformations are parametrized by $(\phi, \xi, \delta t_i)$. The variation of the metric (2.56) can be expressed as

$$\delta g_{ab} = 2\phi \, g_{ab} + \nabla_a \xi_b + \nabla_b \xi_a + \delta t_i \, \partial_i \, g_{ab}, \tag{2.58}$$

which is decomposed in a reparametrization (2.7), a Weyl rescaling (2.11), and a contribution from the variations of the moduli parameters. The latter are called Teichmüller deformations and describe changes in the metric which cannot be written as a combination of diffeomorphism and Weyl transformation. Only the last term is written with a delta because the parameters ξ and ϕ are already infinitesimal. There is an implicit sum over i and we have defined

$$\partial_i := \frac{\partial}{\partial t_i}. \tag{2.59}$$

According to the formula (2.55), the volumes $\Omega_{\text{Diff}_0}[g]$ and $\Omega_{\text{Weyl}}[g]$ of the diffeomorphisms connected to the identity and Weyl group are

$$\Omega_{\text{Diff}_0}[g] :- \int d_g \xi, \tag{2.60a}$$

$$\Omega_{\text{Weyl}}[g] := \int d_g \phi. \tag{2.60b}$$

The full diffeomorphism group has one connected component for each element of the modular group Γ_g, according to (2.9): the volume $\Omega_{\text{Diff}}[g]$ of the full group is the volume of the component connected to the identity times the volume Ω_{Γ_g}

$$\Omega_{\text{Diff}}[g] = \Omega_{\text{Diff}_0}[g] \, \Omega_{\Gamma_g}. \tag{2.60c}$$

We have written that the volume depends on g: but, the metric itself is parametrized in terms of the integration variables, and thus the LHS of (2.60) cannot depend on the variable which is integrated over: Ω_{Diff_0} can depend only on ϕ and Ω_{Weyl} only on ξ. But, all measures (2.34b) are invariant under diffeomorphisms, and thus the result cannot depend on ξ. Moreover, the measure for vector is invariant under Weyl transformation, which means that Ω_{Diff_0} does not depend on ϕ. This implies that the volumes depend only on the moduli parameters

$$\Omega_{\text{Diff}_0}[g] := \Omega_{\text{Diff}_0}[e^{2\phi}\hat{g}] = \Omega_{\text{Diff}_0}[\hat{g}], \qquad \Omega_{\text{Weyl}}[g] := \Omega_{\text{Weyl}}[\mathcal{L}_\xi \hat{g}] = \Omega_{\text{Weyl}}[\hat{g}]. \tag{2.61a}$$

For this reason, it is also sufficient to take the normalization factor Ω_{gauge} to have the same dependence:

$$\Omega_{\text{gauge}}[g] := \Omega_{\text{gauge}}[\hat{g}]. \tag{2.61b}$$

These volumes are also discussed in Sect. 2.3.4.

Computation: Equation (2.61a)

$$\Omega_{\mathrm{Diff}_0}[\mathrm{e}^{2\phi}\hat{g}] = \int \mathrm{d}_{\mathrm{e}^{2\phi}\mathcal{L}_\xi\hat{g}}\,\xi = \int \mathrm{d}_{\mathrm{e}^{2\phi}\hat{g}}\,\xi = \int \mathrm{d}_{\hat{g}}\,\xi = \Omega_{\mathrm{Diff}_0}[\hat{g}],$$

$$\Omega_{\mathrm{Weyl}}[\mathcal{L}_\xi\hat{g}] = \int \mathrm{d}_{\mathrm{e}^{2\phi}\mathcal{L}_\xi\hat{g}}\,\phi = \int \mathrm{d}_{\mathrm{e}^{2\phi}\hat{g}}\,\phi = \Omega_{\mathrm{Weyl}}[\hat{g}].$$

Remark 2.12 (Free-Field Measure for the Liouville Mode) The explicit measure (2.60b) of the Liouville mode is complicated since the inner product contains an exponential of the field:

$$|\delta\phi|^2 = \int \mathrm{d}^2\sigma \sqrt{g}\,\delta\phi^2 = \int \mathrm{d}^2\sigma \sqrt{\hat{g}}\,\mathrm{e}^{2\phi}\delta\phi^2. \tag{2.62}$$

It has been proposed by David–Distler–Kawai [5, 13], and later checked explicitly [8, 9, 24], how to rewrite the measure in terms of a free measure weighted by an effective action. The latter is identified with the Liouville action (Sect. 2.3.3).

In principle, we could follow the standard Faddeev–Popov procedure by inserting a delta function for the gauge fixing condition

$$F_{ab} := g_{ab} - \hat{g}_{ab}^{(f,\phi)}(t), \tag{2.63}$$

with $\hat{g}_{ab}^{(f,\phi)}(t)$ defined in (2.56). However, we will take a detour to take the opportunity to study in detail manipulations of path integrals and to understand several aspects of the geometry of Riemann surfaces. In any case, several points are necessary even when going the short way, but less apparent.

In order to make use of the factorization (2.40) of the integration measure, the variation (2.58) is decomposed into its trace (first term) and traceless parts (last two terms) (2.37)

$$\delta g_{ab} = 2\tilde{\Lambda}\,g_{ab} + (P_1\xi)_{ab} + \delta t_i\,\mu_{iab}, \tag{2.64}$$

where[4]

$$(P_1\xi)_{ab} = \nabla_a\xi_b + \nabla_b\xi_a - g_{ab}\nabla_c\xi^c, \tag{2.65a}$$

$$\mu_{iab} = \partial_i g_{ab} - \frac{1}{2}\,g_{ab}\,g^{cd}\partial_i g_{cd}, \tag{2.65b}$$

$$\tilde{\Lambda} = \Lambda + \frac{1}{2}\,\delta t_i\,g^{ab}\partial_i g_{ab}, \qquad \Lambda = \phi + \frac{1}{2}\,\nabla_c\xi^c. \tag{2.65c}$$

---

[4]For comparison, Polchinski [29] defines P_1 with an overall factor $1/2$.

The objects μ_i are called *Beltrami differentials* and correspond to traceless Teichmüller deformations (the factor of $1/2$ comes from the symmetrization of the metric
indices). The decomposition emphasizes which variations are independent from
each other. In particular, changes to the trace of the metric due to a diffeomorphism
generated by ξ or a modification of the moduli parameters can be compensated by
a Weyl rescaling.

One can use (2.40) to replace the integration over g_{ab} by one over the gauge
parameters ξ and ϕ and over the moduli t_i since they contain all the information
about the metric:

$$Z_g = \int d^{M_g} t \, d_g \tilde{\Lambda} \, d_g (P_1 \xi) \, \Omega_{\text{gauge}}[g]^{-1} \, Z_m[g]. \tag{2.66}$$

It is tempting to perform the change of variables

$$(P_1 \xi, \tilde{\Lambda}) \longrightarrow (\xi, \phi) \tag{2.67}$$

such that

$$d_g (P_1 \xi) \, d_g \tilde{\Lambda} \stackrel{?}{=} d_g \xi \, d_g \phi \, \Delta_{\text{FP}}[g], \tag{2.68}$$

where $\Delta_{\text{FP}}[g]$ is the Jacobian of the transformation

$$\Delta_{\text{FP}}[g] = \det \frac{\partial (P_1 \xi, \tilde{\Lambda})}{\partial (\xi, \phi)} = \det \begin{pmatrix} P_1 & 0 \\ \star & 1 \end{pmatrix} = \det P_1. \tag{2.69}$$

But, one needs to be more careful:

1. The variations involving $P_1 \xi$ and δt_i are not orthogonal and, as a consequence,
   the measure does not factorize.
2. P_1 has zero-modes, i.e. vectors such that $P_1 \xi = 0$, which causes the determinant
   to vanish, $\det P_1 = 0$.

A rigorous analysis will be performed in Sect. 2.3.2 and will lead to additional
factors in the path integral.

Next, if the actions and measures were invariant under diffeomorphisms and
Weyl transformations (which amounts to replace g by $\hat{g}$ everywhere), it would be
possible to factor out the integrations over the gauge parameters and to cancel the
corresponding infinite factors thanks to the normalization $\Omega_{\text{gauge}}[g]$. A new problem
arises because the measures are not Weyl invariant as explained above and one
should be careful when replacing the metric (Sect. 2.3.3).

2.3.2   Reparametrizations and Analysis of P_1

The properties of the operator P_1 are responsible for both problems preventing a direct factorization of the measure; for this reason, it is useful to study it in more detail.

The operator P_1 is an object which takes a vector v to a symmetric traceless 2-tensor T, see (2.65a). Conversely, its adjoint $P_1^\dagger$ can be defined from the scalar product (2.34c)

$$(T, P_1 v)_g = (P_1^\dagger T, v)_g, \tag{2.70}$$

and takes symmetric traceless tensors to vectors. In components, one finds

$$(P_1^\dagger T)_a = -2\nabla^b T_{ab}. \tag{2.71}$$

The Riemann–Roch theorem relates the dimension of the kernels of both operators [26]:

$$\dim \ker P_1^\dagger - \dim \ker P_1 = -3\chi_g = 6g - 6. \tag{2.72}$$

Teichmüller Deformations
We first need to characterize Teichmüller deformations, the variations of moduli parameters which lead to transformations of the metric independent from diffeomorphisms and Weyl rescalings. This means that the different variations must be orthogonal for the inner product (2.34).

First, the deformations must be traceless, otherwise they can be compensated by a Weyl transformation. The traceless metric variations δg which cannot be generated by a vector field ξ are perpendicular to $P_1\xi$ (otherwise, the former would be a linear combination of the latter):

$$(\delta g, P_1 \xi)_g = 0 \quad \Longrightarrow \quad (P_1^\dagger \delta g, \xi)_g = 0. \tag{2.73}$$

Since ξ is arbitrary, this means that the first argument vanishes

$$P_1^\dagger \delta g = 0. \tag{2.74}$$

Metric variations induced by a change in the moduli t_i are in the kernel of $P_1^\dagger$

$$\delta g \in \ker P_1^\dagger. \tag{2.75}$$

Elements of $\ker P_1^\dagger$ are called *quadratic differentials* and a basis (not necessarily orthonormal) of $\ker P_1^\dagger$ is denoted as

$$\ker P_1^\dagger = \text{Span}\{\phi_i\}, \qquad i = 1, \ldots, \dim \ker P_1^\dagger \tag{2.76}$$

(these should not be confused with the Liouville field). The dimension of $\ker P_1^\dagger$ is in fact equal to the dimension of the moduli space (2.51):

$$\dim_{\mathbb{R}} \ker P_1^\dagger = M_g = \begin{cases} 0 & g = 0, \\ 2 & g = 1, \\ 6g - 6 & g > 1. \end{cases} \tag{2.77}$$

The last two terms in the variation (2.64) of δg_{ab} are not orthogonal. Let us introduce the projector on the complement space of $\ker P_1^\dagger$

$$\Pi := P_1 \frac{1}{P_1^\dagger P_1} P_1^\dagger. \tag{2.78}$$

The moduli variations can then be rewritten as

$$\delta t_i \, \mu_i = \delta t_i \, (1 - \Pi)\mu_i + \delta t_i \, \Pi \mu_i = \delta t_i \, (1 - \Pi)\mu_i + \delta t_i \, P_1 \zeta_i. \tag{2.79}$$

The ζ_i exist because $\Pi \mu_i \in \operatorname{Im} P_1$, and they read

$$\zeta_i := \frac{1}{P_1^\dagger P_1} P_1^\dagger \mu_i. \tag{2.80}$$

The first term can be decomposed on the quadratic differential basis (2.76)

$$(1 - \Pi)\mu_i = \phi_j (M^{-1})_{jk} (\phi_k, \mu_i)_g, \tag{2.81}$$

where

$$M_{ij} := (\phi_i, \phi_j)_g. \tag{2.82}$$

Ultimately, the variation (2.64) becomes

$$\delta g_{ab} = (P_1 \tilde{\xi})_{ab} + 2\tilde{\Lambda} \, g_{ab} + Q_{iab} \, \delta t_i. \tag{2.83}$$

where

$$\tilde{\xi} = \xi + \zeta_i \delta t_i, \qquad Q_{iab} = \phi_{jab} \, (M^{-1})_{jk} (\phi_k, \mu_i)_g. \tag{2.84}$$

Correspondingly, the norm of the variation splits in three terms since each variation is orthogonal to the others:

$$|\delta g|_g^2 = |\delta \tilde{\Lambda}|_g^2 + |P_1 \tilde{\xi}|_g^2 + |Q_i \delta t_i|_g^2. \tag{2.85}$$

Since the norm is decomposed as a sum, the measure factorizes:

$$d_g g_{ab} = d_g \tilde{\Lambda} \, d_g (P_1 \tilde{\xi}) \, d_g (Q_i \delta t_i). \tag{2.86}$$

One can then perform a change of coordinates

$$(\tilde{\xi}, \tilde{\Lambda}, Q_i \delta t_i) \longrightarrow (\xi, \Lambda, \delta t_i), \tag{2.87}$$

where Λ was defined in (2.65c). The goal of this transformation is to remove the dependence in the moduli from the measures on the Weyl factor and vector fields, and to recover a finite-dimensional integral over the moduli:

$$d_g \tilde{\Lambda} \, d_g (P_1 \tilde{\xi}) \, d_g (Q_i \delta t_i) = d^{M_g} t \, d_g \Lambda \, d_g (P_1 \xi) \, \frac{\det(\phi_i, \mu_j)_g}{\sqrt{\det(\phi_i, \phi_j)_g}}, \tag{2.88}$$

where the determinants correspond to the Jacobian. The role of the determinant in the denominator is to ensure a correct normalization when the basis is not orthonormal (in particular, it ensures that the Jacobian is independent of the basis). Plugging this result in (2.28) gives the partition function as

$$Z_g = \int_{\mathcal{T}_g} d^{M_g} t \, \frac{1}{\Omega_{\text{gauge}}[\hat{g}]} \int d_g \Lambda \, d_g (P_1 \xi) \, \frac{\det(\phi_i, \mu_j)_g}{\sqrt{\det(\phi_i, \phi_j)_g}} \, Z_m[g]. \tag{2.89}$$

The t_i are integrated over the Teichmüller space $\mathcal{T}_g$ defined by (2.49) because the vectors ξ generate only reparametrizations connected to the identity, and thus the remaining freedom lies in $\text{Met}(\Sigma_g)/G_0$. Next, we study how to perform the changes of variables to remove P_1 from the measure.

Conformal Killing Vectors

In this section, we focus on the $d_g \Lambda \, d_g (P_1 \xi)$ part of the measure and we make contact with the rest at the end.

Infinitesimal reparametrizations generated by a vector field ξ^a produce only transformations close to the identity. For this reason, integrating over all possible vector fields yields the volume (2.60a) of the component of the diffeomorphism group connected to the identity:

$$\int d_g \xi = \Omega_{\text{Diff}_0}[\hat{g}]. \tag{2.90}$$

Remember that the volume depends only on the moduli, but obviously not on ξ (integrated over) nor ϕ (the inner product (2.34b) is invariant). But, due to the existence of zero-modes, one gets an integration over a subset of all vector fields, and this complicates the program, as we discuss now.

Zero-modes $\xi^{(0)}$ of P_1 are called *conformal Killing vectors* (CKV)

$$\xi^{(0)} \in \mathcal{K}_g := \ker P_1 \tag{2.91}$$

and satisfy the conformal Killing equation (see also Sect. 5.1):

$$(P_1 \xi^{(0)})_{ab} = \nabla_a \xi_b^{(0)} + \nabla_b \xi_a^{(0)} - g_{ab} \nabla_c \xi^{(0)c} = 0. \tag{2.92}$$

CKVs correspond to reparametrizations which can be absorbed by a change of the conformal factor. They should be removed from the ξ integration in order to not double-count the corresponding metrics. The dimension of the zero-modes CKV space depends on the genus [26]:

$$\mathsf{K}_g := \dim_{\mathbb{R}} \mathcal{K}_g = \dim_{\mathbb{R}} \ker P_1 = \begin{cases} 6 & g = 0, \\ 2 & g = 1, \\ 0 & g > 1. \end{cases} \tag{2.93}$$

The associated transformations will be interpreted later (Chap. 5). The groups generated by the CKVs are

$$g = 0: \quad \mathcal{K}_0 - \mathrm{SL}(2, \mathbb{C}), \qquad g = 1: \quad \mathcal{K}_1 = \mathrm{U}(1) \times \mathrm{U}(1). \tag{2.94}$$

Note that the first group is non-compact while the second is compact.

A general vector ξ can be separated into a zero-mode part and its orthogonal complement ξ':

$$\xi = \xi^{(0)} + \xi', \tag{2.95}$$

such that

$$(\xi^{(0)}, \xi')_g = 0 \tag{2.96}$$

for the inner product (2.34b). Because zero-modes are annihilated by P_1, the correct change of variables in the partition function (2.66) maps to ξ' only:

$$(P_1 \xi, \Lambda) \longrightarrow (\xi', \phi). \tag{2.97}$$

Integrating over ξ at this stage would double-count the CKV (since they are already described by the ϕ integration). The appropriate Jacobian reads

$$d_g \wedge d_g (P_1 \xi) = d_g \phi \, d_g \xi' \, \Delta_{\mathrm{FP}}[g], \tag{2.98}$$

where the Faddeev–Popov determinant is

$$\Delta_{\text{FP}}[g] = \det' \frac{\partial(P_1\xi, \Lambda)}{\partial(\xi', \phi)} = \det' P_1 = \sqrt{\det' P_1 P_1^\dagger}, \tag{2.99}$$

the prime on the determinant indicating that the zero-modes are excluded. This brings the partition function (2.89) to the form

$$Z_g = \int_{\mathcal{T}_g} d^{M_g} t \, \Omega_{\text{gauge}}[\hat{g}]^{-1} \int d_g\phi \, d_g\xi' \, \frac{\det(\phi_i, \mu_j)_g}{\sqrt{\det(\phi_i, \phi_j)_g}} \, \Delta_{\text{FP}}[g] Z_m[g]. \tag{2.100}$$

Computation: Equation (2.98)

The Jacobian can be evaluated directly:

$$\Delta_{\text{FP}}[g] = \det' \frac{\partial(P_1\xi, \Lambda)}{\partial(\xi', \phi)} = \det' \begin{pmatrix} P_1 & 0 \\ \frac{1}{2}\nabla & 1 \end{pmatrix} = \det' P_1. \tag{2.101}$$

As a consequence of $\det' P_1^\dagger = \det' P_1$, the Jacobian can be rewritten as:

$$\sqrt{\det' P_1^\dagger P_1} = \det' P_1. \tag{2.102}$$

It is instructive to derive this result also by manipulating the path integral. Considering small variations of the fields, one has

$$1 = \int d_g\delta\Lambda \, d_g(P_1\delta\xi) \, e^{-|\delta\Lambda|_g^2 - |P_1\delta\xi'|_g^2}$$

$$= \Delta_{\text{FP}}[g] \int d_g\delta\phi \, d_g\delta\xi' \, e^{-|\delta\phi + \frac{1}{2}\nabla_c\delta\xi^c|_g^2 - |P_1\delta\xi'|_g^2}$$

$$= \Delta_{\text{FP}}[g] \int d_g\delta\phi \, d_g\delta\xi' \, e^{-|\delta\phi|_g^2 - (\delta\xi', P_1^\dagger P_1\delta\xi')_g}$$

$$= \Delta_{\text{FP}}[g] \left(\det' P_1^\dagger P_1 \right)^{-1/2}.$$

That the expression is equal to 1 follows from the normalization of symmetric tensors and scalars (2.34) (the measures appearing in the path integral (2.89) arises without any factor). The third equality holds because the measure is invariant under translations of the fields, and we used the definition of the adjoint.

The volume of the group generated by the vectors orthogonal to the CKV is denoted as

$$\Omega'_{\text{Diff}_0}[g] := \Omega'_{\text{Diff}_0}[\hat{g}] = \int d_g\xi'. \qquad (2.103)$$

As explained in the beginning of this section, one should extract the volume of the full Diff_0 group, not only the volume $\Omega'_{\text{Diff}_0}[g]$. Since the two sets of vectors are orthogonal, we can expect the measures, and thus the volumes, to factorize. However, a Jacobian can and does arise: its role is to take into account the normalization of the zero-modes. Denoting by ψ_i a basis (not necessarily orthonormal) for the zero-modes

$$\ker P_1 = \text{Span}\{\psi_i\}, \qquad i = 1, \ldots, K_g, \qquad (2.104)$$

the change of variables

$$\xi' \longrightarrow \xi \qquad (2.105)$$

reads

$$d_g\xi' = \frac{1}{\sqrt{\det(\psi_i, \psi_j)_g}} \frac{d_g\xi}{\Omega_{\text{ckv}}[g]}, \qquad (2.106)$$

where $\Omega_{\text{ckv}}[g]$ is the volume of the CKV group. The determinant is necessary when the basis is not orthonormal. The relation between the gauge volumes is then

$$\Omega_{\text{Diff}_0}[g] = \sqrt{\det(\psi_i, \psi_j)_g} \, \Omega_{\text{ckv}}[g] \, \Omega'_{\text{Diff}_0}[g]. \qquad (2.107)$$

Note that the CKV volume is given in (2.111) and depends only on the topology but not on the metric. By using arguments similar to the ones which lead to (2.61a), one can expect that each term is independently invariant under Weyl rescaling: this is indeed true (Sect. 2.3.3).

Computation: Equation (2.106)
Let us expand $\xi^{(0)}$ on the zero-mode basis

$$\xi^{(0)} = \alpha_i\psi_i, \qquad (2.108)$$

where the α_i are real numbers, such that one can write the changes of variables

$$\xi \longrightarrow (\xi', \alpha_i). \qquad (2.109)$$

The Jacobian is computed from

$$1 = \int d\xi \, e^{-|\xi|_g^2} = J \int d\xi^{(0)} \, d\xi' \, e^{-|\xi'|_g^2 - |\xi^{(0)}|_g^2}$$

$$= J \int \prod_i d\alpha_i \, e^{-\alpha_i \alpha_j (\psi_i, \psi_j)_g} \int d\xi' \, e^{-|\xi'|_g^2}$$

$$= J \left(\det(\psi_i, \psi_j)_g \right)^{-1/2}.$$

Note that the integration over the α_i is a standard finite-dimensional integral. This gives

$$d\xi = \sqrt{\det(\psi_i, \psi_j)_g} \, d\xi' \prod_i d\alpha_i. \qquad (2.110)$$

Since nothing depends on the α_i, they can be integrated over as in (2.53), giving the volume of the CKV group

$$\Omega_{\text{ckv}}[g] = \int \prod_i d\alpha_i. \qquad (2.111)$$

Replacing the integration over ξ' thanks to (2.106), the path integral becomes

$$Z_g = \int_{\mathcal{T}_g} d^{M_g} t \, \Omega_{\text{gauge}}[\hat{g}]^{-1}$$

$$\times \int d_g \phi \, d_g \xi \, \frac{\det(\phi_i, \mu_j)_g}{\sqrt{\det(\phi_i, \phi_j)_g}} \frac{\Omega_{\text{ckv}}[g]^{-1}}{\sqrt{\det(\psi_i, \psi_j)_g}} \Delta_{\text{FP}}[g] \, Z_m[g]. \quad (2.112)$$

Since the matter action and measure, and the Liouville measure are invariant under reparametrizations, one can perform a change of variables

$$(f^* \hat{g}, f^* \phi, f^* \Psi) \longrightarrow (\hat{g}, \phi, \Psi) \qquad (2.113)$$

such that everything becomes independent of f (or equivalently ξ). Since the measure for ξ is Weyl invariant, it is possible to separate it from the rest of the expression, which yields an overall factor of $\Omega_{\text{Diff}_0}[g]$. This brings the partition function to the form

$$Z_g = \int_{\mathcal{T}_g} d^{M_g} t \, \frac{\Omega_{\text{Diff}_0}[\hat{g}]}{\Omega_{\text{gauge}}[\hat{g}]} \int d_g \phi \, \frac{\det(\phi_i, \mu_j)_g}{\sqrt{\det(\phi_i, \phi_j)_g}} \frac{\Omega_{\text{ckv}}[g]^{-1}}{\sqrt{\det(\psi_i, \psi_j)_g}} \Delta_{\text{FP}}[g] \, Z_m[g],$$

$$(2.114)$$

where the same symbol is used for the metric

$$g_{ab} := g_{ab}^{(\phi)} = e^{2\phi}\hat{g}_{ab}. \tag{2.115}$$

Since the expression is invariant under the full diffeomorphism group $\mathrm{Diff}(\Sigma_g)$ and not just under its component $\mathrm{Diff}_0(\Sigma_g)$, one needs to extract the volume of the full diffeomorphism group before cancelling it with the normalization factor. Otherwise, there is still an over-counting the configurations. Using the relation (2.60c) leads to

$$Z_g = \frac{1}{\Omega_{\Gamma_g}} \int_{\mathcal{T}_g} d^{M_g} t \, \frac{\Omega_{\mathrm{Diff}}[\hat{g}]}{\Omega_{\mathrm{gauge}}[\hat{g}]} \int d_g\phi \, \frac{\det(\phi_i,\mu_j)_g}{\sqrt{\det(\phi_i,\phi_j)_g}} \, \frac{\Omega_{\mathrm{ckv}}[g]^{-1}}{\sqrt{\det(\psi_i,\psi_j)_g}} \Delta_{\mathrm{FP}}[g] \, Z_m[g]. \tag{2.116}$$

The volume Ω_{Γ_g} can be factorized outside the integral because it depends only on the genus and not on the metric. Finally, using the relation (2.52), one can replace the integration over the Teichmüller space by an integration over the moduli space

$$Z_g = \int_{\mathcal{M}_g} d^{M_g} t \, \frac{\Omega_{\mathrm{Diff}}[\hat{g}]}{\Omega_{\mathrm{gauge}}[\hat{g}]} \int d_g\phi \, \frac{\det(\phi_i,\mu_j)_g}{\sqrt{\det(\phi_i,\phi_j)_g}} \, \frac{\Omega_{\mathrm{ckv}}[g]^{-1}}{\sqrt{\det(\psi_i,\psi_j)_g}} \Delta_{\mathrm{FP}}[g] \, Z_m[g]. \tag{2.117}$$

2.3.3 Weyl Transformations and Quantum Anomalies

The next question is whether the integrand depends on the Liouville mode ϕ such that the Weyl volume can be factorized out. While the matter action has been chosen to be Weyl invariant—see the condition (2.19)—the measures cannot be defined to be Weyl invariant. This means that there is a *Weyl* (or conformal) *anomaly*, i.e. a violation of the Weyl invariance due to quantum effects. Since the techniques needed to derive the results of this section are outside the scope of this book, we simply state the results.

It is possible to show that the Weyl anomaly reads [11, p. 929][5]

$$\frac{\Delta_{\mathrm{FP}}[e^{2\phi}\hat{g}]}{\sqrt{\det(\phi_i,\phi_j)_{e^{2\phi}\hat{g}}}} = e^{\frac{c_{gh}}{6}S_L[\hat{g},\phi]} \frac{\Delta_{\mathrm{FP}}[\hat{g}]}{\sqrt{\det(\hat{\phi}_i,\hat{\phi}_j)_{\hat{g}}}} \tag{2.118a}$$

$$Z_m[e^{2\phi}\hat{g}] = e^{\frac{c_m}{6}S_L[\hat{g},\phi]} Z_m[\hat{g}], \tag{2.118b}$$

---

[5]The relation is written for Z_m since the action is invariant and is not affected by the anomaly.

where S_L is the Liouville action

$$S_L[\hat{g}, \phi] := \frac{1}{4\pi} \int d^2\sigma \sqrt{\hat{g}} \left(\hat{g}^{ab} \partial_a \phi \partial_b \phi + \hat{R}\phi \right), \qquad (2.119)$$

where $\hat{R}$ is the Ricci scalar of the metric $\hat{g}_{ab}$. These relations require to introduce counter-terms, discussed further in Sect. 2.3.4. The coefficients c_m and c_{gh} are the central charges respectively of the matter and ghost systems, with

$$c_{gh} = -26. \qquad (2.120)$$

This value will be derived in Sect. 7.2.

The inner products between ϕ_i and μ_j, and between the ψ_i, and the CKV volume are independent of ϕ [26, sec. 14.2.2, 11, p. 931]

$$\det(\phi_i, \mu_j)_{e^{2\phi}\hat{g}} = \det(\hat{\phi}_i, \hat{\mu}_j)_{\hat{g}}, \qquad \det(\psi_i, \psi_j)_{e^{2\phi}\hat{g}} = \det(\psi_i, \psi_j)_{\hat{g}},$$

$$\Omega_{\text{ckv}}[e^{2\phi}\hat{g}] = \Omega_{\text{ckv}}[\hat{g}]. \qquad (2.121)$$

Remark 2.13 (Weyl and Gravitational Anomalies) The Weyl anomaly translates into a non-zero trace of the quantum energy–momentum tensor

$$\langle g^{\mu\nu} T_{\mu\nu} \rangle = \frac{c}{12} R, \qquad (2.122)$$

where c is the central charge of the theory. The Weyl anomaly can be traded for a gravitational anomaly, which means that diffeomorphisms are broken at the quantum level [20].

Inserting (2.118) in (2.117) yields

$$Z_g = \int_{\mathcal{M}_g} d^{M_g} t \, \frac{\Omega_{\text{Diff}}[\hat{g}]}{\Omega_{\text{gauge}}[\hat{g}]} \frac{\det(\phi_i, \hat{\mu}_j)_{\hat{g}}}{\sqrt{\det(\phi_i, \phi_j)_{\hat{g}}}} \frac{\Omega_{\text{ckv}}[\hat{g}]^{-1}}{\sqrt{\det(\psi_i, \psi_j)_{\hat{g}}}} \Delta_{\text{FP}}[\hat{g}] \, Z_m[\hat{g}]$$

$$\times \int d_g\phi \, e^{-\frac{c_L}{6} S_L[\hat{g}, \phi]}, \qquad (2.123)$$

with the Liouville central charge

$$c_L := 26 - c_m. \qquad (2.124)$$

The critical "dimension" is defined to be the value of the matter central charge c_m such that the Liouville central charge cancels

$$c_L = 0 \quad \Longrightarrow \quad c_m = 26. \qquad (2.125)$$

If the number of non-compact dimensions is D, it means that the central charge (2.21) of the transverse CFT satisfies

$$c_\perp = 26 - D, \tag{2.126}$$

In this case, the integrand does not depend on the Liouville mode (because Ω_{Diff} is invariant under Weyl transformations) and the integration over ϕ can be factored out and yields the volume of the Weyl group (2.60b)

$$\int \mathrm{d}_g \phi = \Omega_{\text{Weyl}}[\hat{g}]. \tag{2.127}$$

Then, taking

$$\Omega_{\text{gauge}}[\hat{g}] = \Omega_{\text{Diff}}[\hat{g}] \times \Omega_{\text{Weyl}}[\hat{g}] \tag{2.128}$$

removes the infinite gauge contributions and gives the partition function

$$Z_g = \int_{\mathcal{M}_g} \mathrm{d}^{M_g} t \, \frac{\det(\phi_i, \hat{\mu}_j)_{\hat{g}}}{\sqrt{\det(\phi_i, \phi_j)_{\hat{g}}}} \, \frac{\Omega_{\text{ckv}}[\hat{g}]^{-1}}{\sqrt{\det(\psi_i, \psi_j)_{\hat{g}}}} \, \Delta_{\text{FP}}[\hat{g}] \, Z_m[\hat{g}]. \tag{2.129}$$

2.3.4 Ambiguities, Ultralocality and Cosmological Constant

Different ambiguities remain in the previous computations, starting with the definitions of the measures (2.32) and (2.34), then in obtaining the volume of the diffeomorphism (2.60a) and Weyl (2.60b) groups, and finally in deriving the conformal anomaly (2.118).

These different ambiguities can be removed by renormalizing the worldsheet cosmological constant. This implies that the action

$$S_\mu[g] = \int \mathrm{d}^2\sigma \sqrt{g} \tag{2.130}$$

must be added to the classical Lagrangian, where μ_0 is the bare cosmological constant. This means that Weyl invariance is explicitly broken at the classical level. After performing all the manipulations, μ_0 is determined by removing all ambiguities and enforcing invariance under the Weyl symmetry at the quantum level. This amounts to set the renormalized cosmological constant to zero (since it breaks the Weyl symmetry). The possibility to introduce a counter-term violating a classical symmetry arises because the symmetry itself is broken by a quantum anomaly, so there is no reason to enforce it in the classical action.

We now review each issue separately. First, consider the inner product of a single tensor (2.32): the determinant $\det \gamma_g$ depends on the metric and one should be more careful when fixing the gauge or integrating over all metrics. However, ultralocality implies that the determinant can only be of the form [11, pp. 923]

$$\sqrt{\det \gamma_g} = e^{-\mu_\gamma \, S_\mu[g]}, \tag{2.131}$$

for some $\mu_\gamma \in \mathbb{R}$, since S_μ is the only renormalizable covariant functional depending on the metric but not on its derivatives. The effect is just to redefine the cosmological constant.

Second, the volume of the field space can be defined as the limit $\lambda \to 0$ of a Gaussian integral [11, pp. 931]:

$$\Omega_\Phi = \lim_{\lambda \to 0} \int d_g \Phi \, e^{-\lambda \, (\Phi, \Phi)_g}. \tag{2.132}$$

Due to ultralocality, the Gaussian integral should again be of the form

$$\int d_g \Phi \, e^{-\lambda \, (\Phi, \Phi)_g} = e^{-\mu(\lambda) \, S_\mu[g]}, \tag{2.133}$$

for some constant $\mu(\lambda)$. Hence, the limit $\lambda \to 0$ gives

$$\Omega_\Phi = \int d_g \Phi = e^{-\mu(0) \, S_\mu[g]}, \tag{2.134}$$

which can be absorbed in the cosmological constant. However, the situation is more complicated if $\Phi = \xi, \phi$ since the integration variables also appear in the measure, as it was also discussed before (2.61a). But, in that case, it cannot appear in the expression of the volume in the LHS. Moreover, invariances under diffeomorphisms for both measures, and under Weyl rescalings for the vector measure, imply that the LHS can only depend on the moduli through the background metric $\hat{g}$. The diffeomorphism and Weyl volumes can be written in terms of $e^{-\hat{\mu} \, S_\mu[\hat{g}]}$: since there is no counter-term left (the cosmological constant counter-term is already fixed to cancel the coefficient of $S_\mu[g]$), it is necessary to divide by Ω_{gauge} to cancel the volumes.

Finally, the computation of the Weyl anomaly (2.118) yields divergent terms of the form

$$\lim_{\epsilon \to 0} \frac{1}{\epsilon} \int d^2\sigma \sqrt{g}. \tag{2.135}$$

These divergences are cancelled by the cosmological constant counter-term, see [12, app. 5.A] for more details.

2.3.5 Gauge Fixed Path Integral

As a conclusion of this section, we found that the partition function (2.28) can be written as

$$Z_g = \int_{\mathcal{M}_g} d^{M_g}t \; \frac{\det(\phi_i, \hat{\mu}_j)_{\hat{g}}}{\sqrt{\det(\phi_i, \phi_j)_{\hat{g}}}} \; \frac{\Omega_{ckv}[\hat{g}]^{-1}}{\sqrt{\det(\psi_i, \psi_j)_{\hat{g}}}} \; \Delta_{FP}[\hat{g}] \, Z_m[\hat{g}], \tag{2.136a}$$

$$= \int_{\mathcal{M}_g} d^{M_g}t \; \sqrt{\frac{\det(\phi_i, \hat{\mu}_j)^2_{\hat{g}}}{\det(\phi_i, \phi_j)_{\hat{g}}} \frac{\det' \hat{P}_1^\dagger \hat{P}_1}{\det(\psi_i, \psi_j)_{\hat{g}}} \frac{Z_m[\hat{g}]}{\Omega_{ckv}[\hat{g}]}}. \tag{2.136b}$$

after gauge fixing of the worldsheet diffeomorphisms and Weyl rescalings. It is implicit that the factors for the CKV and moduli are respectively absent for $g > 1$ and $g < 1$. For $g = 0$ the CKV group is non-compact and its volume is infinite. It looks like the partition vanishes, but there are subtleties which will be discussed in Sect. 3.1.3.

Remark 2.14 (Weil–Petersson Metric) When the metric is chosen to be of constant curvature $\hat{R} = -1$, the moduli measure together with the determinants form the Weil–Petersson measure

$$d(WP) = \int_{\mathcal{M}_g} d^{M_g}t \; \frac{\det(\phi_i, \hat{\mu}_j)_{\hat{g}}}{\sqrt{\det(\phi_i, \phi_j)_{\hat{g}}}}. \tag{2.137}$$

In (2.136), the background metric $\hat{g}_{ab}$ is fixed. However, the derivation holds for any choice of $\hat{g}_{ab}$: as a consequence, it makes sense to relax the gauge fixing and allow it to vary while adding gauge symmetries. The first symmetry is background diffeomorphisms:

$$\sigma'^a = \hat{f}^a(\sigma^b), \quad \hat{g}'(\sigma') = f^*\hat{g}(\sigma), \quad \phi'(\sigma') = f^*\phi(\sigma), \quad \Psi'(\sigma') = f^*\Psi(\sigma). \tag{2.138}$$

This symmetry is automatic for $S_m[\hat{g}, \Psi]$ since $S_m[g, \Psi]$ was invariant under (2.5). Similarly, the integration measures are also invariant. A second symmetry is found by inspecting the decomposition (2.56)

$$g_{ab} = f^*\left(e^{2\phi}\hat{g}_{ab}(t)\right), \tag{2.139}$$

which is left invariant under a *background Weyl symmetry* (also called emergent):

$$g'_{ab}(\sigma) = e^{2\omega(\sigma)}g_{ab}(\sigma), \quad \phi'(\sigma) = \phi(\sigma) - \omega(\sigma), \quad \Psi'(\sigma) = \Psi(\sigma). \tag{2.140}$$

Let us stress that it is not related to the Weyl rescaling (2.10) of the metric g_{ab}. The background Weyl rescaling (2.140) is a symmetry even when the physical

Weyl rescaling (2.10) is not. Together, the background diffeomorphisms and Weyl symmetry have three gauge parameters, which is sufficient to completely fix the background metric $\hat{g}$ up to moduli.

In fact, the combination of both symmetries is equivalent to invariance under the physical diffeomorphisms. To prove this statement, consider two metrics g and g' related by a diffeomorphism F and both gauge fixed to pairs $(f, \phi, \hat{g})$ and $(f', \phi', \hat{g}')$:

$$g'_{ab} = F^* g_{ab}, \qquad g'_{ab} = f'^* \big(e^{2\phi'}\hat{g}'_{ab}\big), \qquad g_{ab} = f^*\big(e^{2\phi}\hat{g}_{ab}\big). \qquad (2.141)$$

Then, the gauge fixing parametrizations are related by background symmetries $(\hat{F}, \omega)$ as

$$\hat{F} = f'^{-1} \circ F \circ f, \qquad \phi' = \hat{F}^*(\phi - \omega), \qquad \hat{g}'_{ab} = \hat{F}^*(e^{2\omega}\hat{g}_{ab}). \qquad (2.142)$$

Moreover, this also implies that there is a diffeomorphism $\tilde{f} = F \circ f$ such that g' is gauge fixed in terms of $(\phi, \hat{g})$:

$$g'_{ab} = \tilde{f}^*\big(e^{2\phi}\hat{g}_{ab}\big). \qquad (2.143)$$

Computation: Equation (2.142)

The functions F, f, f', ϕ, ϕ' and the metrics g_{ab}, g'_{ab}, $\hat{g}_{ab}$ and $\hat{g}'_{ab}$ are all fixed and one must find $\hat{F}$ and ω such that the relations (2.141) are compatible. First, one rewrites g'_{ab} in terms of $\hat{g}_{ab}$ and compare with the expression with $\hat{g}'_{ab}$:

$$g'_{ab} = F^* g_{ab} = F^*\big(f^*\big(e^{2\phi}\hat{g}_{ab}\big)\big) = F^*\big(f^*\big(e^{2(\phi-\omega)}e^{2\omega}\hat{g}_{ab}\big)\big)$$
$$= f'^*\big(e^{2\phi'}\hat{g}'_{ab}\big).$$

In the third equality, we have introduced ω because $\hat{g}'_{ab} = \hat{F}^*\hat{g}_{ab}$ is not true in general since there are 3 independent components but $\hat{F}$ has only 2 parameters, so we cannot just define $f' = F \circ f$ and $\phi' = \phi$. This explains the importance of the emergent Weyl symmetry.

Remark 2.15 (Gauge Fixing and Field Redefinition) Although it looks like we are undoing the gauge fixing, this is not exactly the case since the original metric is not used anymore. One can understand the procedure of this section as a field redefinition: the degrees of freedom in g_{ab} are repackaged into two fields $(\phi, \hat{g}_{ab})$ adapted to make some properties of the system more salient. A new gauge symmetry is introduced to maintain the number of degrees of freedom. The latter helps to understand the structure of the action on the background. Finally, in this context, the Liouville action is understood as a Wess–Zumino action, which is defined as the difference between the effective actions evaluated in each metric. Another typical use of this point of view is to rewrite a massive vector field as a massless gauge field together with an axion [30].

Remark 2.16 (Two-Dimensional Gravity) In $2d$ gravity, one does not work in the critical dimension (2.125) and $c_L \neq 0$. Thus, the Liouville mode does not decouple: the conformal anomaly breaks the Weyl symmetry at the quantum level which gives dynamics to gravity, even if it has no degree of freedom classically. As a consequence, one chooses $\Omega_{\text{gauge}} = \Omega_{\text{Diff}}$.

Since the role of the classical Weyl symmetry is not as important as for string theory, it is even not necessary to impose it classically. This leads to consider non-conformal matter [1, 2, 4, 14, 15]. Following the arguments from Sect. 2.1, the existence of the emergent Weyl symmetry (2.140) implies that the total action $S_{\text{grav}}[\hat{g}, \phi] + S_m[\hat{g}, \Psi]$ must be a CFT for a flat background $\hat{g} = \delta$, even if the two actions are not independently CFTs.

2.4 Ghost Action

2.4.1 Actions and Equations of Motion

It is well-known that a determinant can be represented with two anti-commuting fields, called ghosts. The fields carry indices dictated by the map induced by the operator of the Faddeev–Popov determinant: one needs a symmetric and traceless anti-ghost b_{ab} and a vector ghost c^a fields:

$$\Delta_{\text{FP}}[g] = \int d'_g b \, d'_g c \; e^{-S_{\text{gh}}[g,b,c]}, \tag{2.144}$$

where the prime indicates that the ghost zero-modes are omitted. The ghost action is

$$S_{\text{gh}}[g, b, c] := \frac{1}{4\pi} \int d^2\sigma \sqrt{g} \, g^{ab} g^{cd} b_{ac}(P_1 c)_{bd} \tag{2.145a}$$

$$= \frac{1}{4\pi} \int d^2\sigma \sqrt{g} \, g^{ab} \left(b_{ac} \nabla_b c^c + b_{bc} \nabla_a c^c - b_{ab} \nabla_c c^c \right). \tag{2.145b}$$

The ghosts c^a and anti-ghosts b_{ab} are associated respectively to the variations due to the diffeomorphisms ξ^a and to the variations perpendicular to the gauge slice. The normalization of $1/4\pi$ is conventional. In Minkowski signature, the action is multiplied by a factor i.

Since b_{ab} is traceless, the last term of the action vanishes and could be removed. However, this implies to consider traceless variations of the b_{ab} when varying the action (to compute the equations of motion, the energy–momentum tensor, etc.). On the other hand, one can keep the term and consider unconstrained variation of b_{ab} (since the structure of the action will force the variation to have the correct symmetry), which is simpler. A last possibility is to introduce a Lagrange multiplier. These aspects are related to the question of introducing a ghost for the Weyl symmetry, which is described in Sect. 2.4.2.

The equations of motion are

$$(P_1 c)_{ab} = \nabla_a c_b + \nabla_b c_a - g_{ab} \nabla_c c^c = 0, \qquad (P_1^\dagger b)_a = -2\nabla^b b_{ab} = 0. \qquad (2.146)$$

Hence, the classical solutions of b and c are respectively mapped to the zero-modes of the operators $P_1^\dagger$ and P_1, and they are thus associated to the CKV and Teichmüller parameters.

The energy–momentum tensor is

$$T_{ab}^{\text{gh}} = -b_{ac}\nabla_b c^c - b_{bc}\nabla_a c^c + c^c \nabla_c b_{ab} + g_{ab} b_{cd} \nabla^c c^d. \qquad (2.147)$$

Its trace vanishes off-shell (i.e. without using the b and c equations of motion)

$$g^{ab} T_{ab}^{\text{gh}} = 0, \qquad (2.148)$$

which shows that the action (2.145) is invariant under Weyl transformations

$$S_{\text{gh}}[e^{2\omega} g, b, c] = S_{\text{gh}}[g, b, c]. \qquad (2.149)$$

The action (2.145) also has a U(1) global symmetry. The associated conserved charge is called the *ghost number* and counts the number of c ghosts minus the number of b ghosts, i.e.

$$N_{\text{gh}}(b) = -1, \qquad N_{\text{gh}}(c) = 1. \qquad (2.150a)$$

The matter fields are inert under this symmetry:

$$N_{\text{gh}}(\Psi) = 0. \qquad (2.150b)$$

In terms of actions, the path integral (2.136) can be rewritten as

$$Z_g = \int_{\mathcal{M}_g} \mathrm{d}^{M_g} t \, \frac{\det(\phi_i, \hat{\mu}_j)_{\hat{g}}}{\sqrt{\det(\phi_i, \phi_j)_{\hat{g}}}} \, \frac{\Omega_{\text{ckv}}[\hat{g}]^{-1}}{\sqrt{\det(\psi_i, \psi_j)_{\hat{g}}}} \int \mathrm{d}_{\hat{g}} \Psi \, \mathrm{d}'_{\hat{g}} b \, \mathrm{d}'_{\hat{g}} c \, e^{-S_m[\hat{g}, \Psi] - S_{\text{gh}}[\hat{g}, b, c]}.$$

$$(2.151)$$

One can use (2.136) or (2.151) indifferently: the first is more appropriate when using spectral analysis to compute the determinant explicitly, while the second is more natural in the context of CFTs.

2.4.2 Weyl Ghost

Ghosts have been introduced for the reparametrizations (generated by ξ^a) and the traceless part of the metric (the gauge field associated to the transformation): one may wonder why there is not a ghost c_w associated to the Weyl symmetry along with an anti-ghost for the trace of the metric (i.e. the conformal factor). This can be understood from several viewpoints. First, the relation between a metric and its transformation—and the corresponding gauge fixing condition—does not involve any derivative: as such, the Jacobian is trivial. Second, one could choose

$$F_{ab}^\perp = \sqrt{g}\,g_{ab} - \sqrt{\hat{g}}\,\hat{g}_{ab} = 0 \tag{2.152}$$

as a gauge fixing condition instead of (2.63), and the trace component does not appear anywhere. Finally, a local Weyl symmetry is not independent from the diffeomorphisms.

Remark 2.17 (Local Weyl Symmetry) The topic of obtaining a local Weyl symmetry by gauging a global Weyl symmetry (dilatation) is very interesting [16, chap. 15, 19]. Under general conditions, one can express the new action in terms of the Ricci tensor (or of the curvature): this means that the Weyl gauge field and its curvature are composite fields.

Moreover, one finds that local Weyl invariance leads to an off-shell condition while diffeomorphisms give on-shell conditions. This explains why one imposes only Virasoro constraints (associated to reparametrizations) and no constraints for the Weyl symmetry in the covariant quantization.

However, it can be useful to introduce a ghost field c_w for the Weyl symmetry nonetheless. In view of the previous discussion, this field should appear as a Lagrange multiplier which ensures that b_{ab} is traceless. Starting from the action (2.145), one finds

$$S'_{\text{gh}}[g, b, c, c_w] = \frac{1}{4\pi} \int d^2\sigma \sqrt{g}\, g^{ab} \left(b_{ac} \nabla_b c^c + b_{bc} \nabla_a c^c + 2 b_{ab} c_w \right), \tag{2.153}$$

where b_{ab} is not traceless anymore. The ghost c_w is not dynamical since the action does not contain derivatives of it, and it can be integrated out of the path integral to recover (2.145).

The equations of motion for this modified action are

$$\nabla_a c_b + \nabla_b c_a + 2 g_{ab} c_w = 0, \qquad \nabla^a b_{ab} = 0, \qquad g^{ab} b_{ab} = 0. \tag{2.154}$$

Contracting the first equation with the metric gives

$$c_w = -\frac{1}{2} \nabla_a c^a, \tag{2.155}$$

and thus c_w is nothing else than the divergence of the c^a field: the Weyl ghost is a composite field (this makes connection with Remark 2.17)—see also (2.65c). The energy–momentum tensor of the ghosts with action (2.153) is

$$
T_{ab}^{\prime\text{gh}} = - \left(b_{ac}\nabla_b c^c + b_{bc}\nabla_a c^c + 2b_{ab}c_w \right) - \nabla_c(b_{ab}c^c)
$$
$$
+ \frac{1}{2} g_{ab}g^{cd}\left(b_{ce}\nabla_d c^e + b_{de}\nabla_c c^e + 2b_{cd}c_w \right).
\tag{2.156}
$$

The trace of this tensor

$$
g^{ab}T_{ab}^{\prime\text{gh}} = -g^{ab}\nabla_c(b_{ab}c^c)
\tag{2.157}
$$

does not vanish off-shell, but it does on-shell since $g^{ab}b_{ab} = 0$. This implies that the theory is Weyl invariant even if the action is not. It is interesting to contrast this with the trace (2.148) when the Weyl ghost has been integrated out.

The equations of motion (2.146) and energy–momentum tensor (2.147) for the action (2.145) can be easily derived by replacing c_w by its solution in the previous formulas.

> **Computation: Equation (2.156)**
> The first parenthesis comes from varying g^{ab}, the second from the covariant derivatives and the last from the $\sqrt{g}$. The second term comes from
>
> $$
> g^{ab}\left(b_{ac}\delta\nabla_b c^c + b_{bc}\delta\nabla_a c^c \right) = 2g^{ab}b_{ac}\delta\nabla_b c^c = 2g^{ab}b_{ac}\delta\Gamma^c_{\ bd}c^d
> $$
> $$
> = g^{ab}b_{ac}g^{ce}\left(\nabla_b\delta g_{de} + \nabla_d\delta g_{be} - \nabla_e\delta g_{bd} \right)c^d
> $$
> $$
> = b^{ab}\left(\nabla_a\delta g_{bc} + \nabla_c\delta g_{ab} - \nabla_b\delta g_{ac} \right)c^c
> $$
> $$
> = b^{ab}\nabla_c\delta g_{ab}c^c,
> $$
>
> where two terms have cancelled due to the symmetry of b^{ab}. Integrating by part gives the term in the previous equation.

Note that the integration on the Weyl ghost yields a delta function

$$
\int d_g c_w\, e^{-(c_w,\, g^{ab}b_{ab})_g} = \delta\left(g^{ab}b_{ab} \right).
\tag{2.158}
$$

2.4.3 Zero-Modes

The path integral (2.151) excludes the zero-modes of the ghosts. One can expect them to be related to the determinants of elements of $\ker P_1$ and $\ker P_1^\dagger$ with Grassmann coefficients. They can be included after few simple manipulations (see also Appendix C.1.3).

It is simpler to first focus on the b ghost (to avoid the problems related to the CKV). The path integral (2.151) can be rewritten as

$$
Z_g = \int_{\mathcal{M}_g} \mathrm{d}^{\mathsf{M}_g} t \, \frac{\Omega_{\mathrm{ckv}}[\hat{g}]^{-1}}{\sqrt{\det(\psi_i, \psi_j)_{\hat{g}}}} \int \mathrm{d}_{\hat{g}} \Psi \, \mathrm{d}_{\hat{g}} b \, \mathrm{d}_{\hat{g}}' c \prod_{i=1}^{\mathsf{M}_g} (b, \hat{\mu}_i)_{\hat{g}} \, \mathrm{e}^{-S_m[\hat{g}, \Psi] - S_{\mathrm{gh}}[\hat{g}, b, c]}.
$$

$$(2.159)$$

In this expression, c zero-modes are not integrated over, only the b zero-modes are. This is the standard starting point on Riemann surfaces with genus $g \geq 1$. The inner product reads explicitly

$$
(b, \hat{\mu}_i)_{\hat{g}} = \int \mathrm{d}^2 \sigma \sqrt{\hat{g}} \, G_{\perp}^{abcd} b_{ab} \hat{\mu}_{i,cd} = \int \mathrm{d}^2 \sigma \sqrt{\hat{g}} \, g^{ac} g^{bd} b_{ab} \hat{\mu}_{i,cd}. \qquad (2.160)
$$

Computation: Equation (2.159)

Since the zero-modes of b are in the kernel of $P_1^{\dagger}$, it means that the quadratic differentials (2.76) also provide a suitable basis:

$$
b = b_0 + b', \qquad b_0 = b_{0i} \phi_i,
$$

where the b_{0i} are Grassmann-odd coefficients. The first step is to find the Jacobian for the changes of variables $b \to (b', b_{0i})$:

$$
1 = \int \mathrm{d}_{\hat{g}} b \, \mathrm{e}^{-|b|_{\hat{g}}^2} = J \int \mathrm{d}_{\hat{g}} b' \prod_i \mathrm{d} b_{0i} \, \mathrm{e}^{-|b'|_{\hat{g}}^2 - |b_{0i} \phi_i|^2} = J \sqrt{\det(\phi_i, \phi_j)}.
$$

Next, (2.151) has no zero-modes, so one must insert M_g of them at arbitrary positions σ_j^0 to get a non-vanishing result when integrating over $\mathrm{d}^{\mathsf{M}_g} b_{0i}$. The result of the integral is

$$
\int \mathrm{d}^{\mathsf{M}_g} b_{0i} \prod_j b_0(\sigma_j^0) = \int \mathrm{d}^{\mathsf{M}_g} b_{0i} \prod_j [b_{0i} \phi_i(\sigma_j^0)] = \det \phi_i(\sigma_j^0).
$$

The only combination of the ϕ_i which does not vanish is the determinant due to the antisymmetry of the Grassmann numbers. Combining both results leads to

$$
\frac{\mathrm{d}_{\hat{g}} b'}{\sqrt{\det(\phi_i, \phi_j)_{\hat{g}}}} = \frac{\mathrm{d}_{\hat{g}} b}{\det \phi_i(\sigma_j^0)} \prod_{j-1}^{\mathsf{M}_g} b(\sigma_j^0). \qquad (2.161)
$$

The locations positions σ_j^0 are arbitrary (in particular, the RHS does not depend on them since the LHS does not either). Note that more details are provided in Appendix C.1.3.

An even simpler result can be obtained by combining the previous formula with the factor $\det(\phi_i, \hat{\mu}_j)_{\hat{g}}$:

$$d_{\hat{g}}b' \frac{\det(\phi_i, \hat{\mu}_j)_{\hat{g}}}{\sqrt{\det(\phi_i, \phi_j)_{\hat{g}}}} = d_{\hat{g}}b \prod_{j=1}^{M_g}(b, \hat{\mu}_j)_{\hat{g}}. \tag{2.162}$$

This follows from

$$\prod_{j=1}^{M_g} b(\sigma_j^0) = \prod_{j=1}^{M_g}\left[b_{0i}\phi_i(\sigma_j^0)\right] = \det\phi_i(\sigma_j^0)\prod_{j=1}^{M_g}b_{0i},$$

$$\det(\phi_i, \hat{\mu}_j)_{\hat{g}}\prod_{j=1}^{M_g}b_{0i} = \prod_{j=1}^{M_g}\left[b_{0i}(\phi_i, \hat{\mu}_j)_{\hat{g}}\right] = \prod_{j=1}^{M_g}(b_{0i}\phi_i, \hat{\mu}_j)_{\hat{g}} = \prod_{j=1}^{M_g}(b, \hat{\mu}_j)_{\hat{g}}.$$

Note that the previous manipulations are slightly formal: the symmetric traceless fields b_{ab} and $\phi_{i,ab}$ carry indices and there should be a product over the (two) independent components. This is a trivial extension and would just make the notations heavier.

Similar manipulations lead to a new expression which includes also the c zero-mode (but which is not very illuminating):

$$Z_g = \int_{\mathcal{M}_g} d^{M_g}t \, \frac{\Omega_{\text{ckv}}[\hat{g}]^{-1}}{\det\psi_i(\sigma_j^0)} \int d_{\hat{g}}\Psi \, d_{\hat{g}}b \, d_{\hat{g}}c \prod_{j=1}^{K_g^c} \frac{\epsilon_{ab}}{2}c^a(\sigma_j^0)c^b(\sigma_j^0)$$

$$\times \prod_{i=1}^{M_g}(\hat{\mu}_i, b)_{\hat{g}} \, e^{-S_m[\hat{g},\Psi]-S_{\text{gh}}[\hat{g},b,c]}. \tag{2.163}$$

The σ_j^{0a} are $K_g^c = K_g/2$ fixed positions and the integral does not depend on their values. Note that only K_g^c positions are needed because the coordinate is 2-dimensional: fixing 3 points with 2 components correctly gives 6 constraints. Then, $\psi_i(\sigma_j^{0a})$ is a 6-dimensional matrix, with the rows indexed by i and the columns by the pair (a, j).

The expression cannot be simplified further because the CKV factor is infinite for $g = 0$. This is connected to a fact mentioned previously: there is a remaining gauge symmetry which is not taken into account

$$c \longrightarrow c + c_0, \qquad P_1 c_0 = 0. \tag{2.164}$$

A proper account requires to gauge fix this symmetry: the simplest possibility is to insert three or more vertex operators—this topic is discussed in Sect. 3.1.

Finally, note that the same question arises for the b-ghost since one has the symmetry

$$b \longrightarrow b + b_0, \qquad P_1^\dagger b_0 = 0. \tag{2.165}$$

That there is no problem in this case is related to the presence of the moduli.

2.5 Normalization

In the previous sections, the closed string coupling constant g_s did not appear in the expressions. Loops in vacuum amplitudes are generated by splitting of closed strings. By inspecting the amplitudes, it seems that there are $2g$ such splittings (Fig. 2.2), which would lead to a factor g_s^{2g}. However, this is not quite correct: this result holds for a 2-point function. Gluing the two external legs to get a partition function (that is, taking the trace) leads to an additional factor g_s^{-2} (to be determined later), such that the overall factor is g_s^{2g-2}. The fact that it is the appropriate power of the coupling constant can be more easily understood by considering n-point amplitudes (Sect. 3.1). The normalization of the path integral can be completely fixed by unitarity [29].

The above factor has a nice geometrical interpretation. Defining

$$\Phi_0 = \ln g_s \tag{2.166}$$

and remembering the expression (2.4) of the Euler characteristics $\chi_g = 2 - 2g$, the coupling factor can be rewritten as

$$g_s^{2g-2} = e^{-\Phi_0 \chi_g} = \exp\left(-\frac{\Phi_0}{4\pi} \int d^2\sigma \sqrt{g} R\right) = e^{-\Phi_0 S_{EH}[g]}, \tag{2.167}$$

where S_{EH} is the Einstein–Hilbert action. This action is topological in two dimensions. Hence, the coupling constant can be inserted in the path integral simply by shifting the action by the above term. This shows that string theory on a flat

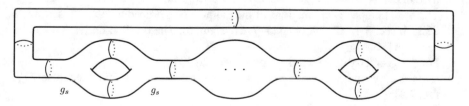

Fig. 2.2 g-loop partition function

target spacetime is completely equivalent to matter minimally coupled to Einstein–Hilbert gravity with a cosmological constant (tuned to impose Weyl invariance at the quantum level). The advantage of describing the coupling power in this fashion is that it directly generalizes to scattering amplitudes and to open strings. The parameter Φ_0 is interpreted as the expectation value of the dilaton. Replacing it by a general field $\Phi(X^\mu)$ is a generalization of the matter non-linear sigma model, but this topic is beyond the scope of this book.

2.6   Summary

In this chapter, we started with a fairly general matter CFT—containing at least D scalar fields X^μ—and explained under which condition it describes a string theory. The most important consequence is that the matter $2d$ QFT must in fact be a $2d$ CFT. We then continued by describing how to gauge fix the integration over the surfaces and we identified the remaining degrees of freedom—the moduli space $\mathcal{M}_g$—up to some residual redundancy—the conformal Killing vector (CKV). Then, we showed how to rewrite the result in terms of ghosts and proved that they are also a CFT. This means that a string theory can be completely described by two decoupled CFTs: a universal ghost CFT and a theory-dependent matter CFT describing the string spacetime embedding and the internal structure. The advantage is that one can forget the path integral formalism altogether and employ only CFT techniques to perform the computations. This point of view will be developed for off-shell amplitudes (Chap. 11) in order to provide an alternative description of how to build amplitudes. It is particularly fruitful because one can also consider matter CFTs which do not have a Lagrangian description. In the next chapter, we describe scattering amplitudes.

2.7   Suggested Readings

Numerous books have been published on the worldsheet string theory. Useful (but not required) complements to this chapter and subsequent ones are [23, 33] for introductory texts and [3, 6, 7, 22, 29] for more advanced aspects.

- The definition of a field measure from a Gaussian integral and manipulations thereof can be found in [18, sec. 15.1, 22.1, 26, chap. 14, 11, 28].
- The most complete explanations of the gauge fixing procedure are [18, sec. 15.1, 22.1, 3, sec. 3.4, 6.2, 29, chap. 5, 7, 21, chap. 5]. The original derivation can be found in [10, 25].
- For the geometry of the moduli space, see [26, 27].
- Ultralocality and its consequences are described in [11, 28] (see also [17, sec. 2.4]).
- The use of a Weyl ghost is shown in [31, sec. 8, 32, sec. 9.2].

References

1. A. Bilal, C. de Lacroix, 2D gravitational Mabuchi action on Riemann surfaces with boundaries. J. High Energy Phys. **2017**(11), 154 (2017). https://doi.org/10.1007/JHEP11(2017)154. arXiv: 1703.10541
2. A. Bilal, L. Leduc, 2D quantum gravity on compact Riemann surfaces with non-conformal matter. J. High Energy Phys. **2017**(01), 089 (2017). https://doi.org/10.1007/JHEP01(2017)089. arXiv: 1606.01901
3. R. Blumenhagen, D. Lüst, S. Theisen, *Basic Concepts of String Theory*, 2013 edn. (Springer, Berlin, 2014)
4. T. Can, M. Laskin, P. Wiegmann, Geometry of quantum hall states: gravitational anomaly and kinetic coefficients. Ann. Phys. **362**, 752–794 (2015). https://doi.org/10.1016/j.aop.2015.02. 013. arXiv: 1411.3105
5. F. David, Conformal field theories coupled to 2-D gravity in the conformal gauge. Mod. Phys. Lett. A **03**(17), 1651–1656 (1988). https://doi.org/10.1142/S0217732388001975
6. P. Deligne, P. Etingof, D.S. Freed, L.C. Jeffrey, D. Kazhdan, J.W. Morgan, D.R. Morrison, E. Witten (eds.), *Quantum Fields and Strings: A Course for Mathematicians. Volume 1* (American Mathematical Society, Providence, 1999)
7. P. Deligne, P. Etingof, D.S. Freed, L.C. Jeffrey, D. Kazhdan, J.W. Morgan, D.R. Morrison, E. Witten (eds.), *Quantum Fields and Strings: A Course for Mathematicians. Volume 2.* (American Mathematical Society, Providence, 1999)
8. E. D'Hoker, Equivalence of Liouville theory and 2-D quantum gravity. Mod. Phys. Lett. A **06**(09), 745–767 (1991). https://doi.org/10.1142/S0217732391000774
9. E. D'Hoker, P.S. Kurzepa, 2D quantum gravity and Liouville theory. Mod. Phys. Lett. A **05**(18), 1411–1421 (1990). https://doi.org/10.1142/S0217732390001608
10. E. D'Hoker, D.H. Phong, Multiloop amplitudes for the Bosonic Polyakov string. Nucl. Phys. B **269**(1), 205–234 (1986). https://doi.org/10.1016/0550-3213(86)90372-X
11. E. D'Hoker, D.H. Phong, The geometry of string perturbation theory. Rev. Mod. Phys. **60**(4), 917–1065 (1988). https://doi.org/10.1103/RevModPhys.60.917
12. P. Di Francesco, P. Mathieu, D. Senechal, *Conformal Field Theory*, 2nd edn. (Springer, Berlin, 1999)
13. J. Distler, H. Kawai, Conformal field theory and 2D quantum gravity. Nucl. Phys. B **321**(2), 509–527 (1989). https://doi.org/10.1016/0550-3213(89)90354-4
14. F. Ferrari, S. Klevtsov, FQHE on curved backgrounds, free fields and large N. J. High Energy Phys. **2014**(12), 086 (2014). https://doi.org/10.1007/JHEP12(2014)086. arXiv: 1410.6802
15. F. Ferrari, S. Klevtsov, S. Zelditch, Gravitational actions in two dimensions and the Mabuchi functional. Nucl. Phys. B **859**(3), 341–369 (2012). https://doi.org/10.1016/j.nuclphysb.2012. 02.003. arXiv: 1112.1352
16. D.Z. Freedman, A. Van Proeyen, *Supergravity* (Cambridge University Press, Cambridge, 2012)
17. H.W. Hamber, *Quantum Gravitation: The Feynman Path Integral Approach* (Springer, Berlin, 2009)
18. B. Hatfield, *Quantum Field Theory of Point Particles and Strings* (Addison Wesley, Boston, 1998)
19. A. Iorio, L. O'Raifeartaigh, I. Sachs, C. Wiesendanger, Weyl-Gauging and conformal invariance. Nucl. Phys. B **495**(1–2), 433–450 (1997). https://doi.org/10.1016/S0550-3213(97)00190-9. arXiv: hep-th/9607110
20. R. Jackiw, Another view on massless matter-gravity fields in two dimensions, Jan 1995
21. M. Kaku, *Introduction to Superstrings and M-Theory* (Springer, Berlin, 1999)
22. E. Kiritsis, *String Theory in a Nutshell* (Princeton University Press, Princeton, 2007)
23. I.D. Lawrie, *A Unified Grand Tour of Theoretical Physics*, 3rd edn. (CRC Press, Boca Raton, 2012)

24. N.E. Mavromatos, J.L. Miramontes,   Regularizing the functional integral in 2D-quantum gravity. Mod. Phys. Lett. A **04**(19), 1847–1853 (1989). https://doi.org/10.1142/S0217732389002082

25. G. Moore, P. Nelson, Measure for moduli the Polyakov string has no nonlocal anomalies. Nucl. Phys. B **266**(1), 58–74 (1986). https://doi.org/10.1016/0550-3213(86)90177-X

26. M. Nakahara, *Geometry, Topology and Physics*, 2nd edn. (Institute of Physics Publishing, Bristol, 2003)

27. P. Nelson, Lectures on strings and moduli space. Phys. Rep. **149**(6), 337–375 (1987). https://doi.org/10.1016/0370-1573(87)90082-2

28. J. Polchinski, Evaluation of the one loop string path integral. Commun. Math. Phys. **104**(1), 37–47 (1986). https://doi.org/10.1007/BF01210791

29. J. Polchinski, *String Theory: Volume 1, An Introduction to the Bosonic String* (Cambridge University Press, Cambridge, 2005)

30. J. Preskill, Gauge anomalies in an effective field theory. Ann. Phys. **210**(2), 323–379 (1991). https://doi.org/10.1016/0003-4916(91)90046-B

31. G. 't Hooft,   Introduction to string theory, May 2004. https://www.staff.science.uu.nl/~hooft101/lectures/stringnotes.pdf

32. K. Wray, An introduction to string theory, May 2011. https://math.berkeley.edu/~kwray/papers/string_theory.pdf.

33. B. Zwiebach, *A First Course in String Theory*, 2nd edn. (Cambridge University Press, Cambridge, 2009)

Worldsheet Path Integral: Scattering Amplitudes

3

Abstract

In this chapter, we generalize the worldsheet path integral to compute scattering amplitudes, which corresponds to insert vertex operators. The gauge fixing from the previous chapter is generalized to this case. In particular, we discuss the 2-point amplitude on the sphere. Finally, we introduce the BRST symmetry and motivate some properties of the BRST quantization, which will be performed in details later. The formulas in this chapter are all covariant: they will be rewritten in complex coordinates in the next chapter.

3.1 Scattering Amplitudes on Moduli Space

In this section, we describe the scattering of n strings. The momentum representation is more natural for describing interactions, especially in string theory. Therefore, each string is characterized by a state $V_{\alpha_i}(k_i)$ with momentum k_i and some additional quantum numbers α_i ($i = 1, \ldots, n$). We start from the worldsheet path integral (2.28) before gauge fixing

$$ Z_g = \int \frac{\mathrm{d}_g\, g_{ab}}{\Omega_{\mathrm{gauge}}[g]}\, Z_m[g], \qquad Z_m[g] = \int \mathrm{d}_g \Psi\, \mathrm{e}^{-S_m[g,\Psi]}. \qquad (3.1) $$

3.1.1 Vertex Operators and Path Integral

The external states are represented by infinite semi-tubes attached to the surfaces. Under a conformal mapping, the tubes can be mapped to points called *punctures* on the worldsheet. At g-loops, the resulting space is a Riemann surface $\Sigma_{g,n}$ of

H. Erbin, *String Field Theory*, Lecture Notes in Physics 980,
https://doi.org/10.1007/978-3-030-65321-7_3

genus g with n punctures (or marked points). The external states are represented by *integrated vertex operators*

$$V_\alpha(k_i) := \int d^2\sigma \sqrt{g(\sigma)}\, V_\alpha(k; \sigma). \tag{3.2}$$

The vertex operators $V_\alpha(k; \sigma)$ are built from the matter CFT operators and from the worldsheet metric g_{ab}. The functional dependence is omitted to not overload the notation, but one should read $V_\alpha(k; \sigma) := V_\alpha[g, \Psi](k; \sigma)$. The integration over the state positions is necessary because the mapping of the tube to a point is arbitrary. Another viewpoint is that it is needed to obtain an expression invariant under worldsheet diffeomorphisms. The vertex operators described general states that not necessarily on-shell: this restriction will be found later when discussing the BRST invariance of scattering amplitudes (Sect. 3.2.2).

Following Sect. 2.3.5, the Einstein–Hilbert action with boundary term

$$S_{\mathrm{EH}}[g] := \frac{1}{4\pi} \int d^2\sigma \sqrt{g}\, R + \frac{1}{2\pi} \oint ds\, k = \chi_{g,n} \tag{3.3}$$

inserted in the path integral equals the Euler characteristics $\chi_{g,n}$ (the g in $\chi_{g,n}$ denotes the genus). On a surface with punctures, the latter is shifted by the number of punctures (which are equivalent to boundaries or disks) with respect to (2.4)

$$\chi_{g,n} := \chi(\Sigma_{g,n}) = 2 - 2g - n. \tag{3.4}$$

This gives the normalization factor

$$g_s^{-\chi_{g,n}} = e^{-\Phi_0 S_{\mathrm{EH}}[g]}, \qquad \Phi_0 := \ln g_s. \tag{3.5}$$

The correctness factor can be verified by inspection of the Riemann surface for the scattering of n strings at g-loops. In particular, the string coupling constant is by definition the interaction strength for the scattering of 3 strings at tree-level. Moreover, the tree-level 2-point amplitude contains no interaction and should have no power of g_s. This factor can also be obtained by unitarity [16].

By inserting these factors in (2.28), the g-loop n-point scattering amplitude is described by

$$A_{g,n}(\{k_i\})_{\{\alpha_i\}} := \int \frac{d_g\, g_{ab}}{\Omega_{\mathrm{gauge}}[g]}\, d_g \Psi\, e^{-S_m[g,\Psi] - \Phi_0 S_{\mathrm{EH}}[g]}$$

$$\times \prod_{i=1}^{n} \left(\int d^2\sigma_i \sqrt{g(\sigma_i)}\, V_{\alpha_i}(k_i; \sigma_i) \right). \tag{3.6}$$

The σ_i dependence of each $\sqrt{g}$ will be omitted from now on since no confusion is possible. The following equivalent notations will be used:

$$A_{g,n}(\{k_i\})_{\{\alpha_i\}} := A_{g,n}(k_1, \ldots, k_n)_{\alpha_1,\ldots,\alpha_n} := A_{g,n}\big(V_{\alpha_1}(k_1), \ldots, V_{\alpha_n}(k_n)\big).$$
(3.7)

The complete (perturbative) amplitude is found by summing over all genus

$$A_n(k_1, \ldots, k_n)_{\alpha_1,\ldots,\alpha_n} = \sum_{g=0}^{\infty} A_{g,n}(k_1, \ldots, k_n)_{\alpha_1,\ldots,\alpha_n}.$$
(3.8)

We omit a genus-dependent normalization that can be determined from unitarity [16]. Sometimes, it is convenient to extract the factor $e^{-\Phi_0 \chi_{g,n}}$ of the amplitude $A_{g,n}$ to display explicitly the genus expansion, but we will not follow this convention here. Since each term of the sum scales as $A_{g,n} \propto g_s^{2g+n-2}$, this expression clearly shows that worldsheet amplitudes are perturbative by definition: this motivates the construction of a string field theory from which the full non-perturbative S-matrix can theoretically be computed.

Finally, the amplitude (3.6) can be rewritten in terms of correlation functions of the matter QFT integrated over worldsheet metrics

$$A_{g,n}(\{k_i\})_{\{\alpha_i\}} = \int \frac{d_g\, g_{ab}}{\Omega_{\text{gauge}}[g]}\, e^{-\Phi_0 S_{\text{EH}}[g]} \int \prod_{i=1}^{n} d^2\sigma_i \sqrt{g}\, \left\langle \prod_{i=1}^{n} V_{\alpha_i}(k_i; \sigma_i) \right\rangle_{m,g}.$$
(3.9)

The correlation function plays the same role as the partition function in (2.28). This shows that string expressions are integrals of CFT expressions over the space of worldsheet metrics (to be reduced to the moduli space).

We address a last question before performing the gauge fixing: what does (3.6) compute exactly: on-shell or off-shell? Green functions or amplitudes? if amplitudes, the S-matrix or just the interacting part T (amputated Green functions)? The first point is that a path integral over connected worldsheets will compute connected processes. We will prove later, when discussing the BRST quantization, that string states must be on-shell (Sects. 3.2 and 3.2.2) and that it corresponds to setting the Hamiltonian (2.26) to zero

$$H = 0.$$
(3.10)

From this fact, it follows that (2.28) must compute amplitudes since non-amputated Green functions diverge on-shell (due to external propagators). Finally, the question of whether it computes the S-matrix $S = 1 + iT$, or just the interacting part T is subtler. At tree-level, they agree for $n \geq 3$, while $T = 0$ for $n = 2$ and S reduces to the identity. This difficulty (discussed further in Sect. 3.1.2) is thus related to the

question of gauge fixing tree-level 2-point amplitude (Sect. 3.1.3). It has long been
believed that (2.28) computes only the interacting part (amputated Green functions),
but it has been understood recently that this is not correct and that (2.28) computes
the S-matrix.

Remark 3.1 (Scattering Amplitudes in QFT) Remember that the S-matrix is separated as

$$S = 1 + iT, \tag{3.11}$$

where 1 denotes the contribution where all particles propagate without interaction.
The connected components of S and T are denoted by S^c and T^c. The n-point
(connected) scattering amplitudes A_n for $n \geq 3$ can be computed from the Green
functions G_n through the LSZ prescription (amputation of the external propagators)

$$A_n(k_1, \ldots, k_n) = G_n(k_1, \ldots, k_n) \prod_{i=1}^{n} (k_i^2 + m_i^2). \tag{3.12}$$

The path integral computes the Green functions G_n; perturbatively, they are
obtained from the Feynman rules. They include a D-dimensional delta function

$$G_n(k_1, \ldots, k_n) \propto \delta^{(D)}(k_1 + \cdots + k_n). \tag{3.13}$$

The 2-point amputated Green function T_2 computed from the LSZ prescription
vanishes on-shell. For example, considering a scalar field at tree-level, one finds

$$T_2 = G_2(k, k') \left(k^2 + m^2 \right)^2 \sim \left(k^2 + m^2 \right) \delta^{(D)}(k + k') \xrightarrow[k^2 \to -m^2]{} 0, \tag{3.14}$$

since

$$G_2(k, k') = \frac{\delta^{(D)}(k + k')}{k^2 + m^2}. \tag{3.15}$$

Hence, $T_2 = 0$ and the S-matrix (3.11) reduces to the identity component $S_2^c = 1_2$
(which is a connected process). There are several ways to understand this result:

1. The recursive definition of the connected S-matrix S^c from the cluster decomposition principle requires a non-vanishing 2-point amplitude [7, sec. 6.1, 11,
   sec. 5.1.5, 23, sec. 4.3].
2. The 2-point amplitude corresponds to the normalization of the 1-particle states
   (overlap of a particle state with itself, which is non-trivial) [19, chap. 5, 22,
   eq. 4.1.4].

3. A single particle in the far past propagating to the far future without interacting is a connected and physical process [7, p. 133].
4. It is required by the unitarity of the 2-point amplitude [8].

These points indicate that the 2-point amplitude is proportional to the identity in the momentum representation [11, p. 212, 22, eq. 4.3.3 and 4.1.5]

$$A_2(k, k') = 2k^0 \, (2\pi)^{D-1} \delta^{(D-1)}(k - k'). \tag{3.16}$$

The absence of interactions implies that the spatial momentum does not change (the on-shell condition implies that the energy is also conserved). This relation is consistent with the commutation relation of the operators with the Lorentz invariant measure[1]

$$[a(k), a^\dagger(k')] = 2k^0 \, (2\pi)^{D-1} \delta^{(D-1)}(k - k'). \tag{3.17}$$

That this holds for all particles at all loops can be proven using the Källen–Lehmann representation [11, p. 212].

On the other hand, the identity part in (3.11) is absent for $n \geq 3$ for connected amplitudes: $S_n^c = T_n^c$ for $n \geq 3$. This shows that the Feynman rules and the LSZ prescription compute only the interacting part T of the on-shell scattering amplitudes. The reason is that the derivation of the LSZ formula assumes that the incoming and outgoing states have no overlap, which is not the case for the 2-point function. A complete derivation of the S-matrix from the path integral is more involved [9, 11, sec. 5.1.5, 25, sec. 6.7] (see also [1]). The main idea is to consider a superposition of momentum states (here, in the holomorphic representation [25, sec. 5.1, 6.4])

$$\phi(\alpha) = \int d^{D-1}k \, \alpha(k)^* a^\dagger(k). \tag{3.18}$$

They contribute a quadratic piece to the connected S-matrix, and setting them to delta functions, one recovers the above result.

3.1.2   Gauge Fixing: General Case

The Faddeev–Popov gauge fixing of the worldsheet diffeomorphisms and Weyl rescaling (2.15) goes through also in this case if the integrated vertex operators are

---

[1] If the modes are defined as $\tilde{a}(k) = a(k)/\sqrt{2k^0}$ such that $[\tilde{a}(k), \tilde{a}^\dagger(k')] = (2\pi)^{D-1}\delta^{(D-1)}(k-k')$, then one finds $\tilde{A}_2(k, k') = (2\pi)^{D-1}\delta^{(D-1)}(k - k')$.

diffeomorphism and Weyl invariant

$$\delta_\xi V_{\alpha_i}(k_i) = \delta_\xi \int d^2\sigma \sqrt{g} \, V_{\alpha_i}(k_i; \sigma) = 0, \tag{3.19a}$$

$$\delta_\omega V_{\alpha_i}(k_i) = \delta_\omega \int d^2\sigma \sqrt{g} \, V_{\alpha_i}(k_i; \sigma) = 0, \tag{3.19b}$$

with the variations defined in (2.7) and (2.11). Diffeomorphism invariance is straightforward if the states are integrated worldsheet scalars. However, if the states are classically Weyl invariant, they are not necessary so at the quantum level: vertex operators are composite operators, which need to be renormalized to be well-defined at the quantum level. Renormalization introduces a scale that breaks Weyl invariance. Enforcing it to be a symmetry of the vertex operators leads to constraints on the latter. We will not enter in the details since it depends on the matter CFT, and we will assume that the operators $V_{\alpha_i}(k_i)$ are indeed Weyl invariant (see [16, sec. 3.6] for more details). In the rest of this book, we will use CFT techniques developed in Chap. 6. The Einstein–Hilbert action is clearly invariant under both symmetries since it is a topological quantity.

Following the computations from Sect. 2.3 leads to a generalization of (2.136) with the vertex operators inserted for the amplitude (3.6)

$$A_{g,n}(\{k_i\})_{\{\alpha_i\}} = g_s^{-\chi_{g,n}} \int_{\mathcal{M}_g} d^{M_g} t \, \frac{\det(\phi_i, \hat{\mu}_j)_{\hat{g}}}{\sqrt{\det(\phi_i, \phi_j)_{\hat{g}}}} \, \frac{\Omega_{\mathrm{ckv}}[\hat{g}]^{-1}}{\sqrt{\det(\psi_i, \psi_j)_{\hat{g}}}}$$

$$\times \int \prod_{i=1}^{n} d^2\sigma_i \sqrt{\hat{g}} \left\langle \prod_{i=1}^{n} \hat{V}_{\alpha_i}(k_i; \sigma_i) \right\rangle_{m,\hat{g}}. \tag{3.20}$$

The hat on the vertex operators indicates that they are evaluated in the background metric $\hat{g}$.

The next step is to introduce the ghosts: following Sect. 2.4, the generalization of (2.159) is

$$A_{g,n}(\{k_i\})_{\{\alpha_i\}} = g_s^{-\chi_{g,n}} \int_{\mathcal{M}_g} d^{M_g} t \, \frac{\Omega_{\mathrm{ckv}}[\hat{g}]^{-1}}{\sqrt{\det(\psi_i, \psi_j)_{\hat{g}}}} \int d_{\hat{g}} b \, d'_{\hat{g}} c \prod_{i=1}^{M_g}(b, \hat{\mu}_i)_{\hat{g}} \, e^{-S_{\mathrm{gh}}[\hat{g}, b, c]}$$

$$\times \int \prod_{i=1}^{n} d^2\sigma_i \sqrt{\hat{g}} \left\langle \prod_{i=1}^{n} \hat{V}_{\alpha_i}(k_i; \sigma_i) \right\rangle_{m,\hat{g}}. \tag{3.21}$$

For the moment, only the b ghosts come with zero-modes. Then, c zero-modes can be introduced in (3.21)

$$A_{g,n} = g_s^{-\chi_{g,n}} \int_{\mathcal{M}_g} d^{M_g} t \, \frac{\Omega_{\mathrm{ckv}}[\hat{g}]^{-1}}{\det \psi_i(\sigma_j^0)}$$

$$\times \int d_{\hat{g}} b \, d_{\hat{g}} c \prod_{j=1}^{K_g^c} \frac{\epsilon_{ab}}{2} c^a(\sigma_j^0) c^b(\sigma_j^0) \prod_{i=1}^{M_g} (\hat{\mu}_i, b)_{\hat{g}} \, e^{-S_{\mathrm{gh}}[\hat{g}, b, c]}$$

$$\times \int \prod_{i=1}^{n} d^2 \sigma_i \sqrt{\hat{g}} \left\langle \prod_{i=1}^{n} \hat{V}_{\alpha_i}(k_i; \sigma_i) \right\rangle_{m, \hat{g}}, \tag{3.22}$$

by following the same derivation as (2.163). The formulas (3.21) and (3.22) are the correct starting point for all g and n. In particular, the c ghosts are not paired with any vertex (a condition often assumed or presented as mandatory). This fact will help resolve some difficulties for the 2-point function on the sphere.

Remember that there is no CKV and no c zero-mode for $g \geq 2$. For the sphere $g = 0$ and the torus $g = 1$, there are CKVs, indicating that there is a residual symmetry in (3.21) and (3.22), which is the global conformal group of the worldsheet. It can be gauge fixed by imposing conditions on the vertex operators.[2] The simplest gauge fixing condition amounts to fix the positions of K_g^c vertex operators through the Faddeev–Popov trick

$$1 = \Delta(\sigma_j^0) \int d\xi \prod_{j=1}^{K_g^c} \delta^{(2)}(\sigma_j - \sigma_j^{0(\xi)}), \quad \sigma_j^{0(\xi)} = \sigma_j^0 + \delta_\xi \sigma_j^0, \quad \delta_\xi \sigma_j^0 = \xi(\sigma_j^0), \tag{3.23}$$

where ξ is a conformal Killing vector, and the variation of σ was given in (2.7). We find that

$$\Delta(\sigma_j^0) = \det \psi_i(\sigma_j^0). \tag{3.24}$$

A priori, the positions σ_j^0 are not the same as the one appearing in (2.163) (since both sets are arbitrary); however, considering the same positions allows to cancel the factor (3.24) with the same one in (2.163).

---

[2]In fact, it is only important to gauge fix for the sphere because the volume of the group is infinite. On the other hand, the volume of the CKV group for the torus is finite-dimensional such that dividing by Ω_{ckv} is not ambiguous.

Computation: Equation (3.24)
The first step is to compute Δ in (3.23). For this, we decompose the CKV ξ on the basis (2.104)

$$\xi\left(\sigma_j^0\right) = \alpha_i \psi_i\left(\sigma_j^0\right)$$

and write the Gaussian integral

$$1 = \int \prod_{j=1}^{K_g^c} d^2 \delta\sigma_j \, e^{-\sum_j (\delta\sigma_j, \delta\sigma_j)} = \Delta \int \prod_{j=1}^{K_g} d\alpha_i \, e^{-\sum_{j,i,i'} (\alpha_i \psi_i(\sigma_j), \alpha_{i'} \psi_{i'}(\sigma_j))}$$

$$= \Delta \left(\det \psi_i(\sigma_j) \right)^{-1}.$$

Again, we have reduced rigour in order to simplify the manipulations.

After inserting the identity (3.23) into (3.22), one can integrate over K_g^c vertex operator positions to remove the delta functions—at the condition that there are at least K_g^c operators. As a consequence, we learn that the proposed gauge fixing works only for $n \geq 1$ if $g = 1$ or $n \geq 3$ if $g = 0$. This condition is equivalent to

$$\chi_{g,n} = 2 - 2g - n < 0. \tag{3.25}$$

In this case, the factors $\det \psi_i(\sigma_j^0)$ cancel and (3.21) becomes

$$A_{g,n}(\{k_i\})_{\{\alpha_i\}} = g_s^{-\chi_{g,n}} \int_{\mathcal{M}_g} d^{M_g} t \int d_{\hat{g}} b \, d_{\hat{g}} c \prod_{j=1}^{K_g^c} \frac{\epsilon_{ab}}{2} c^a\left(\sigma_j^0\right) c^b\left(\sigma_j^0\right)$$

$$\times \prod_{i=1}^{M_g} (\hat{\mu}_i, b)_{\hat{g}} \, e^{-S_{gh}[\hat{g},b,c]}$$

$$\times \int \prod_{i=K_g^c+1}^{n} d^2\sigma_i \sqrt{\hat{g}} \left\langle \prod_{j=1}^{K_g^c} \hat{V}_{\alpha_j}(k_j; \sigma_j^0) \prod_{i=K_g^c+1}^{n} \hat{V}_{\alpha_i}(k_i; \sigma_i) \right\rangle_{m,\hat{g}}.$$

$$\tag{3.26}$$

The result may be divided by a symmetry factor if the delta functions have solutions for several points [16, sec. 5.3]. Performing the gauge fixing for the other cases (in particular, $g = 0, n = 2$ and $g = 1, n = 0$) is more subtle (Sect. 3.1.3 and [16]).

The amplitude can be rewritten in two different ways. First, the ghost insertions can be rewritten in terms of ghost correlation functions

$$A_{g,n}(\{k_i\})_{\{\alpha_i\}} = g_s^{-\chi_{g,n}} \int_{\mathcal{M}_g} d^{M_g} t \int \prod_{i=K_g^c+1}^{n} d^2\sigma_i \sqrt{\hat{g}}$$

$$\times \left\langle \prod_{j=1}^{K_g^c} \frac{\epsilon_{ab}}{2} c^a(\sigma_j^0) c^b(\sigma_j^0) \prod_{i=1}^{M_g} (\hat{\mu}_i, b)_{\hat{g}} \right\rangle_{gh,\hat{g}}$$

$$\times \left\langle \prod_{j=1}^{K_g^c} \hat{V}_{\alpha_j}(k_j; \sigma_j^0) \prod_{i=K_g^c+1}^{n} \hat{V}_{\alpha_i}(k_i; \sigma_i) \right\rangle_{m,\hat{g}}. \qquad (3.27)$$

This form is particularly interesting because it shows that, before integration over the moduli, the amplitudes factorize. This is one of the main advantages of the conformal gauge, since the original complicated amplitude (3.6) for a QFT on a dynamical spacetime reduces to the product of two correlation functions of QFTs on a fixed curved background. In fact, the situation is even simpler when taking a flat background $\hat{g} = \delta$ since both the ghost and matter sectors are CFTs and one can employ all the tools from two-dimensional CFT (Part I) to perform the computations and mostly forget about the path integral origin of these formulas. This approach is particularly fruitful for off-shell (Chap. 11) and superstring amplitudes (Chap. 17).

Remark 3.2 (Amplitudes in 2d Gravity) The derivation of amplitudes for 2d gravity follows the same procedure, up to two differences: (1) there is an additional decoupled (before moduli and position integrations) gravitational sector described by the Liouville field and (2) the matter and gravitational action are not CFTs if the original matter was not.

A second formula can be obtained by bringing the c ghost on top of the matter vertex operators that are at the same positions

$$A_{g,n}(\{k_i\})_{\{\alpha_i\}} = g_s^{-\chi_{g,n}} \int_{\mathcal{M}_g} d^{M_g} t \int \prod_{i=K_g^c+1}^{n} d^2\sigma_i \sqrt{\hat{g}}$$

$$\times \left\langle \prod_{i=1}^{M_g} \hat{B}_i \prod_{j=1}^{K_g^c} \hat{\mathcal{V}}_{\alpha_j}(k_j; \sigma_j^0) \prod_{i=K_g^c+1}^{n} \hat{V}_{\alpha_i}(k_i; \sigma_i) \right\rangle_{\hat{g}}, \qquad (3.28)$$

and where

$$\hat{\mathcal{V}}_{\alpha_j}(k_j; \sigma_j^0) := \frac{\epsilon_{ab}}{2} c^a(\sigma_j^0) c^b(\sigma_j^0) \hat{V}_{\alpha_j}(k_j; \sigma_j^0), \qquad \hat{B}_i := (\hat{\mu}_i, b)_{\hat{g}}. \qquad (3.29)$$

The operators $\mathcal{V}_{\alpha_i}(k_i; \sigma_j^0)$ (a priori off-shell) are called *unintegrated operators*, by opposition to the integrated operators $V_{\alpha_i}(k_i)$. We will see that both are natural elements of the BRST cohomology.

To stress that the $\hat{B}_i$ insertions are really an element of the measure, it is finally possible to rewrite the previous expression as

$$A_{g,n}(\{k_i\})_{\{\alpha_i\}} = g_s^{-\chi_{g,n}} \int_{\mathcal{M}_g \times \mathbb{C}^{n-K_g^c}} \left\langle \bigwedge_{i=1}^{M_g} \hat{B}_i \, dt_i \prod_{j=1}^{K_g^c} \hat{\mathcal{V}}_{\alpha_i}(k_i; \sigma_j^0) \right.$$

$$\left. \times \prod_{i=K_g^c+1}^{n} \hat{V}_{\alpha_i}(k_i; \sigma_i) \, d^2\sigma_i \sqrt{\hat{g}} \right\rangle_{\hat{g}}. \tag{3.30}$$

The result (3.28) suggests a last possibility for improving the expression of the amplitude. Indeed, the different vertex operators do not appear symmetrically: some are integrated over and other come with c ghosts. Similarly, the two types of integrals have different roles: the moduli are related to geometry, while the positions look like external data (vertex operators). However, punctures can obviously be interpreted as part of the geometry, and one may wonder if it is possible to unify the moduli and positions integrals. It is, in fact, possible to put all vertex operators and integrals on the same footing by considering the amplitude to be defined on the moduli space $\mathcal{M}_{g,n}$ of genus-g Riemann surfaces with n punctures instead of just $\mathcal{M}_g$ [16] (see also Sect. 11.3.1).

3.1.3 Gauge Fixing: 2-Point Amplitude

As discussed at the end of Sect. 3.1.1, it has long been believed that the tree-level 2-point amplitude vanishes. There were two main arguments: there are not sufficiently many vertex operators (1) to fix completely the SL(2, $\mathbb{C}$) invariance or (2) to saturate the number of c ghost zero-modes. Let us review both points and then explain why they are incorrect. We will provide the simplest arguments, referring the reader to the literature [8, 18] for more general approaches.

For simplicity, we consider the flat metric $\hat{g} = \delta$ and an orthonormal basis of CKV. The two weight-$(1, 1)$ matter vertex operators are denoted as $V_k(z, \bar{z})$ and $V_{k'}(z', \bar{z}')$ such that the 2-point correlation function on the sphere reads (see Chaps. 6 and 7 for more details)

$$\langle V_k(z, \bar{z}) V_{k'}(z', \bar{z}') \rangle_{S^2} = \frac{i (2\pi)^D \delta^{(D)}(k + k')}{|z - z'|^4}. \tag{3.31}$$

The numerator comes from the zero-modes $e^{i(k+k')\cdot x}$ for a target spacetime with a Lorentzian signature [3, p. 866, 16] (required to make use of the on-shell condition).

Review of the Problem

The tree-level amplitude (3.20) for $n = 2$ reads

$$A_{0,2}(k, k') = \frac{C_{S_2}}{\text{Vol}\,\mathcal{K}_{0,0}} \int d^2z\,d^2z' \left\langle V_k(z, \bar{z}) V_{k'}(z', \bar{z}') \right\rangle_{S^2},\tag{3.32}$$

where $\mathcal{K}_{0,n}$ is the CKV group of $\Sigma_{0,n}$, the sphere with n punctures. In particular, the group of the sphere without puncture is $\mathcal{K}_{0,0} = \text{PSL}(2, \mathbb{C})$. The normalization of the amplitude is $C_{S_2} = 8\pi\alpha'^{-1}$ for $g_s = 1$ [16, 21]. Since there are two insertions, the symmetry can be partially gauge fixed by fixing the positions of the two punctures to $z = 0$ and $z' = \infty$. In this case, the amplitude (3.32) becomes

$$A_{0,2}(k, k') = \frac{C_{S^2}}{\text{Vol}\,\mathcal{K}_{0,2}} \left\langle V_k(\infty, \infty) V_{k'}(0, 0) \right\rangle_{S^2},\tag{3.33}$$

where $\mathcal{K}_{0,2} = \mathbb{R}_+^* \times U(1)$ is the CKV group of the 2-punctured sphere—containing dilatations and rotations.[3] Since the volume of this group is infinite $\text{Vol}\,\mathcal{K}_{0,2} = \infty$, it looks like $A_{0,2} = 0$. However, this forgets that the 2-point correlation function (3.31) contains a D-dimensional delta function. The on-shell condition implies that the conservation of the momentum $k + k' = 0$ is automatic for one component, such that the numerator in (3.33) contains a divergent factor $\delta(0)$

$$A_{0,2}(k, k') = (2\pi)^{D-1}\delta^{(D-1)}(k \mid k') \frac{C_{S_2} 2\pi i\,\delta(0)}{\text{Vol}\,\mathcal{K}_{0,2}}.\tag{3.34}$$

Hence, (3.33) is of the form $A_{0,2} = \infty/\infty$ and one should be careful when evaluating it.

The second argument relies on a loophole in the understanding of the gauge fixed amplitude (3.28). The result (3.28) is often summarized by saying that one can go from (3.20) to (3.28) by replacing $\mathcal{K}_g^c$ integrated vertices $\int V$ by unintegrated vertices $c\bar{c}V$ in order to saturate the ghost zero-modes and to obtain a non-zero result. For $g = 0$, this requires 3 unintegrated vertices. But, since there are only two operators in (3.32), this is impossible and the result must be zero. However, this is also incorrect because it is always possible to insert 6 c zero-modes, as shown by the formulas (2.163) and (3.27). Indeed, they are part of how the path integral measure is defined and do not care of the matter operators. The question is whether they can be attached to vertex operators (for aesthetic reasons or more pragmatically to get natural states of the BRST cohomology). To find the correct result with ghosts requires to start with (3.27) and to see how this can be simplified when there are only two operators.

---

[3]The subgroup and the associated measure depend on the locations of the two punctures.

Computation of the Amplitude

In this section, we compute the 2-point amplitude from (3.33)

$$A_{0,2}(k, k') = \frac{C_{S^2}}{\text{Vol}\,\mathcal{K}_{0,2}} \langle V_k(\infty, \infty) V_{k'}(0, 0) \rangle_{S^2}.$$
(3.35)

The volume of $\mathcal{K}_{0,2}$ reads (by writing a measure invariant under rotations and dilatations, but not translations nor special conformal transformations) [4,5]

$$\text{Vol}\,\mathcal{K}_{0,2} = \int \frac{d^2 z}{|z|^2} = 2 \int_0^{2\pi} d\sigma \int_0^\infty \frac{dr}{r},$$
(3.36)

by doing the change of variables $z = r e^{i\sigma}$. Since the volume is infinite, it must be regularized. A first possibility is to cut off a small circle of radius ϵ around $r = 0$ and $r = \infty$ (corresponding to removing the two punctures at $z = 0, \infty$). A second possibility consists in performing the change of variables $r = e^\tau$ and to add an imaginary exponential

$$\text{Vol}\,\mathcal{K}_{0,2} = 4\pi \int_0^\infty \frac{dr}{r} = 4\pi \int_{-\infty}^\infty d\tau = 4\pi \lim_{\varepsilon \to 0} \int_{-\infty}^\infty d\tau\, e^{i\varepsilon\tau} = 4\pi \times 2\pi \lim_{\varepsilon \to 0} \delta(\varepsilon),$$
(3.37)

such that the regularized volume reads

$$\text{Vol}_\varepsilon\,\mathcal{K}_{0,2} = 8\pi^2\,\delta(\varepsilon).$$
(3.38)

In fact, τ can be interpreted as the Euclidean worldsheet time on the cylinder since r corresponds to the radial direction of the complex plane.

Since the worldsheet is an embedding into the target spacetime, both must have the same signature. As a consequence, for the worldsheet to be also Lorentzian, the formula (3.37) must be analytically continued as $\varepsilon = -iE$ and $\tau = it$ such that

$$\text{Vol}_{M,E}\,\mathcal{K}_{0,2} = 8\pi^2 i\,\delta(E),$$
(3.39)

where the subscript M reminds that one considers the Lorentzian signature. Inserting this expression in (3.34) and taking the limit $E \to 0$, it looks like the two $\delta(0)$ will cancel. However, we need to be careful about the dimensions. Indeed, the worldsheet time τ and energy E are dimensionless, while the spacetime time and energy are not. Thus, it is not quite correct to cancel directly both $\delta(0)$ since they do not have the same dimensions. In order to find the correct relation between the integrals in (3.37) and of the zero-mode in (3.31), we can look at the mode expansion for the scalar field (removing the useless oscillators)

$$X^0(z, \bar{z}) = x^0 + \frac{i}{2} \alpha' k^0 \ln |z|^2 = x^0 + i\alpha' k^0 \tau,$$
(3.40)

where the second equality follows by setting $z = e^\tau$. After analytic continuation $k^0 = -\mathrm{i}k_M^0$, $X^0 = \mathrm{i}X_M^0$, $x^0 = \mathrm{i}x_M^0$, and $\tau = \mathrm{i}t$, we find [26, p. 186]

$$X_M^0 = x_M^0 + \alpha' k_M^0 t. \tag{3.41}$$

This indicates that the measure of the worldsheet time in (3.39) must be rescaled by $1/\alpha' k_M^0$ such that

$$\mathrm{Vol}_M \, \mathcal{K}_{0,2} \longrightarrow \frac{8\pi^2 \mathrm{i}\,\delta(0)}{\alpha' k_M^0} = \frac{C_{S_2}\, 2\pi \mathrm{i}\,\delta(0)}{2 k_M^0}. \tag{3.42}$$

This is equivalent to rescale E by $\alpha' k^0$ and to use $\delta(ax) = a^{-1}\delta(x)$.

Ultimately, the 2-point amplitude becomes (removing the subscript on k^0)

$$A_{0,2}(k, k') = 2k^0 (2\pi)^{D-1} \delta^{(D-1)}(\boldsymbol{k} + \boldsymbol{k}') \tag{3.43}$$

and matches the QFT formula (3.16). We see that taking into account the scale of the coordinates is important to reproduce this result.

The computation displayed here presents some ambiguities because of the regularization. However, this ambiguity can be fixed from unitarity of the scattering amplitudes. A more general version of the Faddeev–Popov gauge fixing has been introduced in [8] to avoid dealing altogether with infinities. It is an interesting question whether these techniques can be extended to compute the tree-level 1- and 0-point amplitudes on the sphere. In most cases, the 1-point amplitude is expected to vanish since 1-point correlation functions of primary operators other than the identity vanish in unitary CFTs.[4] The 0-point function corresponds to the sphere partition function: the saddle point approximation to leading order allows to relate it to the spacetime action evaluated on the classical solution ϕ_0, $Z_0 \sim e^{-S[\phi_0]/\hbar}$. Since the normalization is not known and because $S[\phi_0]$ is expected to be infinite, only comparison between two spacetimes should be meaningful (à la Gibbons–Hawking–York [15, sec. 4.1]). In particular, for Minkowski spacetime we find naively

$$Z_0 \sim \frac{\delta^{(D)}(0)}{\mathrm{Vol}\,\mathcal{K}_0}, \tag{3.44}$$

which is not well-defined. This question has not yet been investigated.

---

[4] The integral over the zero-mode gives a factor $\delta^{(D)}(k)$ that implies $k = 0$. At zero momentum, the time scalar X^0 is effectively described by unitary CFT. However, there can be some subtleties when considering marginal operator.

Expression with Ghosts

There are different ways to rewrite the 2-point amplitude in terms of ghosts. In all cases, one correctly finds the 6 insertions necessary to get a non-vanishing result since, by definition, it is always possible to rewrite the Faddeev–Popov determinant in terms of ghosts. A first approach is to insert $1 = \int d^2 z\, \delta^{(2)}(z)$ inside (3.32) to mimic the presence of a third operator. This is equivalent to use the identity

$$\langle 0|\, c_{-1}\bar{c}_{-1}c_0\bar{c}_0 c_1\bar{c}_1 \,|0\rangle = 1 \tag{3.45}$$

inside (3.33), leading to

$$A_{0,2}(k, k') = \frac{C_{S^2}}{\text{Vol}\,\mathcal{K}_{0,2}} \,\langle \mathscr{V}_k(\infty, \infty)c_0\bar{c}_0\, \mathscr{V}_{k'}(0, 0)\rangle_{S^2}\,, \tag{3.46}$$

where $\mathscr{V}_k(z, \bar{z}) = c\bar{c}V_k(z, \bar{z})$. This shows that (3.16) can also be recovered using the correct insertions of ghosts. The presence of $c_0\bar{c}_0$ can be expected from string field theory since they appear in the kinetic term (10.115).

The disadvantage of this formula is to still contain the infinite volume of the dilatation group. It is also possible to introduce ghosts for the more general gauge fixing presented in [8]. An alternative approach has been proposed in [18].

3.2   BRST Quantization

The symmetries of a Lagrangian dictate the possible terms that can be considered. This continues to hold at the quantum level, and the counter-terms introduced by renormalization are constrained by the symmetries. However, if the path integral is gauge fixed, the original symmetry is no more available for this purpose. Fortunately, one can show that there is a global symmetry (with anti-commuting parameters) remnant of the local symmetry: the BRST symmetry. It ensures consistency of the quantum theory. It also provides a direct access to the physical spectrum.

The goal of this section is to provide a general idea of the BRST quantization for the worldsheet path integral. A more detailed CFT analysis and the consequence for string theory are given in Chap. 8. The reader is assumed to have some familiarity with the BRST quantization in field theory—a summary is given in Appendix C.2.

3.2.1   BRST Symmetry

The partition function (2.159) is not most suitable to display the BRST symmetry. The first step is to restore the dependence in the original metric g_{ab} by introducing a delta function

$$
Z_g = \int_{\mathcal{M}_g} \frac{\mathrm{d}^{M_g}t}{\Omega_{\mathrm{ckv}}[g]} \int \mathrm{d}_g g_{ab}\, \mathrm{d}_g \Psi\, \mathrm{d}_g b\, \mathrm{d}'_g c\, \delta\!\left(\sqrt{g}\, g_{ab} - \sqrt{\hat{g}}\, \hat{g}_{ab}\right)
$$

$$
\times \prod_{i=1}^{M_g} (\phi_i, b)_g\, e^{-S_m[g,\Psi]-S_{\mathrm{gh}}[g,b,c]}.
\tag{3.47}
$$

Note that it is necessary to use the traceless gauge fixing condition (2.152) as it will become clear. The delta function is Fourier transformed in an exponential, thanks to an auxiliary bosonic field

$$
Z_g = \int_{\mathcal{M}_g} \frac{\mathrm{d}^{M_g}t}{\Omega_{\mathrm{ckv}}[g]} \int \mathrm{d}_g g_{ab}\, \mathrm{d}_g B^{ab}\, \mathrm{d}_g \Psi\, \mathrm{d}_g b\, \mathrm{d}'_g c
$$

$$
\times \prod_{i=1}^{M_g} (\phi_i, b)_g\, e^{-S_m[g,\Psi]-S_{\mathrm{gh}}[g,\hat{g},B]-S_{\mathrm{gh}}[g,b,c]},
\tag{3.48}
$$

where the gauge fixing action reads

$$
S_{\mathrm{gf}}[g,\hat{g},B] = -\frac{i}{4\pi} \int \mathrm{d}^2\sigma\, B^{ab}\!\left(\sqrt{g}\, g_{ab} - \sqrt{\hat{g}}\, \hat{g}_{ab}\right).
\tag{3.49}
$$

Varying the action with respect to the auxiliary field B_{ab}, called the Nakanishi–Lautrup field, produces the gauge fixing condition.

The BRST transformations are

$$
\begin{aligned}
\delta_\epsilon g_{ab} &= i\epsilon\, \mathcal{L}_c g_{ab}, & \delta_\epsilon \Psi &= i\epsilon\, \mathcal{L}_c \Psi, \\
\delta_\epsilon c^a &= i\epsilon\, \mathcal{L}_c c^a, & \delta_\epsilon b_{ab} &= \epsilon\, B_{ab}, & \delta_\epsilon B_{ab} &= 0,
\end{aligned}
\tag{3.50}
$$

where ϵ is a Grassmann parameter (anti-commuting number) independent of the position. If the traceless gauge fixing (2.152) is not used, then B_{ab} is not traceless: in that case, the variation $\delta_\epsilon b_{ab}$ will generate a trace, which is not consistent. Since the transformations act on the matter action S_m as a diffeomorphism with vector ϵc^a, it is obvious that it is invariant by itself. It is easy to show that the transformations (3.50) leave the total action invariant in (3.48). The invariance of the measure is given in [16].

Remark 3.3 (BRST Transformations with Weyl Ghost) One can also consider the action (2.153) with the Weyl ghost. In this case, the transformation law of the metric is modified and the Weyl ghost transforms as a scalar

$$
\delta_\epsilon g_{ab} = i\epsilon\, \mathcal{L}_c g_{ab} + i\epsilon\, g_{ab} c_w, \qquad \delta_\epsilon c_w = i\epsilon\, \mathcal{L}_c c_w.
\tag{3.51}
$$

The second term in $\delta_\epsilon g_{ab}$ is a Weyl transformation with parameter ϵc_w. Moreover, b_{ab} and B_{ab} are not symmetric traceless.

The equation of motion for the auxiliary field is

$$B_{ab} = i\, T_{ab} := i\big(T_{ab}^m + T_{ab}^{gh}\big), \tag{3.52}$$

where the RHS is the total energy–momentum tensor (matter plus ghosts). Integrating it out imposes the gauge condition $g_{ab} = \hat{g}_{ab}$ and yields the modified BRST transformations

$$\delta_\epsilon \Psi = i\epsilon\, \mathcal{L}_c \Psi, \qquad \delta_\epsilon c^a = i\epsilon\, \mathcal{L}_c c^a, \qquad \delta_\epsilon b_{ab} = i\epsilon\, T_{ab}. \tag{3.53}$$

Without starting with the path integral (3.48) with auxiliary field, it would have been difficult to guess the transformation of the b ghost. Since c^a is a vector, one can also write

$$\delta_\epsilon c^a = \epsilon\, c^b \partial_b c^a. \tag{3.54}$$

Associated to this symmetry is the BRST current j_B^a and the associated conserved BRST charge Q_B

$$Q_B = \int d\sigma\, j_B^0. \tag{3.55}$$

The charge is nilpotent

$$Q_B^2 = 0, \tag{3.56}$$

and through the presence of the c ghost in the BRST transformation, the BRST charge has ghost number one

$$N_{gh}(Q_B) = 1. \tag{3.57}$$

Variations of the matter fields can be written as

$$\delta_\epsilon \Psi = i\, [\epsilon Q_B, \Psi]_\pm. \tag{3.58}$$

Note that the energy–momentum tensor is BRST exact

$$T_{ab} = [Q_B, b_{ab}]. \tag{3.59}$$

3.2.2 BRST Cohomology and Physical States

Physical state $|\psi\rangle$ is the element of the *absolute* cohomology of the BRST operator

$$|\psi\rangle \in \mathcal{H}(Q_B) := \frac{\ker Q_B}{\mathrm{Im}\, Q_B}, \tag{3.60}$$

or, more explicitly, closed but non-exact states

$$Q_B |\psi\rangle = 0, \qquad \nexists |\chi\rangle : |\psi\rangle = Q_B |\chi\rangle. \tag{3.61}$$

The adjective "absolute" is used to distinguish it from two other cohomologies (relative and semi-relative) defined below. Two states of the cohomology differing by an exact state represent identical physical states

$$|\psi\rangle \sim |\psi\rangle + Q_B |\Lambda\rangle. \tag{3.62}$$

This equivalence relation, translated in terms of spacetime fields, corresponds to spacetime gauge transformations. In particular, it contains the (linearized) reparametrization invariance of the spacetime metric in the closed string sector, and for the open string sector, it contains Yang–Mills symmetries. We will find that it corresponds to the gauge invariance of free string field theory (Chap. 10).

However, physical states satisfy two additional constraints (remember that b_{ab} is traceless symmetric)

$$\int \mathrm{d}\sigma\, b_{ab} |\psi\rangle = 0. \tag{3.63}$$

These conditions are central to string (field) theory, so they will appear regularly in this book. For this reason, it is useful to provide first some general motivations and to refine the analysis later since the CFT language will be more appropriate. Moreover, these two conditions will naturally emerge in string field theory.

In order to introduce some additional terminology, let us define the following quantities:[5]

$$b^+ := \int \mathrm{d}\sigma\, b_{00}, \qquad b^- := \int \mathrm{d}\sigma\, b_{01}. \tag{3.64}$$

---

[5]The objects $b^\pm$ are zero-modes of the b ghost fields. They correspond (up to a possible irrelevant factor) to the modes $b_0^\pm$ in the CFT formulation of the ghost system (7.132)

The *semi-relative* and *relative* cohomologies $\mathcal{H}^-(Q_B)$ and $\mathcal{H}^0(Q_B)$ are defined as[6]

$$\mathcal{H}^-(Q_B) = \mathcal{H}(Q_B) \cap \ker b^-, \qquad \mathcal{H}^0(Q_B) = \mathcal{H}^-(Q_B) \cap \ker b^+. \qquad (3.65)$$

The first constraint arises as a consequence of the topology of the closed string worldsheet: the spatial direction is a circle, which implies that the theory must be invariant under translations along the σ direction (the circle is invariant under rotation). However, choosing a parametrization implies to fix an origin for the spatial direction: this is equivalent to a gauge fixing condition. As usual, this implies that the corresponding generator P_σ of worldsheet spatial translations (2.26) must annihilate the states

$$P_\sigma |\psi\rangle = 0. \qquad (3.66)$$

This is called the *level-matching condition*. Using (3.59), this can be rewritten as

$$P_\sigma |\psi\rangle = \int d\sigma \, T_{01} |\psi\rangle = \int d\sigma \, \{Q_B, b_{01}\} |\psi\rangle = Q_B \int d\sigma \, b_{01} |\psi\rangle, \qquad (3.67)$$

since $Q_B |\psi\rangle = 0$ for a state $|\psi\rangle$ in the cohomology. The simplest way to enforce this condition is to set the state on which Q_B acts to zero[7]

$$b^- |\psi\rangle = 0, \qquad (3.68)$$

which is equivalent to one of the conditions in (3.63).

The second condition does not follow as simply. The Hilbert space can be decomposed according to b^+ as

$$\mathcal{H}^- := \mathcal{H}_\downarrow \oplus \mathcal{H}_\uparrow, \qquad \mathcal{H}_\downarrow := \mathcal{H}^0 := \mathcal{H}^- \cap \ker b^+. \qquad (3.69)$$

Indeed, b^+ is a Grassmann variable and generates a 2-state system. In the ghost sector, the two Hilbert spaces are generated from the ghost vacua $|\downarrow\rangle$ and $|\uparrow\rangle$ obeying

$$b^+ |\downarrow\rangle = 0, \qquad b^+ |\uparrow\rangle = |\downarrow\rangle. \qquad (3.70)$$

---

[6]The BRST cohomologies described in this section are slightly different from the ones used in the rest of this book. To distinguish them, indices are written as superscripts in this section and as subscripts otherwise.

[7]The reverse is not true. We will see in Sect. 3.2.2 the relation between the two conditions in more details.

The action of the BRST charge on states $|\psi_\downarrow\rangle \in \mathcal{H}_\downarrow$ and $|\psi_\uparrow\rangle \in \mathcal{H}_\uparrow$ follows from these relations and from the commutation relation (3.59)

$$Q_B |\psi_\downarrow\rangle = H |\psi_\uparrow\rangle, \qquad Q_B |\psi_\uparrow\rangle = 0, \qquad (3.71)$$

where H is the worldsheet Hamiltonian defined in (2.26). To prove this relation, start first with $H |\psi_\uparrow\rangle$, and then use (3.59)) to get the LHS of the first condition; then apply Q_B to get the second condition (using that Q_B commutes with H, and b^+ with any other operators building the states). For $H \neq 0$, the state $|\psi_\downarrow\rangle$ is not in the cohomology and $|\psi_\uparrow\rangle$ is exact. Thus, the exact and closed states are

$$\text{Im}\, Q_B = \left\{ |\psi_\uparrow\rangle \in \mathcal{H}_\uparrow \mid H |\psi_\uparrow\rangle \neq 0 \right\}, \qquad (3.72a)$$

$$\ker Q_B = \left\{ |\psi_\uparrow\rangle \in \mathcal{H}_\uparrow \right\} \cup \left\{ |\psi_\downarrow\rangle \in \mathcal{H}_\downarrow \mid H |\psi_\downarrow\rangle = 0 \right\}. \qquad (3.72b)$$

This implies that eigenstates of H in the cohomology satisfy the on-shell condition

$$H |\psi\rangle = 0. \qquad (3.73)$$

This is consistent with the fact that scattering amplitudes involve on-shell states. In this case, $|\psi_\uparrow\rangle$ is not exact and is thus a member of the cohomology $\mathcal{H}(Q_B)$, as well as $|\psi_\downarrow\rangle$ since it becomes close. But, the Hilbert space $\mathcal{H}_\uparrow$ must be rejected for two reasons: there would be an apparent doubling of states and scattering amplitudes would behave badly. The first problem arises because one can show that the cohomological subspaces of each space are isomorphic: $\mathcal{H}_\downarrow(Q_B) \simeq \mathcal{H}_\uparrow(Q_B)$. Hence, keeping both subspaces would lead to a doubling of the physical states. For the second problem, consider an amplitude where one of the external states is built from $|\psi_\uparrow\rangle$: the amplitude vanishes if the states are off-shell since the state $|\psi_\uparrow\rangle$ is exact, but it does not vanish on-shell [16, ch. 4]. This means that it must be proportional to $\delta(H)$. But, general properties in QFT forbid such dependence in the amplitude (only poles and cuts are allowed, except if $D = 2$). Projecting out the states in $\mathcal{H}_\uparrow$ is equivalent to require

$$b^+ |\psi\rangle = 0 \qquad (3.74)$$

for physical states, which is the second condition in (3.63).

In fact, this condition can be obtained very similarly as the $b^- = 0$ condition: using the expression of H (2.26) and the commutation relation (3.59), (3.73) is equivalent to

$$Q_B \int d\sigma\, b_{00} |\psi\rangle = 0. \qquad (3.75)$$

Hence, imposing (3.74) allows to automatically ensure that (3.73) holds.

Since the on-shell characters (3.73) of the BRST states and of the BRST symmetry are intimately related to the construction of the worldsheet integral, one can expect difficulty for going off-shell.

3.3   Summary

In this chapter, we derived general formulas for string scattering amplitudes. The general BRST formalism has been summarized. Moreover, we gave general motivations for restricting the absolute cohomology to the smaller relative cohomology. In Chap. 8, a more precise derivation of the BRST cohomology is worked out. It also includes a proof of the no-ghost theorem: the ghosts and the negative-norm states (in Minkowski signature) are unphysical particles and should not be part of the physical states. This theorem asserts that it is indeed the case. It will also be the occasion to recover the details of the spectrum in various cases.

3.4   Suggested Readings

- The delta function approach to the gauge fixing is described in [16, sec. 3.3, 13, sec. 15.3.2], with a more direct computation in [12].
- The most complete references for scattering amplitudes in the path integral formalism are [4, 16].
- Computation of the tree-level 2-point amplitude [8,18] (for discussions of 2-point function, see [4, p. 936–7, 3, 5, 6, 17, p. 863–4]).
- The BRST quantization of string theory is discussed in [2, 14, 16, chap. 4]. For a general discussion, see [10, 20, 23]. The use of an auxiliary field is considered in [24, sec. 3.2].

References

1. J. Collins, A New Approach to the LSZ Reduction Formula (2019). arXiv: 1904.10923
2. B. Craps, K. Skenderis, Comments on BRST quantization of strings. J. High Energy Phys. **2005**(05), 001 (2005). https://doi.org/10.1088/1126-6708/2005/05/001. arXiv: hep-th/0503038
3. P. Deligne, P. Etingof, D.S. Freed, L.C. Jeffrey, D. Kazhdan, J.W. Morgan, D.R. Morrison, E. Witten (edd.), *Quantum Fields and Strings: A Course for Mathematicians. Volume 2* (American Mathematical Society, Providence, 1999)
4. E. D'Hoker, D.H. Phong, The geometry of string perturbation theory. Rev. Modern Phys. **60**(4), 917–1065 (1988). https://doi.org/10.1103/RevModPhys.60.917
5. H. Dorn, H.-J. Otto, Two and three-point functions in Liouville theory. Nuclear Phys. B **429**(2), 375–388 (1994). https://doi.org/10.1016/0550-3213(94)00352-1. arXiv: hep-th/9403141
6. H. Dorn, H.-J. Otto, Some Conclusions for Noncritical String Theory Drawn from Two- and Three-Point Functions in the Liouville Sector (1995). arXiv: hep-th/9501019
7. A. Duncan, *The Conceptual Framework of Quantum Field Theory*, 1st edn. (Oxford University Press, Oxford, 2012)

8. H. Erbin, J. Maldacena, D. Skliros, Two-point string amplitudes. J. High Energy Phys. **2019**(7), 139 (2019). https://doi.org/10.1007/JHEP07(2019)139. arXiv: 1906.06051
9. L.D. Faddeev, *Introduction to Functional Methods*. Conf. Proc. C7507281 (1975), pp. 1–40. https://doi.org/10.1142/9789814412674_0001
10. M. Henneaux, C. Teitelboim, *Quantization of Gauge Systems* (Princeton University Press, Princeton, 1994)
11. C. Itzykson, J.-B. Zuber, *Quantum Field Theory* (Dover Publications, New York, 2006)
12. E. Kiritsis, *String Theory in a Nutshell* (Princeton University Press, Princeton, 2007)
13. I.D. Lawrie, *A Unified Grand Tour of Theoretical Physics*, 3rd edn. (CRC Press, Boca Raton, 2012)
14. P. Mansfield, Nilpotent BRST invariance of the interacting Polyakov string. Nuclear Phys. B **283**(Suppl. C), 551–576 (1987). https://doi.org/10.1016/0550-3213(87)90286-0
15. E. Poisson, *A Relativist's Toolkit: The Mathematics of Black-Hole Mechanics*, 1st edn. (Cambridge University Press, Cambridge, 2007)
16. J. Polchinski, *String Theory: Volume 1, An Introduction to the Bosonic String* (Cambridge University Press, Cambridge, 2005)
17. N. Seiberg, Notes on quantum Liouville theory and quantum gravity. Progr. Theoret. Phys. Suppl. **102**, 319–349 (1990). https://doi.org/10.1143/PTPS.102.319
18. S. Seki, T. Takahashi, Two-point string amplitudes revisited by operator formalism. Phys. Lett. B **800**, 135078 (2020). https://doi.org/10.1016/j.physletb.2019.135078. arXiv: 1909.03672
19. M. Srednicki, *Quantum Field Theory*. English 1st edn. (Cambridge University Press, Cambridge, 2007)
20. J.W. van Holten, Aspects of BRST quantization. Lect. Notes Phys. **659**, 99–166 (2002). arXiv: hep-th/0201124
21. T. Weigand, *Introduction to String Theory* (Springer, Berlin, 2012). http://www.thphys.uni-heidelberg.de/~weigand/Skript-strings11-12/Strings.pdf.
22. S. Weinberg, *The Quantum Theory of Fields, Volume 1: Foundations* (Cambridge University Press, Cambridge, 2005)
23. S. Weinberg, *The Quantum Theory of Fields, Volume 2: Modern Applications* (Cambridge University Press, Cambridge, 2005)
24. P. West, *Introduction to Strings and Branes*, 1st edn. (Cambridge University Press, Cambridge, 2012)
25. J. Zinn-Justin, *Quantum Field Theory and Critical Phenomena*, 4th edn. (Clarendon Press, Oxford, 2002)
26. B. Zwiebach, *A First Course in String Theory*, 2nd edn. (Cambridge University Press, Cambridge, 2009)

Worldsheet Path Integral: Complex Coordinates

4

Abstract

In the two previous chapters, the amplitudes computed from the worldsheet path integrals have been written covariantly for a generic curved background metric. In this chapter, we start to use complex coordinates and finally take the background metric to be flat. This is the usual starting point for computing amplitudes since it allows to make contact with CFTs and to employ tools from complex analysis. We first recall few facts on $2d$ complex manifolds before briefly describing how to rewrite the scattering amplitudes in complex coordinates.

4.1 Geometry of Complex Manifolds

Choosing a flat background metric simplifies the computations. However, we have seen in Sect. 2.3 that there is a topological obstruction to get a globally flat metric. The solution is to work with coordinate patches $(\sigma^0, \sigma^1) = (\tau, \sigma)$ such that the background metric $\hat{g}_{ab}$ is flat in each patch (conformally flat gauge):

$$ds^2 = g_{ab}d\sigma^a d\sigma^b = e^{2\phi(\tau,\sigma)}(d\tau^2 + d\sigma^2), \tag{4.1}$$

or

$$g_{ab} = e^{2\phi}\delta_{ab}, \qquad \hat{g}_{ab} = \delta_{ab}. \tag{4.2}$$

To simplify the notations, we remove the dependence in the flat metric and the hat for quantities (like the vertex operators) expressed in the background metric when no confusion is possible.

© Springer Nature Switzerland AG 2021
H. Erbin, *String Field Theory*, Lecture Notes in Physics 980,
https://doi.org/10.1007/978-3-030-65321-7_4

Introducing complex coordinates

$$z = \tau + i\sigma, \qquad \bar{z} = \tau - i\sigma, \tag{4.3a}$$

$$\tau = \frac{z + \bar{z}}{2}, \qquad \sigma = \frac{z - \bar{z}}{2i}, \tag{4.3b}$$

the metric reads[1]

$$ds^2 = 2g_{z\bar{z}}dzd\bar{z} = e^{2\phi(z,\bar{z})}|dz|^2. \tag{4.4}$$

The metric and its inverse can also be written in components:

$$g_{z\bar{z}} = \frac{e^{2\phi}}{2}, \qquad g_{zz} = g_{\bar{z}\bar{z}} = 0, \tag{4.5a}$$

$$g^{z\bar{z}} = 2e^{-2\phi}, \qquad g^{zz} = g^{\bar{z}\bar{z}} = 0. \tag{4.5b}$$

Equivalently, the non-zero components of the background metric are

$$\hat{g}_{z\bar{z}} = \frac{1}{2}, \qquad \hat{g}^{z\bar{z}} = 2. \tag{4.6}$$

An oriented two-dimensional manifold is a complex manifold: this means that there exists a complex structure, such that the transition functions and changes of coordinates between different patches are holomorphic at the intersection of the two patches:

$$w = w(z), \qquad \bar{w} = \bar{w}(\bar{z}). \tag{4.7}$$

For such a transformation, the Liouville mode transforms as

$$e^{2\phi(z,\bar{z})} = \left|\frac{\partial w}{\partial z}\right|^2 e^{2\phi(w,\bar{w})} \tag{4.8}$$

such that

$$ds^2 = e^{2\phi(w,\bar{w})}|dw|^2. \tag{4.9}$$

This shows also that a conformal structure (2.12) induces a complex structure since the transformation law of ϕ is equivalent to a Weyl rescaling.

---

[1] In Sect. 6.1, we provide more details on the relation between the worldsheet (viewed as a cylinder or a sphere) and the complex plane.

The integration measures are related as

$$d^2\sigma := d\tau \, d\sigma = \frac{1}{2} d^2 z, \qquad d^2 z := dz \, d\bar{z}. \tag{4.10}$$

Due to the factor of 2 in the expression, the delta function $\delta^{(2)}(z)$ also gets a factor of 2 with respect to $\delta^{(2)}(\sigma)$

$$\delta^{(2)}(z) = \frac{1}{2} \delta^{(2)}(\sigma). \tag{4.11}$$

Then, one can check that

$$\int d^2 z \, \delta^{(2)}(z) = \int d^2\sigma \, \delta^{(2)}(\sigma) = 1. \tag{4.12}$$

The basis vectors (derivatives) and one-forms can be found using the chain rule:

$$\partial_z = \frac{1}{2}(\partial_\tau - i\partial_\sigma), \qquad \partial_{\bar{z}} = \frac{1}{2}(\partial_\tau + i\partial_\sigma), \tag{4.13a}$$

$$dz = d\tau + id\sigma, \qquad d\bar{z} = d\tau - id\sigma. \tag{4.13b}$$

The Levi-Civita (completely antisymmetric) tensor is normalized by

$$\epsilon_{01} = \epsilon^{01} = 1. \tag{4.14a}$$

$$\epsilon_{z\bar{z}} = \frac{i}{2}, \qquad \epsilon^{z\bar{z}} = -2i, \tag{4.14b}$$

remembering that it transforms as a density. Integer indices run over local frame coordinates.

The different tensors can be found from the tensor transformation law. For example, the components of a vector V^a in both systems are related by

$$V^z = V^0 + iV^1, \qquad V^{\bar{z}} = V^0 - iV^1 \tag{4.15}$$

such that

$$V = V^0 \partial_0 + V^1 \partial_1 = V^z \partial_z + V^{\bar{z}} \partial_{\bar{z}}. \tag{4.16}$$

For holomorphic coordinate transformations (4.7), the components of the vector do not mix:

$$V^w = \frac{\partial w}{\partial z} V^z, \qquad V^{\bar{w}} = \frac{\partial \bar{w}}{\partial \bar{z}} V^{\bar{z}}. \tag{4.17}$$

This implies that the tangent space of the Riemann surface is decomposed into holomorphic and anti-holomorphic vectors:[2]

$$T\Sigma_g \simeq T\Sigma_g^+ \oplus T\Sigma_g^- , \tag{4.18a}$$

$$V^z \partial_z \in T\Sigma_g^+ , \qquad V^{\bar{z}} \partial_{\bar{z}} \in T\Sigma_g^- , \tag{4.18b}$$

as a consequence of the existence of a complex structure. Similarly, the components of a 1-form ω—which is the only non-trivial form on Σ_g—can be written in terms of the real coordinates as

$$\omega_z = \frac{1}{2}(\omega_0 - i\omega_1), \qquad \omega_{\bar{z}} = \frac{1}{2}(\omega_0 + i\omega_1) \tag{4.19}$$

such that

$$\omega = \omega_0 d\sigma^0 + \omega_1 d\sigma^1 = \omega_z dz + \omega_{\bar{z}} d\bar{z}. \tag{4.20}$$

Hence, a 1-form is decomposed into complex $(1, 0)$- and $(0, 1)$-forms:

$$T^*\Sigma_g \simeq \Omega^{1,0}(\Sigma_g) \oplus \Omega^{0,1}(\Sigma_g), \tag{4.21a}$$

$$\omega_z dz \in \Omega^{1,0}(\Sigma_g), \qquad \omega_{\bar{z}} d\bar{z} \in \Omega^{0,1}(\Sigma_g), \tag{4.21b}$$

since both components will not mixed under holomorphic changes of coordinates (4.7). Finally, the metric provides an isomorphism between $T\Sigma_g^+$ and $\Omega^{0,1}(\Sigma_g)$, and between $T\Sigma_g^-$ and $\Omega^{1,0}(\Sigma_g)$, since it can be used to lower/raise an index while converting it from holomorphic to anti-holomorphic, or conversely:

$$V_z = g_{z\bar{z}} V^{\bar{z}}, \qquad V_{\bar{z}} = g_{z\bar{z}} V^z. \tag{4.22}$$

This can be generalized further by considering components with more indices: all anti-holomorphic indices can be converted to holomorphic indices thanks to the metric:

$$T^{\overbrace{z \cdots z}^{q+ + p-}}_{\underbrace{z \cdots z}_{p+ + q-}} = (g^{z\bar{z}})^{p-} (g_{z\bar{z}})^{q-} T^{\overbrace{z \cdots z}^{q+}\,\overbrace{\bar{z} \cdots \bar{z}}^{q-}}_{\underbrace{z \cdots z}_{p+}\,\underbrace{\bar{z} \cdots \bar{z}}_{p-}}. \tag{4.23}$$

---

[2]However, at this stage, each component can still depend on both z and $\bar{z}$: $V^z = V^z(z, \bar{z})$ and $V^{\bar{z}} = V^{\bar{z}}(z, \bar{z})$.

Hence, it is sufficient to study (p, q)-tensors with p upper and q lower holomorphic indices. In this case, the transformation rule under (4.7) reads

$$T^{\overbrace{w \cdots w}^{q}}_{\underbrace{w \cdots w}_{p}} = \left(\frac{\partial w}{\partial z}\right)^{n} T^{\overbrace{z \cdots z}^{q}}_{\underbrace{z \cdots z}_{p}}, \qquad n := q - p. \tag{4.24}$$

The number $n \in \mathbb{Z}$ is called the helicity or rank.[3] The set of helicity-n tensors is denoted by $\mathcal{T}^{n}$.

The first example is vectors (or equivalently 1-forms): $V^{z} \in \mathcal{T}^{1}$, $V_{z} \in \mathcal{T}^{-1}$. The second most useful case is traceless symmetric tensors, which are elements of $\mathcal{T}^{\pm 2}$. Consider a traceless symmetric tensor $T^{ab} = T^{ba}$ and $g_{ab}T^{ab} = 0$: this implies $T^{01} = T^{10}$ and $T^{00} = -T^{11}$ in real coordinates. The components in complex coordinates are

$$T^{zz} = 2(T^{00} + iT^{01}) \in \mathcal{T}^{2}, \qquad T^{\bar{z}\bar{z}} = 2(T^{00} - iT^{01}) \in \mathcal{T}^{-2}, \qquad T^{z}_{z} = 0. \tag{4.25}$$

Note that

$$T_{zz} = g_{z\bar{z}}g_{z\bar{z}}T^{\bar{z}\bar{z}} = \frac{1}{2}(T^{00} - iT^{01}), \tag{4.26}$$

and $T^{z}_{z} = g_{zz}T^{zz} \in \mathcal{T}^{0}$ corresponds to the trace.

Computation: Equation (4.25)

$$T^{zz} = \left(\frac{\partial z}{\partial \tau}\right)^{2} T^{00} + \left(\frac{\partial z}{\partial \sigma}\right)^{2} T^{11} + 2\frac{\partial z}{\partial \tau}\frac{\partial z}{\partial \sigma} T^{01} = T^{00} - T^{11} + 2i\, T^{01}.$$

Stokes' theorem in complex coordinates follows directly from (B.10):

$$\int d^{2}z\, (\partial_{z}v^{z} + \partial_{\bar{z}}v^{\bar{z}}) = -i \oint (dz\, v^{\bar{z}} - d\bar{z}v^{z}) = -2i \oint_{\partial R} (v_{z}dz - v_{\bar{z}}d\bar{z}), \tag{4.27}$$

where the integration contour is anti-clockwise. To obtain this formula, note that $d^{2}x = \frac{1}{2}d^{2}z$ and $\epsilon_{z\bar{z}} = i/2$, such that the factor $1/2$ cancels between both sides.

---

[3] In fact, it is even possible to consider $n \in \mathbb{Z} + 1/2$ to describe spinors.

4.2   Complex Representation of Path Integral

In the previous section, we have found that tensors of a given rank are naturally decomposed into different subspaces thanks to the complex structure of the manifold. Accordingly, complex coordinates are natural and one can expect most objects in string theory to split similarly into holomorphic and anti-holomorphic sectors (or left- and right-moving). This will be particularly clear using the CFT language (Chap. 6). The main difficulty for this program is due to the matter zero-modes. In this section, we focus on the path integral measure and expression of the ghosts.

There is, however, a subtlety in displaying explicitly the factorization: the notion of "holomorphicity" depends on the metric (because the complex structure must be compatible with the metric for a Hermitian manifold). Since the metric depends on the moduli which are integrated over in the path integral, it is not clear that there is a consistent holomorphic factorization. We will not push the question of achieving a global factorization further (but see Remark 4.1) to focus instead on the integrand. The latter is local (in moduli space) and there is no ambiguity.

The results of the previous section indicate that the basis of Killing vectors (2.104) and quadratic differentials (2.76) split into holomorphic and anti-holomorphic components:

$$\psi_i(z, \bar{z}) = \psi_i^z \partial_z + \psi_i^{\bar{z}} \partial_{\bar{z}}, \qquad \phi_i(z, \bar{z}) = \phi_{i,zz}(dz)^2 + \phi_{i,\bar{z}\bar{z}}(d\bar{z})^2. \qquad (4.28)$$

Similarly, the operators P_1 (2.65a) and $P_1^\dagger$ (2.71) also split:

$$(P_1\xi)_{zz} = 2\nabla_z\xi_z = \partial_z\xi^{\bar{z}}, \qquad (P_1\xi)_{\bar{z}\bar{z}} = 2\nabla_{\bar{z}}\xi_{\bar{z}} = \partial_{\bar{z}}\xi^z, \qquad (4.29a)$$

$$(P_1^\dagger T)_z = -2\nabla^z T_{zz} = -4\,\partial_{\bar{z}}T_{zz}, \qquad (P_1^\dagger T)_{\bar{z}} = -2\nabla^{\bar{z}}T_{\bar{z}\bar{z}} = -4\,\partial_z T_{\bar{z}\bar{z}} \qquad (4.29b)$$

for arbitrary vector ξ and traceless symmetric tensor T (in the background metric). As a consequence, the components of Killing vectors and quadratic differentials are holomorphic or anti-holomorphic as a function of z:

$$\psi^z = \psi^z(z), \qquad \psi^{\bar{z}} = \psi^{\bar{z}}(\bar{z}), \qquad \phi_{zz} = \phi_{zz}(z), \qquad \phi_{\bar{z}\bar{z}} = \phi_{\bar{z}\bar{z}}(\bar{z}), \qquad (4.30)$$

such that it makes sense to consider a complex basis instead of the previous real basis:

$$\ker P_1 = \mathrm{Span}\{\psi_K(z)\} \oplus \mathrm{Span}\{\bar{\psi}_K(\bar{z})\}, \qquad K = 1, \ldots, \mathrm{K}_g^c, \qquad (4.31a)$$

$$\ker P_1^\dagger = \mathrm{Span}\{\phi_I(z)\} \oplus \mathrm{Span}\{\bar{\phi}_I(\bar{z})\}, \qquad I = 1, \ldots, \mathrm{M}_g^c. \qquad (4.31b)$$

The last equation can inspire to search for a similar rewriting of the moduli parameters. In fact, the moduli space itself is a complex manifold and can be endowed with complex coordinates [7, 8]:

$$m_I = t_{2I-1} + it_{2I}, \qquad \bar{m}_I = t_{2I-1} - it_{2I}, \qquad I = 1, \ldots, \mathrm{M}_g^c \qquad (4.32)$$

with the integration measure

$$d^{\mathrm{M}_g} t = d^{2\mathrm{M}_g^c} m. \qquad (4.33)$$

The last ingredient to rewrite the vacuum amplitudes (2.136) is to obtain the determinants. The inner products of vector and traceless symmetric fields also factorize

$$(T_1, T_2) = 2 \int d^2\sigma \sqrt{\hat{g}} \, \hat{g}^{ac} g^{bd} T_{1,ab} T_{2,cd} = 4 \int d^2z \left(T_{1,zz} T_{2,\bar{z}\bar{z}} + T_{1,\bar{z}\bar{z}} T_{2,zz} \right), \qquad (4.34a)$$

$$(\xi_1, \xi_2) = \int d^2\sigma \sqrt{\hat{g}} \, \hat{g}_{ab} \xi^a \xi^b = \frac{1}{4} \int d^2z \left(\xi_1^z \xi_2^{\bar{z}} + \xi_1^{\bar{z}} \xi_2^z \right). \qquad (4.34b)$$

All inner products are evaluated in the flat background metric. For (anti-)holomorphic fields, only one term survives in each integral: since each field appears twice in the determinants (ϕ_i, ϕ_j) and (ϕ_i, ϕ_j), the final expression is a square, which cancels against the squareroot in (2.136). The remaining determinant involves the Beltrami differential (2.65b):

$$\mu_{izz} = \partial_i \bar{g}_{zz}, \qquad \mu_{i\bar{z}\bar{z}} = \partial_i \bar{g}_{\bar{z}\bar{z}} \qquad (4.35)$$

($\bar{g}_{zz} = 0$ in our coordinates system, but its variation under a shift of moduli is not zero). The basis can be changed to a complex basis such that the determinant of inner products between Beltrami and quadratic differentials is a modulus squared. All together, the different formulas lead to the following rewriting of the vacuum amplitude :

$$Z_g = \int_{\mathcal{M}_g} d^{2\mathrm{M}_g^c} m \, \frac{|\det(\phi_I, \mu_J)|^2}{|\det(\phi_I, \bar{\phi}_J)|} \, \frac{\det' P_1^\dagger P_1}{|\det(\psi_I, \bar{\psi}_J)|} \, \frac{Z_m[\delta]}{\Omega_{\mathrm{ckv}}[\delta]}, \qquad (4.36)$$

where the absolute values are to be understood with respect to the basis of P_1 and $P_1^\dagger$, for example $|f(m_I)|^2 := f(m_I) f(\bar{m}_I)$.

The same reasoning can be applied to the ghosts. The c and b ghosts are respectively a vector and a symmetric traceless tensor, both with two independent components: it is customary to define

$$c := c^z, \qquad \bar{c} := c^{\bar{z}}, \qquad b := b_{zz}, \qquad \bar{b} := b_{\bar{z}\bar{z}}. \qquad (4.37)$$

In that case, the action (2.145) reads

$$S_{\text{gh}}[g, b, c] = \frac{1}{2\pi} \int d^2z \left(b\partial_{\bar{z}}c + \bar{b}\partial_z\bar{c} \right).$$ (4.38)

The action is the sum of two holomorphic and anti-holomorphic contributions and it is independent of $\phi(z, \bar{z})$ as expected. In fact, the equations of motion are

$$\partial_z c = 0, \qquad \partial_z b = 0, \qquad \partial_{\bar{z}}\bar{c} = 0, \qquad \partial_{\bar{z}}\bar{b} = 0,$$ (4.39)

such that b and c (resp. $\bar{b}$ and $\bar{c}$) are holomorphic (anti-holomorphic) functions. Then, the integration measure is simply

$$\bigwedge_{i=1}^{M_g} B_i \, dt_i = \bigwedge_{I=1}^{M_g^c} B_I \bar{B}_I \, dm_I \wedge \bar{m}_I, \qquad B_I := (\mu_I, b).$$ (4.40)

Note that B_I does not contain $\bar{b}(\bar{z})$, it is built only from $b(z)$.

Finally, the vacuum amplitude (2.163) reads

$$Z_g = \int_{\mathcal{M}_g} d^{2M_g^c} m \, \frac{\Omega_{\text{ckv}}[\delta]^{-1}}{|\det \psi_I(z_j^0)|^2}$$

$$\times \int d(b, \bar{b}) \, d(c, \bar{c}) \prod_{j=1}^{K_g^c} c(z_j^0)\bar{c}(\bar{z}_j^0) \prod_{I=1}^{M_g^c} |(\mu_I, b)|^2 \, e^{-S_{\text{gh}}[b,c]} \, Z_m[\delta].$$

(4.41)

The c insertions are separated in holomorphic and anti-holomorphic components because, at the end, only the zero-modes contribute. The measures are written as $d(b, \bar{b})$ and $d(c, \bar{c})$ because proving that they factorize is difficult (Remark 4.1).

Remark 4.1 (Holomorphic Factorization) It was proven in [1, 3, 4] (see [7, sec. 9, 5, sec. VII, 9, sec. 3] for reviews) that the ghost and matter path integrals can be globally factorized, up to a factor due to zero-modes. Such a result is suggested by the factorization of the inner products, which imply a factorization of the measures: the caveat is due to the zero-mode determinants and matter measure. Interestingly, the factorization is possible only in the critical dimension (2.125).

4.3 Summary

In this chapter, we have introduced complex notations for the fields, path integral and moduli space.

4.4 Suggested Readings

- Good references for this chapter are [2, 5–8].
- Geometry of complex manifolds is discussed in [2, sec. 6.2, 6, chap. 14, 5].

References

1. A.A. Belavin, V.G. Knizhnik, Algebraic geometry and the geometry of quantum strings. Phys. Lett. B **168**(3), 201–206 (1986). https://doi.org/10.1016/0370-2693(86)90963-9
2. R. Blumenhagen, D. Lüst, S. Theisen, *Basic Concepts of String Theory*, 2013 edn. (Springer, Berlin, 2014)
3. J.B. Bost, T. Jolicoeur, A holomorphy property and the critical dimension in string theory from an index theorem. Phys. Lett. B **174**(3), 273–276 (1986). https://doi.org/10.1016/0370-2693(86)91097-X
4. R. Catenacci, M. Cornalba, M. Martellini, C. Reina, Algebraic geometry and path integrals for closed strings. Phys. Lett. B **172**(3), 328–332 (1986). https://doi.org/10.1016/0370-2693(86)90262-5
5. E. D'Hoker, D.H. Phong, The geometry of string perturbation theory. Rev. Mod. Phys. **60**(4), 917–1065 (1988). https://doi.org/10.1103/RevModPhys.60.917
6. M. Nakahara, *Geometry, Topology and Physics*, 2nd edn. (Institute of Physics Publishing, Bristol, 2003)
7. P. Nelson, Lectures on strings and moduli space. Phys. Rep. **149**(6), 337–375 (1987). https://doi.org/10.1016/0370-1573(87)90082-2
8. J. Polchinski, *String Theory: Volume 1, An Introduction to the Bosonic String* (Cambridge University Press, Cambridge, 2005)
9. D.P. Skliros, E.J. Copeland, P.M. Saffin, Highly excited strings I: generating function. Nucl. Phys. B **916**, 143–207 (2017). https://doi.org/10.1016/j.nuclphysb.2016.12.022. arXiv:1611.06498

Conformal Symmetry in D Dimensions

5

Abstract

Starting with this chapter, we discuss general properties of conformal field theories (CFT). The goal is not to be exhaustive, but to provide a short introduction and to gather the concepts and formulas that are needed for string theory. However, the subject is presented as a standalone topic such that it can be of interest for a more general public.

The conformal group in any dimension is introduced in this chapter. The specific case $D = 2$, which is the most relevant for the current book, is developed in the following chapters.

5.1 CFT on a General Manifold

In this chapter and in the next one, we discuss CFTs as QFTs living on a spacetime $\mathcal{M}$, independently from string theory (there is no reference to a target spacetime). As such, we will use spacetime notations together with some simplifications: coordinates are written as x^μ with $\mu = 0, \ldots, D - 1$ and time is written as $x^0 = t$ ($x^0 = \tau$) in Lorentzian (Euclidean) signature.

Given a metric $g_{\mu\nu}$ on a D-dimensional manifold $\mathcal{M}$, the conformal group $\text{CISO}(\mathcal{M})$ is the set of coordinate transformations (called conformal symmetries or isometries)

$$x^\mu \longrightarrow x'^\mu = x'^\mu(x) \tag{5.1}$$

which leaves the metric invariant up to an overall scaling factor:

$$g_{\mu\nu}(x) \longrightarrow g'_{\mu\nu}(x') = \frac{\partial x^\rho}{\partial x'^\mu} \frac{\partial x^\sigma}{\partial x'^\nu} g_{\rho\sigma}(x) = \Omega(x')^2 g_{\mu\nu}(x'). \tag{5.2}$$

© Springer Nature Switzerland AG 2021
H. Erbin, *String Field Theory*, Lecture Notes in Physics 980,
https://doi.org/10.1007/978-3-030-65321-7_5

This means that angles between two vectors u and v are left invariant under the transformation:

$$\frac{u \cdot v}{|u|\,|v|} = \frac{u' \cdot v'}{|u'|\,|v'|}. \tag{5.3}$$

It is often convenient to parametrize the scale factor by an exponential

$$\Omega := e^{\omega}. \tag{5.4}$$

Considering an infinitesimal transformation

$$\delta x^{\mu} = \xi^{\mu}, \tag{5.5}$$

the condition (5.2) becomes the conformal Killing equation

$$\delta g_{\mu\nu} = \mathcal{L}_{\xi} g_{\mu\nu} = \nabla_{\mu}\xi_{\nu} + \nabla_{\nu}\xi_{\mu} = \frac{2}{d} g_{\mu\nu} \nabla_{\rho}\xi^{\rho}, \tag{5.6}$$

such that the scale factor is

$$\Omega^2 = 1 + \frac{2}{d} \nabla_{\rho}\xi^{\rho}. \tag{5.7}$$

The vector fields ξ satisfying this equation are called conformal Killing vectors (CKV). Conformal transformations form a global subgroup of the diffeomorphism group: the generators of the transformations do depend on the coordinates, but the parameters do not (for an internal global symmetry, both the generators and the parameters do not depend on the coordinates).

The conformal group contains the isometry group ISO($\mathcal{M}$) of $\mathcal{M}$ as a subgroup, corresponding to the case $\Omega = 1$:

$$\text{ISO}(\mathcal{M}) \subset \text{CISO}(\mathcal{M}). \tag{5.8}$$

These transformations also preserve distances between points. The corresponding generators of infinitesimal transformations are called Killing vectors and satisfies the Killing equation

$$\delta g_{\mu\nu} = \mathcal{L}_{\xi} g_{\mu\nu} = \nabla_{\mu}\xi_{\nu} + \nabla_{\nu}\xi_{\mu} = 0. \tag{5.9}$$

They form a subalgebra of the CKV algebra.

An important point is to be made for the relation between infinitesimal and finite transformations: with spacetime symmetries it often happens that the first cannot be exponentiated into the second. The reason is that the (conformal) Killing vectors may be defined only locally, i.e. they are well-defined in a given domain but have singularities outside. When this happens, they do not lead to an invertible transformation, which cannot be an element of the group. These notions are sometimes confused in physics and the term of "group" is used instead of "algebra". We shall be careful in distinguishing both concepts.

Remark 5.1 (Isometries of $\mathcal{M} \subset \mathbb{R}^{p,q}$) In order to find the conformal isometries of a manifold $\mathcal{M}$ which is a subset of $\mathbb{R}^{p,q}$ defined in (5.10), it is sufficient to restrict the transformations of $\mathbb{R}^{p,q}$ to the subset $\mathcal{M}$ [4]. In the process, not all global transformations generically survive. On the other hand, the algebra of local (infinitesimal) transformations for $\mathcal{M}$ and $\mathbb{R}^{p,q}$ are identical since $\mathcal{M}$ is locally like $\mathbb{R}^{p,q}$.

5.2 CFT on Minkowski Space

In this section, we consider the case where $\mathcal{M} = \mathbb{R}^{p,q}$ ($D = p + q$) and where $g = \eta$ is the flat metric with signature (p, q):

$$\eta = \mathrm{diag}(\underbrace{-1, \ldots, -1}_{q}, \underbrace{1, \ldots, 1}_{p}). \tag{5.10}$$

The conformal Killing equation becomes

$$\left(\eta_{\mu\nu}\Delta + (D - 2)\partial_\mu \partial_\nu\right)\partial \cdot \epsilon = 0, \tag{5.11}$$

where Δ is the D-dimensional Beltrami–Laplace operator for the metric $\eta_{\mu\nu}$. The case $D = 2$ is relegated to the next chapter. For $D > 2$, one finds the following transformations:

$$\text{translation:} \quad \xi^\mu = a^\mu, \tag{5.12a}$$

$$\text{rotation \& boost:} \quad \xi^\mu = \omega^\mu{}_\nu x^\nu, \tag{5.12b}$$

$$\text{dilatation:} \quad \xi^\mu = \lambda x^\mu, \tag{5.12c}$$

$$\text{SCT:} \quad \xi^\mu = b^\mu x^2 - 2b \cdot x\, x^\mu, \tag{5.12d}$$

where $\omega_{\mu\nu}$ is antisymmetric. The rotations include Lorentz transformations and SCT means "special conformal transformation".

All parameters $\{a^\mu, \omega_{\mu\nu}, \lambda, b^\mu\}$ are constant. The generators are respectively denoted by $\{P_\mu, J_{\mu\nu}, D, K_\mu\}$. The finite translations and rotations form the Poincaré group $SO(p, q)$, while the conformal group can be shown to be $SO(p + 1, q + 1)$:

$$\mathrm{ISO}(\mathbb{R}^{p,q}) = SO(p, q), \qquad \mathrm{CISO}(\mathbb{R}^{p,q}) = SO(p + 1, q + 1). \tag{5.13}$$

The dimension of this group is

$$\dim SO(p+1, q+1) = \frac{1}{2}(p+q+2)(p+q+1).$$ (5.14)

5.3   Suggested Readings

- References on higher-dimensional CFTs are [1–5].

References

1. P.D. Francesco, P. Mathieu, D. Senechal, *Conformal Field Theory*, 2nd edn. (Springer, Berlin, 1999)
2. J.D. Qualls, Lectures on Conformal Field Theory (2015) arXiv: 1511.04074
3. S. Rychkov, *EPFL Lectures on Conformal Field Theory in D>= 3 Dimensions* (Springer, Berlin, 2016). arXiv: 1601.05000
4. M. Schottenloher, *A Mathematical Introduction to Conformal Field Theory*, 2nd edn. (Springer, Berlin, 2008)
5. D. Simmons-Duffin, *TASI Lectures on the Conformal Bootstrap* (A Caltech Library Service, Pasadena, 2016). arXiv: 1602.07982

Conformal Field Theory on the Plane

6

Abstract

Starting with this chapter, we focus on two-dimensional Euclidean CFTs on the complex plane (or equivalently the sphere). We start by describing the geometry of the sphere and the relation to the complex plane and to the cylinder, in order to make contact with the string worldsheet. Then, we discuss classical CFTs and the Witt algebra obtained by classifying the conformal isometries of the complex plane. Then, we describe quantum CFTs and introduce the operator formalism. This last section is the most important for this book as it includes information on the operator product expansion, Hilbert space, Hermitian and BPZ conjugations.

As described at the beginning of Chap. 5, we use spacetime notations for the coordinates, but follow otherwise the normalization for the worldsheet. In particular, integrals are normalized by 2π. However, the spatial coordinate on the cylinder is still written as σ to avoid confusions: $x^\mu = (\tau, \sigma)$.

6.1 The Riemann Sphere

6.1.1 Map to the Complex Plane

The Riemann sphere Σ_0, which is diffeomorphic to the unit sphere S^2, has genus $g = 0$ and is thus the simplest Riemann surface. Its most straightforward description is obtained by mapping it to the extended[1] complex plane $\bar{\mathbb{C}}$ (also denoted $\hat{\mathbb{C}}$), which is the complex plane $z \in \mathbb{C}$ to which the point at infinity $z = \infty$ is added:

$$\bar{\mathbb{C}} = \mathbb{C} \cup \{\infty\}. \tag{6.1}$$

---

[1] This qualification will often be omitted.

© Springer Nature Switzerland AG 2021
H. Erbin, *String Field Theory*, Lecture Notes in Physics 980,
https://doi.org/10.1007/978-3-030-65321-7_6

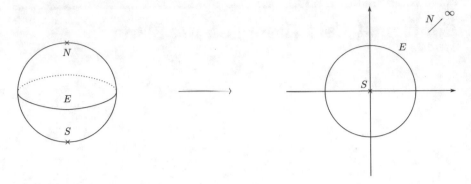

Fig. 6.1 Map from the Riemann sphere to the complex plane. The south and north poles are denoted by the letter S and N, and the equatorial circle by E

One speaks about "*the* point at infinity" because all the points at infinity (i.e. the points z such that $|z| \to \infty$)

$$\lim_{r \to \infty} r\, e^{i\theta} := \infty \tag{6.2}$$

are identified (the limit is independent of θ).

The identification can be understood by mapping (say) the south pole to the origin of the plane and the north pole to infinity[2] (Fig. 6.1) through the stereographic projection

$$z = e^{i\phi} \cot \frac{\theta}{2}, \tag{6.3}$$

where (θ, ϕ) are angles on the sphere. Any circle on the sphere is mapped to a circle in the complex plane. Conversely, the Riemann sphere can be viewed as a compactification of the complex plane.

Introducing Cartesian coordinates (x, y) related to the complex coordinates by[3]

$$z = x + iy, \qquad \bar{z} = x - iy, \tag{6.4a}$$

$$x = \frac{z + \bar{z}}{2}, \qquad y = \frac{z - \bar{z}}{2i}, \tag{6.4b}$$

---

[2]Note that the points are distinguished in order to write the map, but they have nothing special by themselves (i.e. they are not punctures).

[3]General formulas can be found in Sect. 4.1 by replacing (τ, σ) with (x, y). In most cases, the conformal factor is set to zero ($\phi = 0$) in this chapter.

the metric reads

$$ds^2 = dx^2 + dy^2 = dz d\bar{z}. \tag{6.5}$$

The relations between the derivatives in the two coordinate systems are easily found:

$$\partial := \partial_z = \frac{1}{2}(\partial_x - i\partial_y), \qquad \bar{\partial} := \partial_{\bar{z}} = \frac{1}{2}(\partial_x + i\partial_y). \tag{6.6}$$

The indexed form will be used when there is a risk of confusion. If the index is omitted, then the derivative acts directly to the field next to it, for example,

$$\partial\phi(z_1)\partial\phi(z_2) := \partial_{z_1}\partial_{z_2}\phi(z_1)\phi(z_2). \tag{6.7}$$

Generically, the meromorphic and anti-meromorphic parts of an object will be denoted without and with a bar, see (6.55) for an example.

The extended complex plane $\tilde{\mathbb{C}}$ can be covered by two coordinate patches $z \in \mathbb{C}$ and $w \in \mathbb{C}$. In the first, the point at infinity (north pole) is removed, in the second, the origin (south pole) is removed. On the overlap, the transition function is

$$w = \frac{1}{z}. \tag{6.8}$$

This description avoids to work with the infinity: studying the behaviour of $f(z)$ at $z = \infty$ is equivalent to study $f(1/w)$ at $w = 0$.

Since any two-dimensional metric is locally conformally equivalent to the flat metric, it is sufficient to work with this metric in each patch. This is particularly convenient for the Riemann sphere since one patch covers it completely except for one point.

6.1.2 Relation to the Cylinder: String Theory

The worldsheet of a closed string propagating in spacetime is locally topologically a cylinder $\mathbb{R} \times S^1$ of circumference L. In this section, we show that the cylinder can also be mapped to the complex plane—and thus to the Riemann sphere—after removing two points. Since the cylinder has a clear physical interpretation in string theory, it is useful to know how to translate the results from the plane to the cylinder.

It makes also sense to define two-dimensional models on the cylinder independently of a string theory interpretation since the compactification of the spatial direction from $\mathbb{R}$ to S^1 regulates the infrared divergences. Moreover, it leads to a natural definition of a "time" and of a Hamiltonian on the Euclidean plane.

Denoting the worldsheet coordinates in Lorentzian signature by (t, σ) with[4]

$$t \in \mathbb{R}, \qquad \sigma \in [0, L), \qquad \sigma \sim \sigma + L, \tag{6.9}$$

the metric reads

$$ds^2 = -dt^2 + d\sigma^2 = -d\sigma^+ d\sigma^-, \tag{6.10}$$

where the light-cone coordinates

$$d\sigma^\pm = dt \pm d\sigma \tag{6.11}$$

have been introduced. It is natural to perform a Wick rotation from the Lorentzian time t to the Euclidean time

$$\tau = it, \tag{6.12}$$

and the metric becomes

$$ds^2 = d\tau^2 + d\sigma^2. \tag{6.13}$$

It is convenient to introduce the complex coordinates

$$w = \tau + i\sigma, \qquad \bar{w} = \tau - i\sigma \tag{6.14}$$

for which the metric is

$$ds^2 = dw d\bar{w}. \tag{6.15}$$

Note that the relation to Lorentzian light-cone coordinates is

$$w = i(t + \sigma) = i\sigma^+, \qquad \bar{w} = i(t - \sigma) = i\sigma^-. \tag{6.16}$$

Hence, an (anti-)holomorphic function of w ($\bar{w}$) depends only on σ^+ (σ^-) before the Wick rotation: this leads to the identification of the left- and right-moving sectors with the holomorphic and anti-holomorphic sectors of the theory.

The cylinder can be mapped to the complex plane through

$$z = e^{2\pi w/L}, \qquad \bar{z} = e^{2\pi \bar{w}/L}, \tag{6.17}$$

---

[4]Consistently with the comments at the beginning of Chap. 5, the Lorentzian worldsheet time is denoted by t instead of τ_M.

and the corresponding metric is

$$ds^2 = \left(\frac{L}{2\pi}\right)^2 \frac{dz d\bar{z}}{|z|^2}. \tag{6.18}$$

A conformal transformation brings this metric to the flat metric (6.5). The conventions for the various coordinates and maps vary in the different textbooks. We have gathered in Table A.1 the three main conventions and which references use which.

The map from the cylinder to the plane is found by sending the bottom end (corresponding to the infinite past $t \to -\infty$) to the origin of the plane, and the top end (infinite future $t \to \infty$) to the infinity. Since the cylinder has two boundaries (its two ends) the map excludes the point $z = 0$ and $z = \infty$ and one really obtains the space $\bar{\mathbb{C}} - \{0, \infty\} = \mathbb{C}^*$. This space can, in turn, be mapped to the 2-punctured Riemann sphere $\Sigma_{0,2}$.

The physical interpretation for the difference between Σ_0 and $\Sigma_{0,2}$ is simple: since one considers the propagation of a string, it means that the worldsheet corresponds to an amplitude with two external states, which are the mapped to the sphere as punctures (Fig. 6.2, Sect. 3.1.1). Removing the external states (yielding the tree-level vacuum amplitude) corresponds to gluing half-sphere (caps) at each end of the cylinder (Fig. 6.3). Then, it can be mapped to the Riemann sphere without punctures. As a consequence, the properties of tree-level string theory are found by studying the matter and ghost CFTs on the Riemann sphere. Scattering amplitudes are computed through correlation functions of appropriate operators on the sphere. This picture generalizes to higher-genus Riemann surfaces. Moreover,

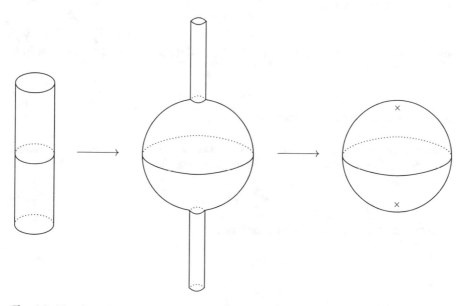

Fig. 6.2 Map from the cylinder to the sphere with two tubes, to the 2-punctured sphere $\Sigma_{0,2}$

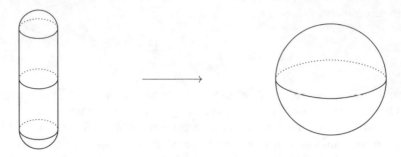

Fig. 6.3 Map from the cylinder with two caps (half-spheres) to the Riemann sphere Σ_0

since local properties of the CFT (e.g. the spectrum of operators) are determined by the conformal algebra, they will be common to all surfaces.

Mathematically, a difference between Σ_0 and $\Sigma_{0,2}$ had to be expected since the sphere has a positive curvature (and $\chi = -2$) but the cylinder is flat (with $\chi = 0$). Punctures contribute negatively to the curvature (and thus positively to the Euler characteristics).

Remark 6.1 The coordinate z is always used as a coordinate on the complex plane, but the corresponding metric may be different—compare (6.5) and (6.18). As explained previously, this does not matter since the theory is insensitive to the conformal factor.

6.2   Classical CFTs

In this section, we consider an action $S[\Psi]$ which is conformally invariant. We first identify and discuss the properties of the conformal algebra and group, before explaining how a CFT is defined.

6.2.1   Witt Conformal Algebra

Since the Riemann sphere is identified with the complex plane, they share the same conformal group and algebra. Consider the metric (6.5)

$$ds^2 = dz d\bar{z}, \tag{6.19}$$

then, any meromorphic change of coordinates

$$z \longrightarrow z' = f(z), \qquad \bar{z} \longrightarrow \bar{z}' = \bar{f}(\bar{z}) \tag{6.20}$$

is a conformal transformation since the metric becomes

$$ds^2 = dz'd\bar{z}' = \left|\frac{df}{dz}\right|^2 dz d\bar{z}. \tag{6.21}$$

However, only holomorphic functions which are globally defined on $\bar{\mathbb{C}}$ are elements of the group. At the algebra level, any holomorphic function $f(z)$ regular in a domain D gives a well-defined transformation in this domain D. Hence, the algebra is infinite-dimensional. On the other hand, $f(z)$ is only meromorphic on $\mathbb{C}$ generically: it cannot be exponentiated to a group element. We first characterize the algebra and then obtain the conditions to promote the local transformations to global ones.

Since the transformations are defined only locally, it is sufficient to consider an infinitesimal transformation

$$\delta z = v(z), \qquad \delta \bar{z} = \bar{v}(\bar{z}), \tag{6.22}$$

where $v(z)$ is a meromorphic vector field on the Riemann sphere. Indeed, the conformal Killing equation (5.6) in $D = 2$ is equivalent to the Cauchy–Riemann equations:

$$\bar{\partial} v = 0, \qquad \partial \bar{v} = 0. \tag{6.23}$$

The vector field admits a Laurent series

$$v(z) = \sum_{n \in \mathbb{Z}} v_n z^{n+1}, \qquad \bar{v}(\bar{z}) = \sum_{n \in \mathbb{Z}} \bar{v}_n \bar{z}^{n+1}, \tag{6.24}$$

and the v_n and $\bar{v}_n$ are to be interpreted as the parameters of the transformation. A basis of vectors (generators) is

$$\ell_n = -z^{n+1} \partial_z, \qquad \bar{\ell}_n = -\bar{z}^{n+1} \partial_{\bar{z}}, \qquad n \in \mathbb{Z}. \tag{6.25}$$

One can check that each set of generators satisfies the Witt algebra

$$[\ell_m, \ell_n] = (m - n)\ell_{m+n}, \qquad [\bar{\ell}_m, \bar{\ell}_n] = (m - n)\bar{\ell}_{m+n}, \qquad [\ell_m, \bar{\ell}_n] = 0. \tag{6.26}$$

Since there are two commuting copies of the Witt algebra, it is natural to extend the ranges of the coordinates from $\mathbb{C}$ to $\mathbb{C}^2$ and to consider z and $\bar{z}$ as independent variables. In particular, this gives a natural action of the product algebra over $\mathbb{C}^2$. This procedure will be further motivated when studying CFTs since the holomorphic and anti-holomorphic parts will generally split, and it makes sense to study them separately. Ultimately, physical quantities can be extracted by imposing the condition $\bar{z} = z^*$ at the end (the star is always reserved for the complex conjugation, the bar will generically denote an independent variable). In that case, the two algebras are also related by complex conjugation.

Note that the variation of the metric (B.6) under a meromorphic change of coordinates (6.22) becomes

$$\delta g_{z\bar{z}} = \partial v + \bar{\partial}\bar{v}, \qquad \delta g_{zz} = \delta g_{\bar{z}\bar{z}} = 0. \tag{6.27}$$

6.2.2 PSL(2, $\mathbb{C}$) Conformal Group

The next step is to determine the globally defined vectors and to study the associated group.

First, the conditions for a vector $v(z)$ to be well-defined at $z = 0$ are

$$\lim_{|z|\to 0} v(z) < \infty \quad \Longrightarrow \quad \forall n < -1 : \quad v_n = 0. \tag{6.28}$$

The behaviour at $z = \infty$ can be investigated thanks to the map $z = 1/w$

$$v(1/w) = \frac{dz}{dw} \sum_n v_n w^{-n-1}, \tag{6.29}$$

where the additional derivative arises because v is a vector. Then, the regularity conditions at $z = \infty$ are

$$\lim_{|z|\to\infty} v(z) = \lim_{|w|\to 0} \frac{dz}{dw} v(1/w) = - \lim_{|w|\to 0} \frac{v(1/w)}{w^2} < \infty$$

$$\Longrightarrow \forall n > 1 : \quad v_n = 0. \tag{6.30}$$

As a result, the globally defined generators are

$$\{\ell_{-1}, \ell_0, \ell_1\} \cup \{\bar{\ell}_{-1}, \bar{\ell}_0, \bar{\ell}_1\}, \tag{6.31}$$

where

$$\ell_{-1} = -\partial_z, \qquad \ell_0 = -z\partial_z, \qquad \ell_1 = -z^2\partial_z. \tag{6.32}$$

It is straightforward to check that they form two copies of the $\mathfrak{sl}(2, \mathbb{C})$ algebra

$$[\ell_0, \ell_{\pm 1}] = \mp\ell_{\pm 1}, \qquad [\ell_1, \ell_{-1}] = 2\ell_0. \tag{6.33}$$

The global conformal group is sometimes called Möbius group:

$$\mathrm{PSL}(2, \mathbb{C}) := \mathrm{SL}(2, \mathbb{C})/\mathbb{Z}_2 \sim \mathrm{SO}(3, 1), \tag{6.34}$$

where the additional division by $\mathbb{Z}_2$ is clearer when studying an explicit representation. It corresponds with ker P_1 defined in (2.91):

$$\mathcal{K}_0 = \mathrm{PSL}(2, \mathbb{C}). \tag{6.35}$$

A matrix representation of $\mathrm{SL}(2, \mathbb{C})$ is

$$g = \begin{pmatrix} a & b \\ c & d \end{pmatrix}, \qquad a, b, c, d \in \mathbb{C}, \qquad \det g = ad - bc = 1, \tag{6.36}$$

which shows that this group has six real parameters

$$\mathsf{K}_0 := \dim \mathrm{SL}(2, \mathbb{C}) = 6. \tag{6.37}$$

The associated transformation on the complex plane reads

$$f_g(z) = \frac{az + b}{cz + d}. \tag{6.38}$$

The quotient by $\mathbb{Z}_2$ is required since changing the sign of all parameters does not change the transformation. These transformations have received different names: Möbius, projective, homographic, linear fractional transformations...

Holomorphic vector fields are then of the form

$$v(z) = \beta + 2\alpha z + \gamma z^2, \qquad \bar{v}(\bar{z}) - \bar{\beta} + 2\bar{\alpha}\bar{z} + \bar{\gamma}\bar{z}^2, \tag{6.39}$$

where

$$a = 1 + \alpha, \qquad b = \beta, \qquad c = -\gamma, \qquad d = 1 - \alpha. \tag{6.40}$$

The finite transformations associated with (5.12) are

$$\text{translation:} \qquad f_g(z) = z + a, \qquad\qquad a \in \mathbb{C}, \tag{6.41a}$$

$$\text{rotation:} \qquad f_g(z) = \zeta\, z, \qquad\qquad |\zeta| = 1, \tag{6.41b}$$

$$\text{dilatation:} \qquad f_g(z) = \lambda\, z, \qquad\qquad \lambda \in \mathbb{R}, \tag{6.41c}$$

$$\text{SCT:} \qquad f_g(z) = \frac{z}{cz + 1}, \qquad\qquad c \in \mathbb{C}. \tag{6.41d}$$

Investigation leads to the following association between the generators and transformations:

- translation: ℓ_{-1} and $\bar{\ell}_{-1}$;
- dilatation (or radial translation): $(\ell_0 + \bar{\ell}_0)$;

- rotation (or angular translation): $i(\ell_0 - \bar{\ell}_0)$;
- special conformal transformation: ℓ_1 and $\bar{\ell}_1$.

The inversion defined by

$$\text{inversion:} \qquad I^+(z) := I(z) := \frac{1}{z} \tag{6.42a}$$

is not an element of $\mathrm{SL}(2, \mathbb{C})$. However, the inversion with a minus sign

$$I^-(z) := -I(z) = I(-z) = -\frac{1}{z} \tag{6.42b}$$

is a $\mathrm{SL}(2, \mathbb{C})$ transformation.

A useful transformation is the circular permutation of $(0, 1, \infty)$:

$$g_{\infty,0,1}(z) = \frac{1}{1-z}. \tag{6.43}$$

6.2.3 Definition of a CFT

A CFT is characterized by its set of (composite) fields (also called operators) $\mathcal{O}(z, \bar{z})$ which correspond to any local expression constructed from the fields Ψ appearing in the Lagrangian and of their derivatives.[5] For example, in a scalar field theory, the simplest operators are of the form $\partial^m \phi^n$.

Among the operators, two particular categories are distinguished according to their transformation laws:

- primary operator:

$$\forall f \text{ meromorphic} : \qquad \mathcal{O}(z, \bar{z}) = \left(\frac{\mathrm{d}f}{\mathrm{d}z}\right)^h \left(\frac{\mathrm{d}\bar{f}}{\mathrm{d}\bar{z}}\right)^{\bar{h}} \mathcal{O}'\big(f(z), \bar{f}(\bar{z})\big), \tag{6.44}$$

- quasi-primary (or $\mathrm{SL}(2, \mathbb{C})$ primary) operator:

$$\forall f \in \mathrm{PSL}(2, \mathbb{C}) : \qquad \mathcal{O}(z, \bar{z}) = \left(\frac{\mathrm{d}f}{\mathrm{d}z}\right)^h \left(\frac{\mathrm{d}\bar{f}}{\mathrm{d}\bar{z}}\right)^{\bar{h}} \mathcal{O}'\big(f(z), \bar{f}(\bar{z})\big). \tag{6.45}$$

The parameters $(h, \bar{h})$ are the conformal weights of the operator $\mathcal{O}$ (both are independent from each other), and combinations of them give the conformal

---

[5]Not all CFTs admit a Lagrangian description. But, since we are mostly interested in string theories defined from Polyakov's path integral, it is sufficient to study CFTs with a Lagrangian.

dimension Δ and spin s:

$$\Delta := h + \bar{h}, \qquad s := h - \bar{h}. \tag{6.46}$$

The conformal weights correspond to the charges of the operator under ℓ_0 and $\bar{\ell}_0$. We will use "$(h, \bar{h})$ (quasi-)primary" as a synonym of "(quasi-)primary field with conformal weight $(h, \bar{h})$".

Remark 6.2 (Complex Conformal Weights) While we consider $h, \bar{h} \in \mathbb{R}$, and more specifically $h, \bar{h} \geq 0$ for a unitary theory (which is the case of string theory except for the reparametrization ghosts), theories with $h, \bar{h} \in \mathbb{C}$ make perfectly sense. One example is the Liouville theory with complex central charge $c \in \mathbb{C}$ [31, 33] (central charges are defined below, see (6.58)).

Primaries and quasi-primaries are hence operators which have nice transformations, respectively, under the algebra and group. Obviously, a primary is also a quasi-primary. These transformations are similar to those of a tensor with h holomorphic and $\bar{h}$ anti-holomorphic indices (Sect. 4.1). Another point of view is that the object

$$\mathcal{O}(z, \bar{z}) \, dz^h d\bar{z}^{\bar{h}} \tag{6.47}$$

is invariant under local / global conformal transformations.

The notation $f \circ \mathcal{O}$ indicates the complete change of coordinates, including the tensor transformation law and the possible corrections if the operator is not primary.[6] For a primary field, we have

$$f \circ \mathcal{O}(z, \bar{z}) := f'(z)^h \bar{f}'(\bar{z})^{\bar{h}} \, \mathcal{O}'\big(f(z), \bar{f}(\bar{z})\big). \tag{6.48}$$

We stress that it does not correspond to function composition.

Under an infinitesimal transformation

$$\delta z = v(z), \qquad \delta \bar{z} = \bar{v}(\bar{z}), \tag{6.49}$$

a primary operator changes as

$$\delta \mathcal{O}(z, \bar{z}) = (h \, \partial v + v \, \partial) \mathcal{O}(z, \bar{z}) + (\bar{h} \, \bar{\partial} \bar{v} + \bar{v} \, \bar{\partial}) \mathcal{O}(z, \bar{z}). \tag{6.50}$$

The transformation of a non-primary field contains additional terms, see, for example, (6.89).

Remark 6.3 (Higher-Genus Riemann Surfaces) According to Remark 5.1, all Riemann surfaces Σ_g share the same conformal algebra since locally they are all subsets

---

[6]In fact, one has $f \circ \mathcal{O} := f^* \mathcal{O}$ in the notations of Chap. 2.

of $\mathbb{R}^2$. On the other hand, one finds that no global transformations are defined for $g > 1$, and only the subgroup $U(1) \times U(1)$ survives for the torus.

The most important operator in a CFT is the energy–momentum tensor $T_{\mu\nu}$, if it exists as a local operator. According to Sect. 2.1, this tensor is conserved and traceless

$$\nabla^\nu T_{\mu\nu} = 0, \qquad g^{\mu\nu} T_{\mu\nu} = 0. \tag{6.51}$$

The traceless equation in components reads

$$g^{\mu\nu} T_{\mu\nu} = 4\, T_{z\bar{z}} = T_{xx} + T_{yy} = 0 \tag{6.52}$$

which implies that the off-diagonal component vanishes in complex coordinates

$$T_{z\bar{z}} = 0. \tag{6.53}$$

Then, the conservation equation yields

$$\partial_z T_{\bar{z}\bar{z}} = 0, \qquad \partial_{\bar{z}} T_{zz} = 0, \tag{6.54}$$

such that the non-vanishing components T_{zz} and $T_{\bar{z}\bar{z}}$ are, respectively, holomorphic and anti-holomorphic. This motivates the introduction of the notations:

$$T(z) := T_{zz}(z), \qquad \bar{T}(\bar{z}) := T_{\bar{z}\bar{z}}(\bar{z}). \tag{6.55}$$

This is an example of the factorization between the holomorphic and anti-holomorphic sectors.

Currents are local objects and thus one expects to be able to write an infinite number of such currents associated with the Witt algebra. Applying the Noether procedure gives

$$J_v(z) := J_v^{\bar{z}}(z) = -T(z)v(z), \qquad \bar{J}_v(\bar{z}) := J_v^z(\bar{z}) = -\bar{T}(\bar{z})\bar{v}(\bar{z}). \tag{6.56}$$

6.3   Quantum CFTs

The previous section was purely classical. The quantum theory is first defined through the path integral

$$Z = \int d\Psi\, e^{-S[\Psi]}. \tag{6.57}$$

We will also develop an operator formalism. The latter is more general than the path integral and allows to work without reference to path integrals and Lagrangians.

This is particularly fruitful as it extends the class of theories and parameter ranges (e.g. Remark 6.2) which can be studied.

6.3.1 Virasoro Algebra

As discussed in Sect. 2.3.3, field measures in path integrals display a conformal anomaly, meaning that they cannot be defined without introducing a scale. This anomaly can be traded for a gravitational anomaly by introducing counter-terms in the action [9, 13, sec. 3.2, 14, 17, 21, 24]. As a consequence, the Witt algebra (6.26) is modified to its central extension, the Virasoro algebra.[7] The generators in both sectors are denoted by $\{L_n\}$ and $\{\bar{L}_n\}$ and are called Virasoro operators (or modes). The algebra is given by

$$[L_m, L_n] = (m - n)L_{m+n} + \frac{c}{12} m(m - 1)(m + 1)\delta_{m+n}, \tag{6.58a}$$

$$[\bar{L}_m, \bar{L}_n] = (m - n)\bar{L}_{m+n} + \frac{\bar{c}}{12} m(m - 1)(m + 1)\delta_{m+n}, \tag{6.58b}$$

$$[L_m, \bar{L}_n] = 0, \qquad [c, L_m] = 0, \qquad [\bar{c}, \bar{L}_m] = 0, \tag{6.58c}$$

where $c, \bar{c} \in \mathbb{C}$ are the holomorphic and anti-holomorphic central charges. Consistency of the theory on a curved space implies $\bar{c} - c$, but there is otherwise no constraint on the plane [13, 38].

The $\mathfrak{sl}(2, \mathbb{C})$ subalgebra is not modified by the central extension. This means that states are still classified by eigenvalues of $(h, \bar{h})$ of $(L_0, \bar{L}_0)$.

Remark 6.4 In most models relevant for string theory, one finds that the central charges are real, $c, \bar{c} \in \mathbb{R}$. Moreover, unitarity requires them to be positive $c, \bar{c} > 0$, and only reparametrization ghosts do not satisfy this condition. On the other hand, it makes perfect sense to discuss general CFTs for $c, \bar{c} \in \mathbb{C}$ (the Liouville theory is such an example [31, 33]).

6.3.2 Correlation Functions

A n-point correlation function is defined by

$$\left\langle \prod_{i=1}^{n} \mathcal{O}_i(z_i, \bar{z}_i) \right\rangle = \int d\Psi \, e^{-S[\Psi]} \prod_{i=1}^{n} \mathcal{O}_i(z_i, \bar{z}_i), \tag{6.59}$$

---

[7]That the central charge in the Virasoro algebra indicates a diffeomorphism anomaly can be understood from the fact that.

choosing a normalization such that $\langle 1 \rangle = 1$. The path integral defines the time-ordered product (on the cylinder) of the corresponding operators.

Invariance under global transformations leads to strong constraints on the correlation functions. For quasi-primary fields, they transform under SL(2, $\mathbb{C}$) as

$$\left\langle \prod_{i=1}^{n} \mathcal{O}_i(z_i, \bar{z}_i) \right\rangle = \prod_{i=1}^{n} \left(\frac{df}{dz}(z_i) \right)^{h_i} \left(\frac{df}{d\bar{z}}(\bar{z}_i) \right)^{\bar{h}_i} \times \left\langle \prod_{i=1}^{n} \mathcal{O}_i\big(f(z_i), \bar{f}(\bar{z}_i)\big) \right\rangle. \tag{6.60}$$

Considering an infinitesimal variation (6.50) yields a differential equation for the n-point function

$$\delta \left\langle \prod_{i=1}^{n} \mathcal{O}_i(z_i, \bar{z}_i) \right\rangle = \sum_{i=1}^{n} \big(h_i \partial_i v(z_i) + v(z_i) \partial_i + \text{c.c.} \big) \left\langle \prod_{i=1}^{n} \mathcal{O}_i(z_i, \bar{z}_i) \right\rangle = 0, \tag{6.61}$$

where $\partial_i := \partial_{z_i}$ and v is a vector (6.39) of $\mathfrak{sl}(2, \mathbb{C})$. These equations are sufficient to determine completely the forms of the 1-, 2-, and 3-point functions of quasi-primaries:

$$\langle \mathcal{O}_i(z_i, \bar{z}_i) \rangle = \delta_{h_i, 0} \delta_{\bar{h}_i, 0}, \tag{6.62a}$$

$$\langle \mathcal{O}_i(z_i, \bar{z}_i) \mathcal{O}_j(z_j, \bar{z}_j) \rangle = \delta_{h_i, h_j} \delta_{\bar{h}_i, \bar{h}_j} \frac{g_{ij}}{z_{ij}^{2h_i} \bar{z}_{ij}^{2\bar{h}_i}}, \tag{6.62b}$$

$$\langle \mathcal{O}_i(z_i, \bar{z}_i) \mathcal{O}_j(z_j, \bar{z}_j) \mathcal{O}_k(z_k, \bar{z}_k) \rangle = \frac{C_{ijk}}{z_{ij}^{h_i + h_j - h_k} z_{jk}^{h_j + h_k - h_i} z_{ki}^{h_i + h_k - h_j}}$$

$$\times \frac{1}{\bar{z}_{ij}^{\bar{h}_i + \bar{h}_j - \bar{h}_k} \bar{z}_{jk}^{\bar{h}_j + \bar{h}_k - \bar{h}_i} \bar{z}_{ki}^{\bar{h}_i + \bar{h}_k - \bar{h}_j}}, \tag{6.62c}$$

where we have defined

$$z_{ij} = z_i - z_j. \tag{6.63}$$

The coefficients C_{ijk} are called structure constants and the matrix g_{ij} defines a metric (Zamolodchikov metric) on the space of fields. The metric is often taken to be diagonal $g_{ij} = \delta_{ij}$, which amounts to use an orthonormal eigenbasis of L_0 and $\bar{L}_0$. The vanishing of the 1-point function of a non-primary quasi-primary holds only on the plane: for example, the value on the cylinder can be non-zero since the map is not globally defined—see in particular (6.167).

Remark 6.5 (Logarithmic CFTs) Logarithmic CFTs display a set of unusual properties [7, 8, 11, 15, 25]. In particular, the correlation functions are not of the form displayed above. The most striking feature of those theories is that the L_0 operator is non-diagonalisable (but it can be set in the Jordan normal form).

Remark 6.6 (Fake Identity) Usually, the only primary operator with $h = \bar{h} = 0$ is the identity 1. While this is always true for unitary theories, there are non-unitary theories ($c \leq 1$ Liouville theory, SLE, loop models) where there is another field (called the indicator, marking operator, or also fake identity) with $h = \bar{h} = 0$ [1, 5, 16, 19, 27, 31, 33]. The main difference between both fields is that the identity is a degenerate field (it has a null descendant), whereas the other operator with $h = \bar{h} = 0$ is not. Such theories will not be considered in this book. Operators with $h = \hbar = 0$ can also be built by combining several CFTs, and they play a very important role in string theory since they describe on-shell states.

Finally, the 4-point function is determined up to a function of a single variable x and its complex conjugate:

$$\left\langle \prod_{i=1}^{4} \mathcal{O}_i(z_i, \bar{z}_i) \right\rangle = f(x, \bar{x}) \prod_{i<j} \frac{1}{z_{ij}^{(h_i+h_j)-h/3}} \times \text{c.c.},\tag{6.64}$$

where

$$h := \sum_{i=1}^{4} h_i, \qquad \bar{h} := \sum_{i=1}^{4} \bar{h}_i.\tag{6.65}$$

The cross-ratio x is SL$(2, \mathbb{C})$ invariant and reads

$$x := \frac{z_{12}z_{34}}{z_{13}z_{24}}.\tag{6.66}$$

The interpretation is that the SL$(2, \mathbb{C})$ invariance allows to fix 3 of the points to an arbitrary value, and the final result does not depend on this choice.

6.4 Operator Formalism and Radial Quantization

Radial quantization is a convenient description of a CFT on the plane in terms of operators. It relies on the maps given in Sect. 6.1.2:

$$z = e^{\tau+i\sigma} = x + iy.\tag{6.67}$$

Taking the physical spacetime to be the cylinder, every question is rephrased on the complex plane in order to exploit the powerful tools from complex analysis. The term "radial quantization" comes from the fact that time translation of the cylinder

$$\tau \longrightarrow \tau + T\tag{6.68}$$

corresponds to dilatation on the plane

$$z \longrightarrow e^T z. \tag{6.69}$$

Thus, time evolution on the cylinder and radial evolution (from the origin to the complex infinity) are identified. In particular, the Hamiltonian of the system of the plane is

$$H = \frac{2\pi}{L}(L_0 + \bar{L}_0), \tag{6.70}$$

since the RHS is the dilatation operator. The cylinder length L was defined in (6.9). The theory is quantized according to this Hamiltonian. In the string theory language, a state with $H = 0$ is said to be on-shell:

$$\text{on-shell state:} \quad h + \bar{h} = 0. \tag{6.71}$$

6.4.1　Radial Ordering and Commutators

Time-ordering in τ becomes radial ordering in the plane:

$$R\big(A(z)B(w)\big) = \begin{cases} A(z)B(w) & |z| > |w|, \\ (-1)^F \, B(w)A(z) & |w| > |z|, \end{cases} \tag{6.72}$$

where $F = 0$ ($F = 1$) for bosonic (fermionic) operators. Radial ordering will often be kept implicit.

The equal-time (anti-)commutator becomes an equal radius commutator defined by point-splitting:

$$[A(z), B(w)]_{\pm, |z|=|w|} = \lim_{\delta \to 0} \big(A(z)B(w)|_{|z|=|w|+\delta} \pm B(w)A(z)|_{|z|=|w|-\delta}\big). \tag{6.73}$$

If A and B are two operators which can be written as the contour integrals of $a(z)$ and $b(z)$ (corresponding to integral over closed curves on the cylinder)

$$A = \oint_{C_0} \frac{dz}{2\pi i}\, a(z), \qquad B = \oint_{C_0} \frac{dz}{2\pi i}\, b(z), \tag{6.74}$$

then one finds the following commutators:

$$[A, B]_{\pm} = \oint_{C_0} \frac{dw}{2\pi i} \oint_{C_w} \frac{dz}{2\pi i}\, a(z)b(w), \tag{6.75a}$$

$$[A, b(w)]_{\pm} = \oint_{C_w} \frac{dz}{2\pi i}\, a(z)b(w). \tag{6.75b}$$

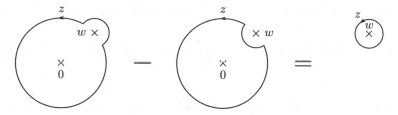

Fig. 6.4 Graphical proof of (6.75)

The contours C_0 and C_w are, respectively, centred around the points 0 and w. For a proof, see Fig. 6.4. Since these are contour integrals in the complex plane, the Cauchy–Riemann formula (B.1) can be used to write the result as soon as one knows the poles of the above expression (ultimately, this amounts to pick the sum of residues). In CFTs, the poles of such expressions are given by operator product expansions (OPE), defined below (Sect. 6.4.2).

Given a conserved current j^μ

$$\partial_\mu j^\mu = \partial j^z + \bar\partial j^z = 2(\partial j_{\bar z} + \bar\partial j_z) = 0, \tag{6.76}$$

the associated conserved charge is defined by

$$Q = \frac{1}{2\pi i} \oint_{C_0} (j_z \mathrm{d}z - j_{\bar z}\mathrm{d}\bar z), \tag{6.77}$$

where C_0 denotes the anti-clockwise contour around $z = 0$ (equivalently the interior of the contour is located to the left). The difference of sign in the second term follows directly from Stokes' theorem (B.14g) (and can be understood as a conjugation of the contour). The additional factor of $1/2\pi$ is consistent with the normalization of spatial integrals in two dimensions. The current components are not necessarily holomorphic and anti-holomorphic at this level, but in practice this will often be the case (and each component is independently conserved), and one writes

$$j(z) := j_z(z), \qquad \bar j(\bar z) := j_{\bar z}(\bar z). \tag{6.78}$$

In this case, the charge also splits into a holomorphic and an anti-holomorphic (left- and right-moving[8]) contributions

$$Q = Q_L + Q_R, \qquad Q_L := \frac{1}{2\pi i} \oint_{C_0} j(z)\mathrm{d}z, \qquad Q_R := -\frac{1}{2\pi i} \oint_{C_0} \bar j(\bar z)\mathrm{d}\bar z. \tag{6.79}$$

---

[8]For charges, we use subscript L and R to distinguish both sectors to avoid introducing a new symbol for the total charge. However, since $Q = Q_L$ in the holomorphic sector, it is often not necessary to distinguish between the two symbols when acting on an operator or a state (however, this is useful for writing mode expansions). We do not write a bar on Q_R because the charges do not depend on the position.

The infinitesimal variation of a field under the symmetry generated by Q reads

$$
\delta_\epsilon \mathcal{O}(z, \bar{z}) = -[\epsilon Q, \mathcal{O}(z, \bar{z})] = -\epsilon \oint_{C_z} \frac{dw}{2\pi i} \, j(w) \mathcal{O}(z, \bar{z}) + \epsilon \oint_{C_z} \frac{d\bar{w}}{2\pi i} \, \bar{j}(\bar{w}) \mathcal{O}(z, \bar{z}).
$$

(6.80)

The contour integrals are easily evaluated once the OPE between the current and the operator is known. This formula gives the infinitesimal variation under the transformation for any field, not only for primaries.

> **Computation: Equation (6.77)**
> In real coordinates, the charge is defined by integrating the time component of the current j^μ over space for fixed time (A.23):
>
> $$
> Q = \frac{1}{2\pi} \int d\sigma \, j^0.
> $$
>
> The first step is to rewrite this formula covariantly. Since the time is fixed on the slice, $d\tau = 0$ and one can write
>
> $$
> Q = \frac{1}{2\pi} \int (d\sigma \, j^0 - d\tau \, j^1) = -\frac{1}{2\pi} \int \epsilon_{\mu\nu} j^\mu \, dx^\nu.
> $$
>
> The last formula is valid for any contour. Moreover, it can be evaluated for complex coordinates:
>
> $$
> Q = -\frac{1}{2\pi} \oint \epsilon_{z\bar{z}} (j^z \, d\bar{z} - j^{\bar{z}} \, dz) = -\frac{i}{4\pi} \oint (j^z \, d\bar{z} - j^{\bar{z}} \, dz)
> $$
>
> $$
> = -\frac{1}{2\pi i} \oint (j_z \, dz - j_{\bar{z}} \, d\bar{z}).
> $$
>
> One finds a contour integral because $\tau = $ cst circles of the cylinder are mapped to $|z| = $ cst contours.

6.4.2 Operator Product Expansions

The operator product expansion (OPE) is a tool used frequently in CFT: it means that when two local operators come close to each other, it is possible to replace their product by a sum of local operators

$$
\mathcal{O}_i(z_i, \bar{z}_i) \mathcal{O}_j(z_j, \bar{z}_j) = \sum_k \frac{c_{ij}^k}{z_{ij}^{h_i + h_j - h_k} \bar{z}_{ij}^{\bar{h}_i + \bar{h}_j - \bar{h}_k}} \mathcal{O}_k(z_j, \bar{z}_j),
$$

(6.81)

where the OPE coefficients c_{ij}^k are some constants and the sum runs over all operators. When $\mathcal{O}_k$ is primary, the coefficients c_{ij}^k are related to the structure constants and the field metric by

$$C_{ijk} = g_{k\ell} c_{ij}^\ell. \tag{6.82}$$

The radius of convergence for the OPE is given by the distance to the nearest operators in the correlation function. The OPE defines an associative algebra (commutative for bosonic operators), and the holomorphic sector forms a subalgebra (called the chiral algebra).

Example 6.1: OPE with the Identity

The OPE of a field $\phi(z)$ with the identity 1 is found by a direct series expansion

$$\phi(z)1 = \sum_{n \in \mathbb{N}} \frac{(z - w)^n}{n!} \partial^n \phi(w). \tag{6.83}$$

Obviously there are no singular terms. ◀

Starting from this point we consider only the holomorphic sector except when stated otherwise. The formula for the OPE (6.81) can be rewritten as

$$A(z)B(w) := \sum_{n=-\infty}^{N} \frac{\{AB\}_n(z)}{(z - w)^n} \tag{6.84}$$

to simplify the manipulations. N is an integer and there are singular terms if $N > 0$. Generally, only the terms singular as $w \to z$ are necessary in the computations (for example, to use the Cauchy–Riemann formula (B.1)): equality up to non-singular terms is denoted by a tilde

$$A(z)B(w) \sim \sum_{n=1}^{N} \frac{\{AB\}_n(z)}{(z - w)^n} =: \overbrace{A(z)B(w)}. \tag{6.85}$$

The RHS of this expression defines the contraction of the operators A and B.

While, most of the time, only singular terms are kept

$$\phi_i(z_i)\phi_j(z_j) \sim \sum_k \theta(h_i + h_j - h_k) \frac{c_{ij}^k}{(z - w)^{h_i + h_j - h_k}} \phi_k(w) \tag{6.86}$$

(with $\theta(x)$ the Heaviside step function), it can happen that one keeps also non-singular terms (the product of two OPE have singular terms coming from non-singular terms multiplying singular terms). Explicit contractions of operators through the OPE are also denoted by a bracket when there are other operators.

For a primary field $\phi(z)$, one finds the OPE with the energy–momentum tensor to be

$$T(z)\phi(w) \sim \frac{h\,\phi(w)}{(z-w)^2} + \frac{\partial\phi(w)}{z-w}, \tag{6.87}$$

where h is the conformal weight of the field. This OPE together with (6.80) for $j(z) = -v(z)T(z)$ correctly reproduces (6.50).

Computation: Equation (6.50)

$$\delta\phi(z) = \oint_{C_z} \frac{dw}{2\pi i}\, v(w)T(w)\phi(z) \sim \oint_{C_z} \frac{dw}{2\pi i}\, v(w)\left(\frac{h\,\phi(z)}{(w-z)^2} + \frac{\partial\phi(z)}{w-z}\right)$$

$$= h\,\partial v(z)\,\phi(z) + v(z)\partial\phi(z).$$

For a non-primary operator, the OPE becomes more complicated (as it is reflected by the transformation property), but the conformal weight can still be identified at the term in z^{-2}. The most important example is the energy–momentum tensor: the central charge is found as the coefficient of the z^{-4} term its OPE with itself:

$$T(z)T(w) \sim \frac{c/2}{(z-w)^4} + \frac{2T(w)}{(z-w)^2} + \frac{\partial T(w)}{z-w}. \tag{6.88}$$

The OPE indicates that the conformal weight of T is $h = 2$. Using (6.80) for $j(z) = -v(z)T(z)$, one finds the infinitesimal variation

$$\delta T = 2\,\partial v\, T + v\,\partial T + \frac{c}{12}\,\partial^3 v. \tag{6.89}$$

The last term vanishes for global transformations: this translates the fact that T is only a quasi-primary. The finite form of this transformation is

$$T'(w) = \left(\frac{dz}{dw}\right)^{-2}\left(T(z) - \frac{c}{12}\,S(w,z)\right) = \left(\frac{dz}{dw}\right)^{-2} T(z) + \frac{c}{12}\,S(z,w), \tag{6.90}$$

where $S(w,z)$ is the Schwarzian derivative

$$S(w,z) = \frac{w^{(3)}}{w'} - \frac{3}{2}\left(\frac{w''}{w'}\right)^2, \tag{6.91}$$

where the derivatives of w are with respect to z. This vanishes if the transformation is in SL(2, $\mathbb{C}$), and it transforms as

$$S(u, z) = S(w, z) + \left(\frac{dw}{dz}\right)^2 S(u, w) \tag{6.92}$$

under successive changes of coordinates.

Computation: Equation (6.89)

$$\delta T(z) = \oint_{C_z} \frac{dw}{2\pi i} v(w) T(w) T(z) \sim \oint_{C_z} \frac{dw}{2\pi i} v(w) \left(\frac{c/2}{(z-w)^4} + \frac{2T(w)}{(z-w)^2} + \frac{\partial T(w)}{z-w}\right)$$

$$= \frac{c}{2 \times 3!} \partial^3 v(z) + 2\partial v(z) T(z) + v(z)\partial T(z).$$

6.4.3 Hermitian and BPZ Conjugation

In this section, we introduce two different notions of conjugations: one is adapted for amplitudes because it defines a unitary Euclidean time evolution, while the second is more natural as an inner product of CFT states. Both can be interpreted as providing a map from in-states to out-states on the cylinder.

Given an operator $\mathcal{O}$, we need to define an operation $\mathcal{O}^\ddagger$—called *Euclidean adjoint* (or simply *adjoint*)—which, after Wick rotation from Euclidean to Lorentzian signature, can be interpreted as the Hermitian adjoint.[9] This is necessary in order to define a Hermitian inner product and to impose reality conditions.

To motivate the definition, consider first the cylinder in Lorentzian signature. Since Hermitian conjugation does not affect the Lorentzian coordinates, the Euclidean time must reverse its sign:

$$t^\dagger = -i\tau^\dagger = t \implies \tau^\dagger = -\tau. \tag{6.93}$$

Hence, an appropriate definition of the Euclidean adjoint is a Hermitian conjugation together with time reversal.[10] Another point of view is that the time evolution operator $U(\tau) := e^{-\tau H}$ is not unitary when H is Hermitian $H^\dagger = H$: the solution is to define a new Euclidean adjoint $U(\tau)^\ddagger := U(-\tau)^\dagger$ such that $U(\tau)$ is unitary for it.

---

[9]In [28], it is denoted by a bar on top of the operator: we avoid this notation since the bar already denotes the anti-holomorphic sector. In [40], it is indicated by a subscript hc. Otherwise, in most of the literature, it has no specific symbol since one directly works with the modes.

[10]The Euclidean adjoint can be used to define an inner product: positive-definiteness of the latter is called *reflection positivity* or OS-positive and is a central axiom of constructive QFT.

Time reversal on the cylinder corresponds to inversion and complex conjugation on the complex plane:

$$z \xrightarrow{\tau \to -\tau} e^{-\tau + i\sigma} = \frac{1}{z^*} = I(\bar{z}), \tag{6.94}$$

where $I(z) = 1/z$ is the inversion (6.42).[11] On the real surface[12] $\bar{z} = z^*$, which leads to the definition of the Euclidean adjoint as follows:

$$\mathcal{O}(z, \bar{z})^{\ddagger} := \left(\bar{I} \circ \mathcal{O}(z, \bar{z}) \right)^{\dagger}, \tag{6.95}$$

where $\bar{I}(z) := 1/\bar{z}$. If $\mathcal{O}$ is quasi-primary, we have

$$\mathcal{O}(z, \bar{z})^{\ddagger} = \left[\frac{1}{\bar{z}^{2h} z^{2\bar{h}}} \mathcal{O}\left(\frac{1}{\bar{z}}, \frac{1}{z} \right) \right]^{\dagger} = \frac{1}{z^{2h} \bar{z}^{2\bar{h}}} \mathcal{O}^{\dagger}\left(\frac{1}{z}, \frac{1}{\bar{z}} \right). \tag{6.96}$$

The last equality shows that Euclidean conjugation is equivalent to take the conjugate of all factors of i but otherwise leaves z and $\bar{z}$ unaffected. The Euclidean adjoint acts by complex conjugation of any c-number and reverses the order of the operators (acting as a transpose):

$$(\lambda \, \mathcal{O}_1 \cdots \mathcal{O}_n)^{\ddagger} = \lambda^* \, \mathcal{O}_n^{\ddagger} \cdots \mathcal{O}_1^{\ddagger}, \qquad \lambda \in \mathbb{C}, \tag{6.97}$$

without any sign.

The second operation, called the *BPZ conjugation*, is useful. It can be defined in two different ways:

$$\mathcal{O}(z, \bar{z})^{t} := I^{\pm} \circ \mathcal{O}(z, \bar{z}) = \frac{(\mp 1)^{h + \bar{h}}}{z^{2h} \bar{z}^{2\bar{h}}} \mathcal{O}\left(\pm \frac{1}{z}, \pm \frac{1}{\bar{z}} \right), \tag{6.98}$$

where $I^{\pm}(z) = \pm 1/z$ is the inversion (6.42). The minus and plus signs are, respectively, more convenient when working with the open and closed strings.[13] The BPZ conjugation does not complex conjugate c-number nor changes the order of the operators:[14]

$$(\lambda \, \mathcal{O}_1 \cdots \mathcal{O}_n)^{t} = \lambda \, \mathcal{O}_1^{t} \cdots \mathcal{O}_n^{t}, \qquad \lambda \in \mathbb{C}. \tag{6.99}$$

---

[11] We do not write "$z^{\dagger}$" because this notation is confusing as one should not complex conjugate the factor of i in the exponential (Sect. 6.2.1).

[12] Remember that $\bar{z}$ is not the complex conjugate of z but an independent variable.

[13] The index t should not be confused with the matrix transpose: it is used in opposition with $\ddagger$ and $\dagger$ to indicate that no complex conjugation is involved.

[14] However, the fields become anti-radially ordered after a BPZ conjugation since it sends z to $1/z$. The radial ordering can be restored by (anti-)commuting the fields, which can introduce additional signs [36]. This problem does not arise when working in terms of the modes.

The identity is invariant under both conjugation

$$1^{\ddagger} = 1^{t} = 1.$$

(6.100)

6.4.4   Mode Expansion

Any field of weight $(h, \bar{h})$ can be expanded in terms of modes $\mathcal{O}_{m,n}$

$$\mathcal{O}(z, \bar{z}) = \sum_{m,n} \frac{\mathcal{O}_{m,n}}{z^{m+h} \bar{z}^{n+\bar{h}}}.$$

(6.101)

Note that the modes $\mathcal{O}_{m,n}$ themselves are operators. The ranges of the two indices are such that

$$m + h \in \mathbb{Z} + \nu, \qquad n + \bar{h} \in \mathbb{Z} + \bar{\nu}, \qquad \nu, \bar{\nu} = \begin{cases} 0 & \text{periodic,} \\ 1/2 & \text{anti-periodic.} \end{cases}$$

(6.102)

The values of ν and $\bar{\nu}$ depend on whether the fields satisfy periodic or anti-periodic boundary conditions on the plane (for half-integer weights, the periodicity is reversed on the cylinder):

$$\mathcal{O}(e^{2\pi i} z, \bar{z}) = e^{2\pi i \nu} \mathcal{O}(z, \bar{z}), \qquad \mathcal{O}(z, e^{2\pi i} \bar{z}) = e^{2\pi i \bar{\nu}} \mathcal{O}(z, \bar{z}).$$

(6.103)

Depending on whether the weights are integers or half-integers, additional terminology is introduced:

- If $h \in \mathbb{Z} + 1/2$, then one can choose anti-periodic (*Neveu–Schwarz* or NS) or periodic (*Ramond* or R) boundary conditions on the cylinder (reversed for the plane):

$$\nu, \bar{\nu} = \begin{cases} 0 & \text{NS} \\ 1/2 & \text{R.} \end{cases}$$

(6.104)

  The indices are half-integers (resp. integers) for the NS (R) sector.
- If $h \in \mathbb{Z}$, periodic (or untwisted) boundary conditions are more natural, but anti-periodic boundary conditions may also be considered:

$$\nu, \bar{\nu} = \begin{cases} 0 & \text{untwisted} \\ 1/2 & \text{twisted.} \end{cases}$$

(6.105)

The modes of untwisted (resp. twisted) fields have integer (half-integers) indices.

The mode expansions have no branch cut (fractional power of z or $\bar{z}$) for periodic fields (bosonic untwisted or fermionic twisted). We will see explicit examples of such operators later in this book.

Under Euclidean conjugation (6.95), the modes are related by

$$(\mathcal{O}^{\ddagger})_{-m,-n} = (\mathcal{O}_{m,n})^{\dagger}. \tag{6.106}$$

In particular, if the operator is Hermitian (under the Euclidean adjoint), the reality condition on the modes relates the negative modes with the conjugated positive modes

$$\mathcal{O}^{\ddagger} = \mathcal{O} \quad \Longrightarrow \quad (\mathcal{O}_{m,n})^{\dagger} = \mathcal{O}_{-m,-n}. \tag{6.107}$$

When no confusion is possible (for Hermitian operators), we will write $\mathcal{O}^{\dagger}_{m,n}$ instead of $(\mathcal{O}_{m,n})^{\dagger}$.

For a holomorphic field $\phi(z)$, the above expansion becomes

$$\phi(z) = \sum_{n \in \mathbb{Z}+h+\nu} \frac{\phi_n}{z^{n+h}}. \tag{6.108}$$

Conversely, the modes are recovered from the field through

$$\phi_n = \oint_{C_0} \frac{\mathrm{d}z}{2\pi \mathrm{i}} z^{n+h-1} \phi(z), \tag{6.109}$$

where the integration is counter-clockwise around the origin.

If the field is Hermitian, then

$$\phi^{\ddagger} = \phi \quad \Longrightarrow \quad (\phi_n)^{\dagger} = \phi_{-n}. \tag{6.110}$$

The operators ϕ_n have a conformal weight of $-n$ (since the weight of z is -1). The BPZ conjugate of the modes is

$$\phi_n^t = (I^{\pm} \circ \phi)_n = (-1)^h (\pm 1)^n \phi_{-n}. \tag{6.111}$$

Computation: Equation (6.111)

$$\phi_n^t = (I^{\pm} \circ \phi)_n = \oint \frac{\mathrm{d}z}{2\pi \mathrm{i}} z^{n+h-1} I^{\pm} \circ \phi(z)$$

$$= \oint \frac{\mathrm{d}z}{2\pi \mathrm{i}} z^{n+h-1} \left(\mp \frac{1}{z^2}\right)^h \phi\left(\pm \frac{1}{z}\right)$$

$$= (\mp 1)^h \oint \frac{dz}{2\pi i} z^{n-h-1} \phi\left(\pm\frac{1}{z}\right)$$

$$= (\mp 1)^h \oint \frac{dw}{2\pi i} \left(\pm\frac{1}{w}\right)^{n-h} w^{-1}\phi(w)$$

$$= (\mp 1)^h (\pm 1)^{n-h} \oint \frac{dw}{2\pi i} w^{-n+h-1}\phi(w),$$

where we have set $w = \pm 1/z$ such that

$$\frac{dz}{z} = \mp\frac{dw}{w^2 z} = -\frac{dw}{w}, \tag{6.112}$$

and the minus sign disappears upon reversing the contour orientation.

The mode expansion of the energy–momentum tensor is

$$T(z) = \sum_{n\in\mathbb{Z}} \frac{L_n}{z^{n+2}}, \qquad L_n = \oint \frac{dz}{2\pi i} T(z) z^{n+1}, \tag{6.113}$$

where one recognizes the Virasoro operators as the modes. In most situations, the Virasoro operators are Hermitian

$$L_n^\dagger = L_{-n}. \tag{6.114}$$

The OPE (6.88) and (6.87) together with (6.75a) help to reconstruct the Virasoro algebra (6.58) and the commutation relations between the L_m and the modes ϕ_n of a weight h primary:

$$[L_m, \phi_n] = \big(m(h-1) - n\big)\phi_{m+n}. \tag{6.115}$$

This easily gives the commutation relation for the complete field:

$$[L_m, \phi(z)] = z^m\big(z\partial + (n+1)h\big)\phi(z). \tag{6.116}$$

We will often use (6.58) and (6.115) for $m = 0$:

$$[L_0, L_{-n}] = nL_{-n}, \qquad [L_0, \phi_{-n}] = n\phi_{-n}. \tag{6.117}$$

This means that both ϕ_n and L_n act as raising operators for L_0 if $n < 0$, and as lowering operators if $n > 0$ (remember that L_0 is the Hamiltonian in the holomorphic sector). When both the holomorphic and anti-holomorphic sectors enter, it is convenient to introduce the combinations

$$L_n^\pm = L_n \pm \bar{L}_n, \tag{6.118}$$

such that L_0^+ is the Hamiltonian.

Finally, every holomorphic current $j(z)$ has a conformal weight $h = 1$ and can be expanded as

$$j(z) = \sum_n \frac{j_n}{z^{n+1}}. \tag{6.119}$$

By definition, the zero-mode is equal to the holomorphic charge

$$Q_L = j_0. \tag{6.120}$$

6.4.5 Hilbert Space

The Hilbert space of the CFT is denoted by $\mathcal{H}$. The $SL(2, \mathbb{C})$ (or *conformal*) *vacuum*[15] $|0\rangle$ is defined by the state which is invariant under the global conformal transformations:

$$L_0|0\rangle = 0, \qquad L_{\pm 1}|0\rangle = 0. \tag{6.121}$$

Expectation value of an operator $\mathcal{O}$ in the $SL(2, \mathbb{C})$ vacuum is denoted as:

$$\langle \mathcal{O} \rangle := \langle 0|\mathcal{O}|0\rangle. \tag{6.122}$$

If the fields are expressed in terms of creation and annihilation operators (which happens e.g. for free scalars, free fermions and ghosts), then the Hilbert space has the structure of a Fock space.

State-Operator Correspondence

The state–operator correspondence identifies every state $|\mathcal{O}\rangle$ of the CFT Hilbert space with an operator $\mathcal{O}(z, \bar{z})$ through

$$|\mathcal{O}\rangle = \lim_{z, \bar{z} \to 0} \mathcal{O}(z, \bar{z})|0\rangle = \mathcal{O}(0, 0)|0\rangle. \tag{6.123}$$

Such a state can be interpreted as an "in" state since it is located at $\tau \to -\infty$ on the cylinder. Focusing now on a holomorphic field $\phi(z)$, the state is defined as

$$|\phi\rangle = \lim_{z \to 0} \phi(z)|0\rangle = \phi(0)|0\rangle. \tag{6.124}$$

---

[15]There are different notions of "vacuum", see (6.129). However, the $SL(2, \mathbb{C})$ vacuum is unique. Indeed, it is mapped to the unique identity operator under the state–operator correspondence (however, there can be other states of weight 0, see Remark 6.6).

For this to make sense, the modes which diverge as $z \to 0$ must annihilate the vacuum. In particular, for a weight h field $\phi(z)$, one finds

$$\forall n \geq -h+1: \quad \phi_n|0\rangle = 0. \tag{6.125}$$

Thus, the ϕ_n for $n \geq -h + 1$ are annihilation operators for the vacuum $|0\rangle$, and conversely the states ϕ_n with $n < -h+1$ are creation operators. As a consequence, the state $|\phi\rangle$ is found by applying the mode $n = -h$ to the vacuum:

$$|\phi\rangle = \phi_{-h}|0\rangle = \oint \frac{dz}{2\pi i} \frac{\phi(z)}{z} |0\rangle. \tag{6.126}$$

Since L_{-1} is the generator of translations on the plane, one finds

$$\phi(z)|0\rangle = e^{zL_{-1}}\phi(0)e^{-zL_{-1}}|0\rangle = e^{zL_{-1}}|\phi\rangle. \tag{6.127}$$

The vacuum $|0\rangle$ is the state associated with the identity 1. Translating the conditions (6.125) to the energy–momentum tensor gives

$$\forall n \geq -1: \quad L_n|0\rangle = 0. \tag{6.128}$$

This is consistent with the definition (6.121) since it includes the $\mathrm{sl}(2, \mathbb{C})$ subalgebra.

If $h < 0$, some of the modes with $n > 0$ do not annihilate the vacuum: (6.117) implies that some states have an energy lower than the one of $|0\rangle$. The state $|\Omega\rangle$ (possibly degenerate) with the lowest energy is called the *energy vacuum*

$$\forall |\phi\rangle \in \mathcal{H}: \quad \langle \Omega|L_0|\Omega\rangle \leq \langle \phi|L_0|\phi\rangle. \tag{6.129}$$

It is obtained by acting repetitively with the modes $\phi_{n>0}$. This vacuum defines a new partition of the non-zero-modes operators into annihilation and creation operators. If there are zero-modes, i.e. $n = 0$ modes, then the vacuum is degenerate since they commute with the Hamiltonian, $[L_0, \phi_0] = 0$ according to (6.117). The partition of the zero-modes into creation and annihilation operators depends on the specific state chosen among the degenerate vacua.

The energy a_Ω of $|\Omega\rangle$, which is also its L_0 eigenvalue

$$L_0|\Omega\rangle := a_\Omega|\Omega\rangle, \tag{6.130}$$

is called *zero-point energy*. Bosonic operators with negative h are dangerous because they lead to an infinite negative energy together with an infinite degeneracy (from the zero-mode).

The conjugate vacuum is defined by BPZ or Hermitian conjugation

$$\langle 0| = |0\rangle^\ddagger = |0\rangle^t \tag{6.131}$$

since both leave the identity invariant. It is also annihilated by the $\mathfrak{sl}(2,\mathbb{C})$ subalgebra:

$$\langle 0|L_0 = 0, \qquad \langle 0|L_{\pm 1} = 0. \tag{6.132}$$

Since there are two kinds of conjugation, two different conjugated states can be defined. They are also called "out" states since they are located at $\tau \to \infty$ on the cylinder (Fig. 6.2).

Euclidean and BPZ Conjugations and Inner Products

The Euclidean adjoint $\langle \mathcal{O}^{\ddagger}|$ of the state $|\mathcal{O}\rangle$ is defined as

$$\langle \mathcal{O}^{\ddagger}| = \lim_{w,\bar{w}\to 0} \langle 0|\mathcal{O}(w,\bar{w})^{\ddagger} = \lim_{w,\bar{w}\to 0} \frac{1}{w^{2h}\bar{w}^{2\bar{h}}} \langle 0|\mathcal{O}\left(\frac{1}{\bar{w}},\frac{1}{w}\right)^{\dagger} \tag{6.133a}$$

$$= \lim_{z,\bar{z}\to\infty} z^{2h}\bar{z}^{2\bar{h}}\langle 0|\mathcal{O}^{\dagger}(z,\bar{z}) \tag{6.133b}$$

$$= \langle 0|I \circ \mathcal{O}^{\dagger}(0,0), \tag{6.133c}$$

where the two coordinate systems are related by $w = 1/\bar{z}$. From this formula, the definition of the adjoint of a holomorphic operator ϕ follows

$$\langle \phi^{\ddagger}| = \lim_{\bar{w}\to 0} \langle 0|\phi(w)^{\ddagger} = \lim_{\bar{w}\to 0} \frac{1}{w^{2h}}\langle 0|\phi^{\dagger}\left(\frac{1}{w}\right) \tag{6.134a}$$

$$= \lim_{z\to\infty} z^{2h}\langle 0|\phi^{\dagger}(z) \tag{6.134b}$$

$$= \langle 0|I \circ \phi^{\dagger}(0). \tag{6.134c}$$

Then, expanding the field in terms of the modes gives

$$\langle \phi^{\ddagger}| = \langle 0|(\phi^{\dagger})_h. \tag{6.135}$$

The BPZ conjugated state is

$$\langle \phi| := \lim_{w\to 0} \langle 0|\phi(w)^t \tag{6.136a}$$

$$= (\pm 1)^h \lim_{z\to\infty} z^{2h}\langle 0|\phi(z) \tag{6.136b}$$

$$= \langle 0|I^{\pm} \circ \phi(0). \tag{6.136c}$$

In terms of the modes, one has

$$\langle \phi| = (\pm 1)^h\langle 0|\phi_h. \tag{6.137}$$

If ϕ is Hermitian, then the relation between both conjugated states corresponds to a reality condition:

$$\langle \phi^{\ddagger}| = (\pm 1)^h \langle \phi|. \tag{6.138}$$

Taking the BPZ conjugation of the conditions (6.125) tells which modes must annihilate the conjugate vacuum:

$$\forall n \leq h - 1: \quad \langle 0|\phi_n = 0, \tag{6.139}$$

and one finds more particularly for the Virasoro operators

$$\forall n \leq 1: \quad \langle 0|L_n = 0. \tag{6.140}$$

This can also be derived directly from (6.136) by requiring that applying an operator on the conjugate vacuum $\langle 0|$ is well-defined.

All conditions taken together mean that the expectation value of the energy–momentum tensor in the conformal vacuum vanishes:

$$\langle 0|T(z)|0\rangle = 0. \tag{6.141}$$

In particular, this means that the energy vacuum $|\Omega\rangle$, if different from $|0\rangle$, has a negative energy.

The Hermitian[16] and BPZ inner products are, respectively, defined by

$$\langle \phi_i^{\ddagger}|\phi_j\rangle = \langle 0|\bar{I} \circ \phi_j(0)\phi_i(0)|0\rangle = \lim_{\substack{z \to \infty \\ w \to 0}} z^{2h_i}\langle 0|\phi_i^{\dagger}(z)\phi_j(w)|0\rangle, \tag{6.142a}$$

$$\langle \phi_i|\phi_j\rangle = \langle 0|I \circ \phi_j(0)\phi_i(0)|0\rangle = (\pm 1)^{h_i} \lim_{\substack{z \to \infty \\ w \to 0}} z^{2h_i}\langle 0|\phi_i(z)\phi_j(w)|0\rangle. \tag{6.142b}$$

These products can be recast as 2-point correlation functions (6.62b) on the sphere:

$$\langle \phi_i|\phi_j\rangle = \langle I \circ \phi_i(0)\phi_j(0)\rangle, \qquad \langle \phi_i^{\ddagger}|\phi_j\rangle = \langle I \circ \phi_i^{\dagger}(0)\phi_j(0)\rangle. \tag{6.143}$$

From the state–operator correspondence, the action of one operator on the in-state can be reinterpreted as the matrix element of this operator using the two external states, or also as a 3-point function:

$$\langle \phi_i|\phi_j(z)|\phi_k\rangle = (\pm 1)^{h_i} \lim_{w \to \infty} w^{2h_i} \langle \phi_i(w)\phi_j(z)\phi_k(0)\rangle. \tag{6.144}$$

---

[16]Depending on the normalization, it can also be anti-Hermitian.

Given a basis of states $\{\phi_i\}$ (i can run over both discrete and continuous indices), the *conjugate* or *dual states* $\{\phi_i^c\}$ are defined by

$$\langle \phi_i^c | \phi_j \rangle = \delta_{ij} \tag{6.145}$$

(the delta function is discrete and/or continuous according to the indices).

Verma Modules

If $\phi(z)$ is a weight h primary, then the associated state $|\phi\rangle$ satisfies:

$$L_0 |\phi\rangle = h |\phi\rangle, \qquad \forall n \geq 1: \quad L_n |\phi\rangle = 0. \tag{6.146}$$

Such a state is also called a highest-weight state. The descendant states are defined by all possible states of the form

$$|\phi_{\{n_i\}}\rangle := \prod_i L_{-n_i} |\phi\rangle, \tag{6.147}$$

where the same L_{-n_i} can appear multiple times and $n_i > 0$. The set of states $\phi_{\{n_i\}}$ is called a *Verma module* $V(h, c)$. One finds that the L_0 eigenvalues of this state is

$$L_0 = h + \sum_i n_i. \tag{6.148}$$

Normal Ordering

The *normal ordering* of an operator with respect to a vacuum corresponds to placing all creation (resp. annihilation) operators of this vacuum on the left (resp. right). From this definition, the expectation value of a normal ordered operator in the vacuum vanishes identically. The main reason for normal ordering is to remove singularities in expectation values.

Given an operator $\phi(z)$, we define two normal orderings:

- The conformal normal order (CNO) $: \mathcal{O} :$ is defined with respect to the conformal vacuum (6.121):

$$\langle 0 |: \mathcal{O} :| 0 \rangle = 0. \tag{6.149}$$

- The energy normal order (ENO) $\overset{\star}{\mathcal{O}}\!\!\star$ is defined with respect to the energy vacuum (6.129):

$$\langle \Omega | \overset{\star}{\mathcal{O}}\!\!\star | \Omega \rangle = 0. \tag{6.150}$$

We first discuss the conformal normal ordering before explaining how to relate it to the energy normal ordering.

Given two operators A and B, the simplest normal ordering amounts to subtract the expectation value:

$$: A(z)B(w) : \overset{?}{=} A(z)B(w) - \langle A(z)B(w) \rangle. \tag{6.151}$$

This is equivalent to defining the products of two operators at coincident points via point-splitting:

$$: A(z)B(z) : \overset{?}{=} \lim_{w \to z} \left(A(z)B(w) - \langle A(z)B(w) \rangle \right). \tag{6.152}$$

While this works well for free fields, this does not generalize for composite or interacting fields.

The reason is that this procedure removes only the highest singularity in the product: it does not work if the OPE has more than one singular term. An appropriate definition is

$$: A(z)B(w) : := A(z)B(w) - \overbrace{A(z)B(w)} = \sum_{n \in \mathbb{N}} (z-w)^n \{AB\}_{-n}(z), \tag{6.153}$$

where the contraction between A and B is defined in (6.85), and the second equality comes from (6.84).

Then, the product evaluated at coincident points is found by taking the limit (in this case the argument is often indicated only at the end of the product)

$$: AB(z) : := : A(z)B(z) : := \lim_{w \to z} : A(z)B(w) : = \{AB\}_0(z). \tag{6.154}$$

Indeed, since all powers of $(z-w)$ are positive in the RHS of (6.153), all terms but the first one disappear. The form of (6.154) shows that the normal order can also be computed with the contour integral

$$: AB(z) : = \oint_{C_z} \frac{dw}{2\pi i} \frac{A(z)B(w)}{z-w}. \tag{6.155}$$

It is common to remove the colons of normal ordering when there is no ambiguity and, in particular, to write

$$AB(z) := : AB(z) :. \tag{6.156}$$

In terms of modes, one has

$$: AB(z) : = \sum_m \frac{: AB :_m}{z^{m+h_A+h_B}}, \tag{6.157a}$$

$$: AB :_m = \sum_{n \le -h_A} A_n B_{m-n} + \sum_{n > -h_A} B_{m-n} A_n. \tag{6.157b}$$

This expression makes explicit that normal ordering is non-commutative and non-associative:

$$: AB(z) : \neq : BA(z) :, \qquad : A(BC)(z) : \neq : (AB)C(z) :. \tag{6.158}$$

The product of normal ordered operators can then be computed using Wick theorem. In fact, one is more interested in the contraction of two such operators in order to recover the OPE between these operators: the product is then derived with (6.153).

If A_i ($i = 1, 2, 3$) are *free fields*, one has

$$A_1(z) : A_2 A_3(w) : = : A_1(z) A_2 A_3(w) : + \overparen{A_1(z) : A_2} A_3(w) :,$$

$$\overparen{A_1(z) : A_2} A_3(w) : = \overparen{A_1(z) A_2}(w) : A_3(w) : + \overparen{A_1(z) A_3}(w) : A_2(w) :. \tag{6.159}$$

If the fields are not free, then the contraction cannot be extracted from the normal ordering. Similarly if there are more fields, then one needs to perform all the possible contractions.

Given two free fields A and B, one has the following identities:

$$A(z) : B(w)^n : = n \, \overparen{A(z) B(w)} : B(w)^{n-1} :, \tag{6.160a}$$

$$A(z) : e^{B(w)} : = \overparen{A(z) B(w)} : e^{B(w)} :, \tag{6.160b}$$

$$: e^{A(z)} :: e^{B(w)} : = \exp\left(\overparen{A(z) B(w)}\right) : e^{A(z)} e^{B(w)} :. \tag{6.160c}$$

The last relation generalizes for a set of n fields A_i:

$$\prod_{i=1}^{n} : e^{A_i} : = : \exp\left(\sum_{i=1}^{n} A_i\right) : \exp \sum_{i<j} \langle A_i A_j \rangle, \tag{6.161a}$$

$$\left\langle \prod_{i=1}^{n} : e^{A_i} : \right\rangle = \exp \sum_{i<j} \langle A_i A_j \rangle. \tag{6.161b}$$

Computation: Equation (6.160b)

$$A(z) : e^{B(w)} : = A(z) \sum_{n} \frac{1}{n!} : B(w)^n : = \overparen{A(z) B(w)} \sum_{n} \frac{1}{(n-1)!} : B(w)^{n-1} :.$$

Computation: Equation (6.160c)

$$: e^{A(z)} :: e^{B(w)} := \sum_{m,n} \frac{1}{m!n!} : A(z)^m :: B(w)^n :$$

$$= \sum_{m,n,k} \frac{k!}{m!n!} \binom{m}{k}\binom{n}{k} \left(\overline{A(z)B(w)}\right)^k : A(z)^{m-k} :: B(w)^{n-k} :$$

$$= \sum_{m,n,k} \frac{1}{k!(m-k)!(n-k)!} \left(\overline{A(z)B(w)}\right)^k : A(z)^{m-k} :: B(w)^{n-k} :.$$

The factorial $k!$ counts the number of possible ways to contract the two operators.

The general properties of normal ordered expressions are identical for both vacua: what differs is the precise computation in terms of the operators (or modes). Hence, the energy normal ordering can be defined in parallel with (6.157), but changing the definitions of creation and annihilation operators:

$$\overset{\star}{\star} AB(z)\overset{\star}{\star} = \sum_m \frac{\overset{\star}{\star} AB(z)\overset{\star}{\star}_n}{z^{m+h_A+h_B}}, \tag{6.162a}$$

$$\overset{\star}{\star} AB\overset{\star}{\star}_m = \sum_{n \leq 0} A_n B_{m-n} + \sum_{n > 0} B_{m-n} A_n. \tag{6.162b}$$

To simplify the definition we assume that A_0 is a creation operator and it is thus included in the first sum (this must be adapted in function of which vacuum state is chosen if the latter is degenerate).

The relation between the normal ordered modes is

$$: AB :_m = \overset{\star}{\star} AB\overset{\star}{\star}_m + \sum_{n=0}^{h_A-1} [B_{m+n}, A_{-n}]. \tag{6.163}$$

Computation: Equation (6.163)

$$: AB :_m = \sum_{n \leq -h_A} A_n B_{m-n} + \sum_{n > -h_A} B_{m-n} A_n$$

$$= \sum_{n \geq h_A} A_{-n} B_{m+n} + \sum_{n > 0} B_{m-n} A_n + \sum_{n=0}^{h_A-1} B_{m+n} A_{-n}$$

$$= \sum_{n\geq 0} A_{-n} B_{m+n} + \sum_{n>0} B_{m-n} A_n + \sum_{n=0}^{h_A-1} [B_{m+n}, A_{-n}]$$

$$= {}^{\star}_{\star} AB {}^{\star}_{\star}{}_m + \sum_{n=0}^{h_A-1} [B_{m+n}, A_{-n}].$$

The choice of the normal ordering for the operators is related to the ordering ambiguity when quantizing the system: when the product of two non-commuting modes appears in the classical composite field, the corresponding quantum operator is ambiguous (generally up to a constant). In practice, one starts with the conformal ordering since it is invariant under conformal transformations and because one can compute with contour integrals. Then, the expression can be translated in the energy ordering using (6.163). But, knowing how the conformal and energy vacua are related, it is often simpler to find the difference between the two orderings by applying the operator on the vacua.

6.4.6   CFT on the Cylinder

According to (6.44), the relation between the field on the cylinder and on the plane is

$$\phi(z) = \left(\frac{L}{2\pi}\right)^h z^{-h} \phi_{\text{cyl}}(w) \tag{6.164}$$

(quantities without indices are on the plane by definition). The mode expansion on the cylinder is

$$\phi_{\text{cyl}} = \left(\frac{2\pi}{L}\right)^h \sum_{n\in\mathbb{Z}} \phi_n e^{-\frac{2\pi}{L} w} = \left(\frac{2\pi}{L}\right)^h \sum_{n\in\mathbb{Z}} \frac{\phi_n}{z^n}. \tag{6.165}$$

Using the finite transformation (6.90) for the energy–momentum tensor T, one finds the relation

$$T_{\text{cyl}}(w) = \left(\frac{2\pi}{L}\right)^2 \left(T(z)z^2 - \frac{c}{24}\right). \tag{6.166}$$

For the L_0 mode, one finds

$$(L_0)_{\text{cyl}} = L_0 - \frac{c}{24}, \tag{6.167}$$

and thus the Hamiltonian is

$$H = (L_0)_{\text{cyl}} + (\bar{L}_0)_{\text{cyl}} = L_0 + \bar{L}_0 - \frac{c + \bar{c}}{24}. \tag{6.168}$$

6.5 Suggested Readings

- The most complete reference on CFTs is [6] but it lacks some recent developments. Two excellent complementary books are [3, 35].

  String theory books generally dedicate a fair amount of pages to CFTs: particularly good summaries can be found in [2, 23, 28, 29].

  Finally, a modern and fully algebraic approach can be found in [31, 32]. Other good reviews are [30, 39].
- There are various other books [18, 20, 22, 26] and reviews [4, 10, 12, 34, 37].
- The maps from the sphere and the cylinder to the complex plane are discussed in [28, sec. 2.6, 6.1].
- Normal ordering is discussed in details in [6, chap. 6] (see also [2, sec. 4.2, 28, sec. 2.2]).
- Euclidean conjugation is discussed in [6, sec. 6.1.1, [28], p. 202–3]. For a comparison of Euclidean and BPZ conjugations, see [[40], sec. 2.2, [36], p. 11].
- Normal ordering and difference between the different definitions are described in [28, chap. 2, 6, sec. 6.5].

References

1. T. Bautista, A. Dabholkar, H. Erbin, Quantum gravity from timelike Liouville theory. J. High Energy Phys. **2019**(10), 284 (2019). https://doi.org/10.1007/JHEP10(2019)284. arXiv: 1905.12689
2. R. Blumenhagen, D. Lüst, S. Theisen, *Basic Concepts of String Theory* (Springer, Berlin, 2014)
3. R. Blumenhagen, E. Plauschinn, *Introduction to Conformal Field Theory: With Applications to String Theory*. Lecture Notes in Physics (Springer, Berlin, 2009). https://www.springer.com/de/book/9783642004490
4. J. Cardy, Conformal field theory and statistical mechanics (2008). arXiv: 0807.3472
5. G. Delfino, J. Viti, On three-point connectivity in two-dimensional percolation. J. Phys. A Math. Theor. **44**(3), 032001 (2011). https://doi.org/10.1088/1751-8113/44/3/032001. arXiv: 1009.1314
6. P. Di Francesco, P. Mathieu, D. Senechal, *Conformal Field Theory*, 2nd edn. (Springer, Berlin, 1999)
7. M. Flohr, On modular invariant partition functions of conformal field theories with logarithmic operators. Int. J. Mod. Phys. A **11**(22), 4147–4172 (1996). https://doi.org/10.1142/S0217751X96001954. arXiv: hep-th/9509166
8. M. Flohr, Bits and pieces in logarithmic conformal field theory. Int. J. Mod. Phys. A **18**(25), 4497–4591 (2003). https://doi.org/10.1142/S0217751X03016859. arXiv: hep-th/0111228
9. K. Fujikawa, U. Lindström, N.K. Nielsen, M. Rocek, P. van Nieuwenhuizen, Regularized BRST-coordinate-invariant measure. Phys. Rev. D **37**(2), 391–405 (1988). https://doi.org/10.1103/PhysRevD.37.391

10. M.R. Gaberdiel, An introduction to conformal field theory (1999). https://doi.org/10.1088/0034-4885/63/4/203. arXiv: hep-th/9910156
11. M.R. Gaberdiel, An algebraic approach to logarithmic conformal field theory. Int. J. Mod. Phys. A **18**(25), 4593–4638 (2003). https://doi.org/10.1142/S0217751X03016860. arXiv: hep-th/0111260
12. P. Ginsparg, Applied conformal field theory (1988). arXiv: hep-th/9108028
13. M.B. Green, J.H. Schwarz, E. Witten, *Superstring Theory: Introduction*, vol. 1. (Cambridge University Press, Cambridge, 1988)
14. E. Guadagnini, Central charge, trace and gravitational anomalies in two dimensions. Phys. Rev. D **38**(8), 2482–2489 (1988). https://doi.org/10.1103/PhysRevD.38.2482
15. V. Gurarie, Logarithmic operators in conformal field theory. Nucl. Phys. B **410**(3), 535–549 (1993). https://doi.org/10.1016/0550-3213(93)90528-W. arXiv: hep-th/9303160
16. D. Harlow, J. Maltz, E. Witten, Analytic continuation of Liouville theory. J. High Energy Phys. **12**, 071 (2011). https://doi.org/10.1007/JHEP12(2011)071. arXiv: 1108.4417
17. M. Hatsuda, P. van Nieuwenhuizen, W. Troost, A. Van Proeyen, The regularized phase space path integral measure for a scalar field coupled to gravity. Nucl. Phys. B **335**(1), 166–196 (1990). https://doi.org/10.1016/0550-3213(90)90176-E
18. M. Henkel, *Conformal Invariance and Critical Phenomena*, 1st edn. (Springer, Berlin, 2010)
19. Y. Ikhlef, J.L. Jacobsen, H. Saleur, Three-point functions in $c \leq 1$ Liouville theory and conformal loop ensembles (v1). Phys. Rev. Lett. **116**(13), 130601 (2016). https://doi.org/10.1103/PhysRevLett.116.130601. arXiv: 1509.03538
20. C. Itzykson, J.-M. Drouffe, *Théorie statistique des champs*, vol. 2. (EDP Sciences, 2000)
21. R. Jackiw, Another view on massless matter-gravity fields in two dimensions (1995). arXiv: hep-th/9501016
22. S.V. Ketov, *Conformal Field Theory* (World Scientific, Singapore, 1995)
23. E. Kiritsis, *String Theory in a Nutshell* (Princeton University Press, Princeton, 2007)
24. M. Knecht, S. Lazzarini, F. Thuillier, Shifting the Weyl anomaly to the chirally split diffeomorphism anomaly in two dimensions. Phys. Lett. B **251**(2), 279–283 (1990). https://doi.org/10.1016/0370-2693(90)90936-Z
25. I.I. Kogan, N.E. Mavromatos, World-sheet logarithmic operators and target space symmetries in string theory. Phys. Lett. B **375**(1–4), 111–120 (1996). https://doi.org/10.1016/0370-2693(96)00195-5. arXiv: hep-th/9512210
26. G. Mussardo, *Statistical Field Theory: An Introduction to Exactly Solved Models in Statistical Physics* (Oxford University Press, Oxford, 2009)
27. M. Picco, R. Santachiara, J. Viti, G. Delfino, Connectivities of Potts Fortuin–Kasteleyn clusters and time-like Liouville correlator. Nucl. Phys. B **875**(3), 719–737 (2013). https://doi.org/10.1016/j.nuclphysb.2013.07.014. arXiv: 1304.6511
28. J. Polchinski, *String Theory: Volume 1, An Introduction to the Bosonic String* (Cambridge University Press, Cambridge, 2005)
29. J. Polchinski, *String Theory, Volume 2: Superstring Theory and Beyond* (Cambridge University Press, Cambridge, 2005)
30. J.D. Qualls, Lectures on conformal field theory (2015). arXiv: 1511.04074
31. S. Ribault, Conformal field theory on the plane (2014). arXiv: 1406.4290
32. S. Ribault, Minimal lectures on two-dimensional conformal field theory. SciPost Phys. Lect. Notes (2018). https://doi.org/10.21468/SciPostPhysLectNotes.1. arXiv: 1609.09523
33. S. Ribault, R. Santachiara, Liouville theory with a central charge less than one. J. High Energy Phys. **2015**(8), 109 (2015). https://doi.org/10.1007/JHEP08(2015)109. arXiv: 1503.02067
34. A.N. Schellekens, Introduction to conformal field theory (2016). http://www.nikhef.nl/~t58/CFT.pdf
35. M. Schottenloher, *A Mathematical Introduction to Conformal Field Theory*, 2nd edn. (Springer, Berlin, 2008)
36. A. Sen, Reality of superstring field theory action. J. High Energy Phys. **2016**(11), 014 (2016). https://doi.org/10.1007/JHEP11(2016)014. arXiv: 1606.03455

37. J. Teschner, A guide to two-dimensional conformal field theory. Les Houches Lect. Notes (2017). https://doi.org/10.1093/oso/9780198828150.003.0002. arXiv: 1708.00680
38. D. Tong, Lectures on string theory (2009). arXiv: 0908.0333
39. X. Yin, Aspects of two-dimensional conformal field theories, in *Proceedings of Theoretical Advanced Study Institute Summer School 2017 "Physics at the Fundamental Frontier" PoS(TASI2017)*, vol. 305 (SISSA Medialab) (2018), p. 003. https://doi.org/10.22323/1.305.0003
40. B. Zwiebach, Closed string field theory: quantum action and the BV master equation. Nucl. Phys. B **390**(1), 33–152 (1993). https://doi.org/10.1016/0550-3213(93)90388-6. arXiv: hep-th/9206084

CFT Systems

7

Abstract

This chapter summarizes the properties of some CFT systems. We focus on the free scalar field and on the first-order bc system (which generalizes the reparametrization ghosts). For the different systems, we first provide an analysis on a general curved background before focusing on the complex plane. This is sufficient to describe the local properties on all Riemann surfaces $g \geq 0$.

7.1 Free Scalar

7.1.1 Covariant Action

The Euclidean action of a free scalar X on a curved background $g_{\mu\nu}$ is

$$S = \frac{c}{4\pi\ell^2} \int d^2x \sqrt{g}\, g^{\mu\nu} \partial_\mu X \partial_\nu X, \tag{7.1}$$

where ℓ is a length scale[1] and

$$\epsilon := \begin{cases} +1 & \text{spacelike} \\ -1 & \text{timelike} \end{cases}, \qquad \sqrt{\epsilon} := \begin{cases} +1 & \text{spacelike} \\ i & \text{timelike} \end{cases} \tag{7.2}$$

denotes the signature of the kinetic term. The field is periodic along σ

$$X(\tau, \sigma) \sim X(\tau, \sigma + 2\pi). \tag{7.3}$$

---

[1] To be identified with the string scale, such that $\alpha' = \ell^2$.

© Springer Nature Switzerland AG 2021
H. Erbin, *String Field Theory*, Lecture Notes in Physics 980,
https://doi.org/10.1007/978-3-030-65321-7_7

The energy–momentum tensor reads

$$T_{\mu\nu} = -\frac{\epsilon}{\ell^2}\left[\partial_\mu X \partial_\nu X - \frac{1}{2}g_{\mu\nu}(\partial X)^2\right],$$
(7.4)

and it is traceless

$$T^\mu_\mu = 0.$$
(7.5)

The equation of motion is

$$\Delta X = 0,$$
(7.6)

where Δ is the Laplacian (A.28).

The simplest method for finding the propagator in flat space is by using the identity (assuming that there is no boundary term)

$$0 = \int dX \frac{\delta}{\delta X(\sigma)}\left(e^{-S[X]}X(\sigma')\right),$$
(7.7)

which yields a differential equation for the propagator:

$$\langle \partial^2 X(\sigma)X(\sigma')\rangle = -2\pi\epsilon\ell^2\delta^{(2)}(\sigma - \sigma').$$
(7.8)

This is easily integrated to

$$\langle X(\sigma)X(\sigma')\rangle = -\frac{\epsilon\ell^2}{2}\ln|\sigma - \sigma'|^2.$$
(7.9)

Computation: Equation (7.9)

By translation and rotation invariance, one has

$$\langle X(\sigma)X(\sigma')\rangle = G(r), \qquad r = |\sigma - \sigma'|.$$
(7.10)

In polar coordinates, the Laplacian reads

$$\Delta G(r) = \frac{1}{r}\partial_r(rG'(r)).$$
(7.11)

Integrating the differential equation (7.8) over $d^2\sigma = r\,dr\,d\theta$ yields

$$-2\pi\epsilon\ell^2 = 2\pi\int_0^r dr'\, r' \times \frac{1}{r'}\partial_{r'}(r'G'(r')) = 2\pi r G'(r).$$
(7.12)

The solution is

$$G'(r) = -\epsilon \ell^2 \ln r \tag{7.13}$$

and the form (7.9) follows by writing

$$\ln r = \frac{1}{2} \ln r^2 = \frac{1}{2} \ln |\sigma - \sigma'|^2. \tag{7.14}$$

The action (7.1) is obviously invariant under constant translations of X:

$$X \longrightarrow X + a, \qquad a \in \mathbb{R}. \tag{7.15}$$

The associated $U(1)$ current[2] is conserved and reads

$$J^\mu := 2\pi i\epsilon \frac{\partial \mathcal{L}}{\partial(\partial_\mu X)} = \frac{i}{\ell^2} g^{\mu\nu} \partial_\nu X, \qquad \nabla_\mu J^\mu = 0. \tag{7.16}$$

On flat space, the charge follows from (A.23):

$$p = \frac{1}{2\pi} \int d\sigma \, J^0 = \frac{i}{2\pi \ell^2} \int d\sigma \, \partial^0 X. \tag{7.17}$$

This charge is called *momentum* because it corresponds to the spacetime momentum in string theory.

Moreover, there is a another *topological current*

$$\tilde{J}^\mu := -i \, \epsilon^{\mu\nu} J_\nu = \frac{1}{\ell^2} \epsilon^{\mu\nu} \partial_\nu X, \tag{7.18}$$

which is identically conserved:

$$\nabla_\mu \tilde{J}^\mu \propto \epsilon^{\mu\nu} [\nabla_\mu, \nabla_\nu] X = 0 \tag{7.19}$$

since $[\nabla_\mu, \nabla_\nu] = 0$ when acting on a scalar field. Note that $\tilde{J}^\mu$ is the Hodge dual of J^μ. The conserved charge is called the *winding number* and reads on flat space:

$$w = \frac{1}{2\pi} \int d\sigma \, \tilde{J}^0 = \frac{1}{2\pi \ell^2} \int_0^{2\pi} d\sigma \, \partial_1 X = \frac{1}{2\pi \ell^2} \big(X(\tau, 2\pi) - X(\tau, 0) \big). \tag{7.20}$$

Remark 7.1 (Normalization of the Current) The definition of the current (7.16) may look confusing. The factor of i is due to the Euclidean signature, see (A.25a), and the

---

[2]The group is $\mathbb{R}$ but the algebra is $\mathfrak{u}(1)$ (since locally there is no difference between the real line and the circle).

factor of 2π comes from the normalization of the spatial integral. We have inserted ϵ in order to interpret the conserved charge p as a component of the momentum contravariant vector in string theory.

To make contact with string theory, consider D scalar fields $X^a(x^\mu)$. Then, the current becomes

$$J_a^\mu = \frac{i}{2\pi\ell^2}\,\eta_{ab}\partial^\mu X^b, \tag{7.21}$$

where the position of the indices is in agreement with the standard form of Noether's formula (A.25a) (a current has indices in opposite locations as the parameters and fields). Since we have $\eta_{00} = -1 = \epsilon_{X_0}$, we find that $J^{0\mu} = \epsilon_{X_0} J_0^\mu$ has no epsilon after replacing the expression (7.17) of J_0^μ.

The transformation $X^a \to X^a + c^a$ is a global translation in target spacetime: the charge p^a is identified with the spacetime momentum. The factor of i indicates that p^a is the Euclidean contravariant momentum vector by comparison with (A.7).

The convention of this section is to always work with quantities which will become contravariant vector to avoid ambiguity.

7.1.2 Action on the Complex Plane

In complex coordinates, the action on flat space reads

$$S = \frac{\epsilon}{2\pi\ell^2} \int dz d\bar{z}\, \partial_z X \partial_{\bar{z}} X, \tag{7.22}$$

giving the equation of motion:

$$\partial_z \partial_{\bar{z}} X = 0. \tag{7.23}$$

This indicates that $\partial_z X$ and $\partial_{\bar{z}} X$ are, respectively, holomorphic and anti-holomorphic such that

$$X(z, \bar{z}) = X_L(z) + X_R(\bar{z}), \tag{7.24}$$

and we will remove the subscripts when there is no ambiguity (for example, when the position dependence is written):

$$X(z) := X_L(z), \qquad X(\bar{z}) := X_R(\bar{z}). \tag{7.25}$$

It looks like $X_L(z)$ and $X_R(\bar{z})$ are unrelated, but this is not the case because of the zero-mode, as we will see below.

The U(1) current is written as

$$J := J_z = \frac{i}{\ell^2} \partial_z X, \qquad \bar{J} := J_{\bar{z}} = \frac{i}{\ell^2} \partial_{\bar{z}} X, \qquad (7.26)$$

where we used the relations $J_z = J^{\bar{z}}/2$ and $J_{\bar{z}} = J^z/2$. The equation of motion implies that the current J is holomorphic, and $\bar{J}$ is anti-holomorphic:

$$\bar{\partial} J = 0, \qquad \partial \bar{J} = 0. \qquad (7.27)$$

The momentum splits into left- and right-moving parts:

$$p = p_L + p_R, \qquad p_L = \frac{1}{2\pi i} \oint dz\, J, \qquad p_R = -\frac{1}{2\pi i} \oint d\bar{z}\, \bar{J}. \qquad (7.28)$$

The components of the topological current (7.18) are related to the ones of the U(1) current:

$$\tilde{J}_z = \frac{i}{\ell^2} \partial_z X = J, \qquad \tilde{J}_{\bar{z}} = -\frac{i}{\ell^2} \partial_{\bar{z}} X = -\bar{J}. \qquad (7.29)$$

As a consequence, the winding number is

$$w = p_L - p_R. \qquad (7.30)$$

Note that we have the relations

$$p_L = \frac{p + w}{2}, \qquad p_R = \frac{p - w}{2}, \qquad (7.31a)$$

$$p^2 + w^2 = p_L^2 + p_R^2, \qquad 2pw = p_L^2 - p_R^2. \qquad (7.31b)$$

The energy–momentum tensor is

$$T := T_{zz} = -\frac{\epsilon}{\ell^2} \partial_z X \partial_z X, \qquad \bar{T} := T_{\bar{z}\bar{z}} = -\frac{\epsilon}{\ell^2} \partial_{\bar{z}} X \partial_{\bar{z}} X, \qquad T_{z\bar{z}} = 0. \qquad (7.32)$$

Since the $\partial_z X$ ($\partial_{\bar{z}} X$) is (anti-)holomorphic, so is $T(z)$ ($\bar{T}(\bar{z})$). Since the energy–momentum tensor, the current and the field itself (up to zero-modes) split into holomorphic and anti-holomorphic components in a symmetric way, it is sufficient to focus on one of the sectors, say the holomorphic one.

The other primary operators of the theory are given by the vertex operators $V_k(z)$:[3]

$$V_k(z, \bar{z}) := \, :e^{i\epsilon k X(z, \bar{z})}: . \tag{7.33}$$

Remark 7.2 In fact, it is possible to introduce more general vertex operators

$$V_{k_L, k_R}(z, \bar{z}) := \, :e^{2i\epsilon\left(k_L X(z) + k_R X(\bar{z})\right)}: , \tag{7.34}$$

but we will not consider them in this book.

Remark 7.3 (Plane and Cylinder Coordinates) The action in w-coordinate (cylinder) takes the same form as a result of the conformal invariance of the scalar field, which in practice results from the cancellation between the determinant and inverse metric. As a consequence, every quantity derived from the classical action (equation of motion, energy–momentum tensor...) will have the same form in both coordinate systems: we will focus on the z-coordinate, writing the w-coordinate expression when it is insightful to compare. This is not anymore the case at the quantum level: anomalies may translate into anomalous tensor transformations such that it is necessary to keep track of the surface on which the tensor is defined. To distinguish between the plane and cylinder quantities, an index "cyl" is added when necessary (by convention, all quantities without further specification are on the plane).

7.1.3 OPE

The OPE between X and itself is directly found from the propagator:

$$X(z)X(w) \sim -\frac{\epsilon\ell^2}{2} \ln(z - w). \tag{7.35}$$

By successive derivations, one finds the OPE between X and ∂X

$$\partial X(z)X(w) \sim -\frac{\epsilon\ell^2}{2} \frac{1}{z - w}, \tag{7.36}$$

and between ∂X with itself

$$\partial X(z)\partial X(w) \sim -\frac{\epsilon\ell^2}{2} \frac{1}{(z - w)^2}. \tag{7.37}$$

The invariance under the permutation of z and w reflects the fact that X is bosonic and that both operators in (7.37) are identical.

---

[3]The ϵ in the exponential is consistent with interpreting X and k as a contravariant vector.

The OPE between ∂X and T allows to verify that the field ∂X is primary with $h = 1$:

$$T(z)\partial X(w) \sim \frac{\partial X(w)}{(z-w)^2} + \frac{\partial(\partial X(w))}{z-w}.$$

(7.38)

The OPE of T with itself gives

$$T(z)T(w) \sim \frac{1}{2}\frac{1}{(z-w)^4} + \frac{2T(w)}{(z-w)^2} + \frac{\partial T(w)}{z-w}$$

(7.39)

which shows that the central charge is

$$c = 1.$$

(7.40)

One finds that the operator $\partial^n X$ has conformal weight

$$h = n$$

(7.41)

since the OPE with T is

$$T(z)\partial^n X(w) \sim \cdots + \frac{n\,\partial^n X(w)}{(z-w)^2} + \frac{\partial(\partial^n X(w))}{z-w},$$

(7.42)

where the dots indicate higher negative powers of $(z-w)$. These states are not primary for $n \geq 2$. Explicitly, for $n = 2$, one finds

$$T(z)\partial^2 X(w) \sim \frac{2\,\partial X(w)}{(z-w)^3} + \frac{2\,\partial^2 X}{(z-w)^2} + \frac{\partial(\partial^2 X(w))}{z-w}.$$

(7.43)

The OPE of a vertex operator with the current J is

$$J(z)V_k(w, \bar{w}) \sim \frac{\ell^2 k}{2}\frac{V_k(w, \bar{w})}{z-w}.$$

(7.44)

This shows that the vertex operators V_k are eigenstates of the U(1) holomorphic current with the eigenvalue given by the momentum (with a normalization of ℓ^2). Then, the OPE with T:

$$T(z)V_k(w, \bar{w}) \sim \frac{h_k V_k(w, \bar{w})}{(z-w)^2} + \frac{\partial V_k(w, \bar{w})}{z-w}$$

(7.45)

together with its anti-holomorphic counterpart show that the V_k are primary operators with weight

$$(h_k, \bar{h}_k) = \left(\frac{\epsilon\ell^2 k^2}{4}, \frac{\epsilon\ell^2 k^2}{4}\right), \qquad \Delta_k = \frac{\epsilon\ell^2 k^2}{2}, \qquad s_k = 0.$$

(7.46)

Note that classically $h_k = 0$ since $\ell \sim \hbar$ [10, p. 81]. The weight is invariant under $k \to -k$. Finally, the OPE between two vertex operators is

$$V_k(z, \bar{z}) V_{k'}(w, , \bar{w}) \sim \frac{V_{k+k'}(w, \bar{w})}{(z - w)^{-\epsilon k k' \ell^2 / 2}}, \tag{7.47}$$

where only the leading term (non-necessarily singular) is displayed. In particular, correlation functions should be computed for $\epsilon k k' < 0$ in order to avoid exponential growth.

Computation: Equation (7.38)

$$T(z) \partial X(w) = -\frac{\epsilon}{\ell^2} \; :\partial X(z) \partial X(z): \; \partial X(w)$$

$$\sim -\frac{2\epsilon}{\ell^2} :\partial X(z) \overbrace{\partial X(z): \partial X}(w) \sim \frac{\partial X(z)}{(z - w)^2}.$$

The result (7.38) follows by Taylor expanding the numerator.

Computation: Equation (7.39)

$$T(z) \partial X(w) = \frac{1}{\ell^4} :\partial X(z) \partial X(z): :\partial X(w) \partial X(w):$$

$$\sim \frac{1}{\ell^4} \Bigg[\; :\overbrace{\partial X(z) \partial X(z): :\overbrace{\partial X(w)} \partial X(w): \; + \; :\overbrace{\partial X(z) \partial X(z): :\partial X(w) \overbrace{\partial X(w)}}:$$

$$+ :\overbrace{\partial X(z) \partial X(z): :\partial X(w)} \partial X(w): + \text{perms} \Bigg]$$

$$\sim 2 \times \frac{1}{4} \frac{1}{(z - w)^4} - 4 \times \frac{1}{2\ell^2} \frac{1}{(z - w)^2} :\partial X(z) \partial X(w):$$

$$\sim \frac{1}{2} \frac{1}{(z - w)^4} - \frac{2}{\ell^2} \frac{1}{(z - w)^2} \Big(:\partial X(w) \partial X(w): + (z - w) :\partial^2 X(w) \partial X(w): \Big).$$

Computation: Equation (7.42)

$$T(z) \partial^n X(w) \sim \partial_w^{n-1} \frac{\partial X(z)}{(z - w)^2}$$

$$\sim n! \frac{\partial X(z)}{(z - w)^{n+1}}$$

$$\sim \frac{n!}{(z-w)^{n+1}} \Big(\cdots + \tfrac{1}{(n-1)!} (z - w)^{n-1} \partial^{n-1} (\partial X(w))$$

$$+ \tfrac{1}{n!} (z - w)^n \partial^n (\partial X(w)) \Big).$$

Computation: Equation (7.44)

Using (6.160b), one has

$$\partial X(z) V_k(w, \bar{w}) \sim i\epsilon k \, \overline{\partial X(z) X}(w) \, V_k(w, \bar{w}) \sim i\epsilon k \left(-\frac{\epsilon \ell^2}{2} \frac{1}{z - w} \right) V_k(w, \bar{w}).$$

Computation: Equation (7.45)

$$T(z) V_k(w, \bar{w}) \sim -\frac{\epsilon}{\ell^2} : \overline{\partial X(z) \partial X(z)} : :e^{i\epsilon k X(w, \bar{w})}:$$

$$\sim \frac{i\epsilon k}{2} \frac{1}{z - w} \, \partial X(z) : e^{i\epsilon k X(w, \bar{w})}: - \frac{\epsilon}{\ell^2} \, \overline{\partial X(z) : \partial X(z) e^{i\epsilon k X(w, \bar{w})}}:$$

$$\sim \frac{i\epsilon k}{2} \frac{1}{z - w} \left(:\partial X(z) \, e^{i\epsilon k X(w, \bar{w})}: + \overline{\partial X(z) : e^{i\epsilon k X(w, \bar{w})}}: \right)$$

$$+ \frac{i\epsilon k}{2} \frac{:\partial X(z) e^{i\epsilon k X(w, \bar{w})}:}{z - w}$$

$$\sim \frac{\epsilon k^2 \ell^2}{4} \frac{V_k(w, \bar{w})}{(z - w)^2} + i\epsilon k \frac{:\partial X(w) e^{ick X(w, \bar{w})}:}{z - w}.$$

In the first line, we consider a single contraction (hence, there is no factor of 2): the reason is that considering the contractions symmetrically and not successively counts twice the first term of the last line. Indeed, there is only one way to generate this term. It is also possible to achieve the same result by expanding the exponential.

Computation: Equation (7.47)

Using (6.160c) and keeping only the leading term, one has

$$V_k(z, \bar{z}) V_{k'}(w, \bar{w}) \sim \exp\left(-kk' \, \overline{X(z, \bar{z}) X}(w, \bar{w}) \right) :e^{i\epsilon k X(z, \bar{z})} e^{i\epsilon k' X(w, \bar{w})}:$$

$$\sim (z - w)^{\epsilon k k' \ell^2 / 2} V_{k+k'}(w, \bar{w}).$$

7.1.4 Mode Expansions

Since ∂X is holomorphic and of weight $h = 1$, it can be expanded as:[4]

$$\partial X = -i\sqrt{\frac{\ell^2}{2}} \sum_{n\in\mathbb{Z}} \alpha_n z^{-n-1}, \qquad \bar\partial X = -i\sqrt{\frac{\ell^2}{2}} \sum_{n\in\mathbb{Z}} \bar\alpha_n \bar z^{-n-1}, \tag{7.48}$$

where an individual mode can be extracted with a contour integral:

$$\alpha_n = i \oint \frac{dz}{2\pi i} z^{n-1} \partial X(z), \qquad \bar\alpha_n = i \oint \frac{dz}{2\pi i} z^{n-1} \bar\partial X(z). \tag{7.49}$$

Integrating this formula gives

$$X(z) = \frac{x_L}{2} - i\sqrt{\frac{\ell^2}{2}} \alpha_0 \ln z + i\sqrt{\frac{\ell^2}{2}} \sum_{n\neq 0} \frac{\alpha_n}{n} z^{-n},$$

$$X(\bar z) = \frac{x_R}{2} - i\sqrt{\frac{\ell^2}{2}} \bar\alpha_0 \ln \bar z + i\sqrt{\frac{\ell^2}{2}} \sum_{n\neq 0} \frac{\bar\alpha_n}{n} \bar z^{-n}. \tag{7.50}$$

The zero-modes are, respectively, α_0 and $\bar\alpha_0$ for ∂X and $\bar\partial X$, and x_L and x_R for X_L and X_R. The meaning of the modes will become clearer in Sect. 7.1.5 where we study the commutation relations.

First, we relate the zero-modes α_0 and $\bar\alpha_0$ to the conserved charges p_L and p_R (7.28) of the U(1) current:

$$p_L = \frac{\alpha_0}{\sqrt{2\ell^2}}, \qquad p_R = \frac{\bar\alpha_0}{\sqrt{2\ell^2}} \tag{7.51}$$

such that

$$X(z) = \frac{x_L}{2} - i\ell^2 p_L \ln z + i\sqrt{\frac{\ell^2}{2}} \sum_{n\neq 0} \frac{\alpha_n}{n} z^{-n}. \tag{7.52}$$

Then, the relations (7.28) and (7.30) allow to rewrite this result in terms of the momentum p and winding w:

$$p = \frac{1}{\sqrt{2\ell^2}} (\alpha_0 + \bar\alpha_0), \qquad w = \frac{1}{\sqrt{2\ell^2}} (\alpha_0 - \bar\alpha_0). \tag{7.53}$$

---

[4]The Fourier expansion is taken to be identical for $\epsilon = \pm 1$ fields since ∂X is contravariant in target space. The difference between the two cases will appear in the commutators.

These relations can be inverted as

$$\alpha_0 = \sqrt{\frac{\ell^2}{2}}\,(p + w), \qquad \bar{\alpha}_0 = \sqrt{\frac{\ell^2}{2}}\,(p - w). \qquad (7.54)$$

In the same sense that there are two momenta p_L and p_R conjugated to x_L and x_R, it makes sense to introduce two coordinates x and q conjugated to p and w. From string theory, the operator x is called the centre-of-mass. The expression (7.54) suggests to write

$$x_L = x + q, \qquad x_R = x - q, \qquad (7.55)$$

and conversely:

$$x = \frac{1}{2}(x_L + x_R), \qquad q = \frac{1}{2}(x_L - x_R). \qquad (7.56)$$

In terms of these new variables, the expansion of the full $X(z, \bar{z})$ reads

$$X(z, \bar{z}) = x - i\frac{\ell^2}{2}\left(p \ln |z|^2 + w \ln \frac{z}{\bar{z}}\right) + i\sqrt{\frac{\ell^2}{2}}\sum_{n \neq 0}\frac{1}{n}(\alpha_n z^{-n} + \bar{\alpha}_n \bar{z}^{-n}). \qquad (7.57)$$

In terms of the coordinates on the cylinder, the part without oscillations becomes

$$X(\tau, \sigma) = x - i\ell^2 p\tau + \ell^2 w\sigma + \cdots \qquad (7.58)$$

Note how the presence of ℓ^2 gives the correct scale to the second term. The mode q does not appear at all, and x is the zero-mode of the complete field $X(z, \bar{z})$. As it is well-known, the physical interpretation of x and p is as the position and momentum of the centre-of-mass of the string.[5] If there is a compact dimension, then w counts the number of times the string winds around it, and q can be understood as the position of the centre-of-mass after a T-duality.[6]

---

[5]In worldsheet Lorentzian signature, this becomes $X(\tau, \sigma) = x + \ell^2 p\tau + \ell^2 w\sigma$ as expected.

[6]T-duality and compact bosons fall outside the scope of this book and we refer the reader to [11, chap. 17, 9, chap. 8] for more details.

Computation: Equation (7.51)

$$p_L = \frac{1}{2\pi i} \oint dz\, J = \frac{i}{\ell^2} \frac{1}{2\pi i} \oint dz\, \partial X = \frac{i}{\ell^2} \frac{1}{2\pi i} \oint dz\, \partial X$$

$$= \frac{1}{\sqrt{2\ell^2}} \frac{1}{2\pi i} \oint dz \sum_n \alpha_n z^{-n-1} = \frac{1}{\sqrt{2\ell^2}} \alpha_0.$$

The computation gives p_R after replacing α_0 by $\bar{\alpha}_0$.

If the scalar field is non-compact but periodic on the cylinder, the periodicity condition

$$X(\tau, \sigma + 2\pi) \sim X(\tau, \sigma) \tag{7.59}$$

translates as

$$X(e^{2\pi i}z, e^{-2\pi i}\bar{z}) \sim X(z, \bar{z}). \tag{7.60}$$

Evaluating the LHS from (7.50) gives a constraint on the zero-modes:

$$X(e^{2\pi i}z, e^{-2\pi i}\bar{z}) = X(z, \bar{z}) - i\sqrt{\frac{\ell^2}{2}} (\alpha_0 - \bar{\alpha}_0), \tag{7.61}$$

which implies

$$\alpha_0 = \bar{\alpha}_0 \implies p_L = p_R = \frac{p}{2}, \qquad w = 0. \tag{7.62}$$

The other cases will not be discussed in this book, but we still use the general notation to make the contact with the literature easier. This also implies that X_L and X_R cannot be periodic independently. Hence, the zero-mode couples the holomorphic and anti-holomorphic sectors together.

The *number operators* $N_n\ \bar{N}_n$ at level $n > 0$ are defined by

$$N_n = \frac{\epsilon}{n} \alpha_{-n}\alpha_n, \qquad \bar{N}_n = \frac{\epsilon}{n} \bar{\alpha}_{-n}\bar{\alpha}_n. \tag{7.63}$$

The modes have been normal ordered. They count the number of excitations at the level n: the factor n^{-1} is necessary because the modes are not canonically normalized. Then, one can build the *level operators*

$$N = \sum_{n>0} n\, N_n. \tag{7.64}$$

They count the number of excitations at level n weighted by the level itself. This corresponds to the total energy due to the oscillations (the higher the level, the more energy it needs to be excited).

The Virasoro operators are

$$L_m = \frac{\epsilon}{2} \sum_n :\alpha_n \alpha_{m-n}: \tag{7.65}$$

For $m \neq 0$, we have

$$m \neq 0: \qquad L_m = \frac{\epsilon}{2} \sum_{n \neq 0, m} :\alpha_n \alpha_{m-n}: + \epsilon\, \alpha_0 \alpha_m, \tag{7.66}$$

there is no ordering ambiguity and the normal order can be removed. In the case of the zero-mode, one finds

$$L_0 = \frac{\epsilon}{2} \sum_n :\alpha_n \alpha_{-n} := N + \frac{\epsilon}{2} \alpha_0^2 = N + \epsilon \ell^2\, p_L^2, \tag{7.67}$$

using (7.64) and (7.51). It is also useful to define $\widehat{L}_0$ which corresponds to L_0 stripped from the zero-mode contribution:

$$\widehat{L}_0 := N. \tag{7.68}$$

Similarly, the anti-holomorphic zero-mode is

$$\bar{L}_0 = \bar{N} + \epsilon \ell^2\, p_R^2, \qquad \widehat{\bar{L}}_0 := \bar{N}, \tag{7.69}$$

such that

$$L_0^+ = N + \bar{N} + \epsilon \ell^2\, (p_L^2 + p_R^2) = N + \bar{N} + \frac{\epsilon \ell^2}{2}\, (p^2 + w^2), \tag{7.70a}$$

$$L_0^- = N - \bar{N} + \epsilon \ell^2\, (p_L^2 - p_R^2) = N - \bar{N} + \epsilon \ell^2\, wp, \tag{7.70b}$$

where $L_0^\pm := L_0 \pm \bar{L}_0$ as defined in (6.118). The last equality of each line follows from (7.31b). The expression of L_0^+ for $N = \bar{N} = 0$ matches the weights (7.46) of the vertex operators for $p_L = p_R = p/2$ (no winding), which will be interpreted below. It is a good place to stress that p_L, p_R, p, and w are operators, while k is a number.

7.1.5 Commutators

The commutators can be computed from (6.75a) knowing the OPE (7.37). The modes of ∂X and $\bar{\partial} X$ satisfy

$$[\alpha_m, \alpha_n] = \epsilon\, m\, \delta_{m+n,0}, \qquad [\bar{\alpha}_m, \bar{\alpha}_n] = \epsilon\, m\, \delta_{m+n,0}, \qquad [\alpha_m, \bar{\alpha}_n] = 0 \qquad (7.71)$$

for all $m, n \in \mathbb{Z}$ (including the zero-modes). The appearance of the factor m in the RHS explains the normalization of the number operator (7.63).

From the commutators of the zero-modes, we directly find the ones for the momentum and winding:

$$[p, w] = [p, p] = [w, w] = 0, \qquad [p, \alpha_n] = [p, \bar{\alpha}_n] = [w, \alpha_n] = [w, \bar{\alpha}_n] = 0. \tag{7.72}$$

The OPE (7.36) yields

$$[x_L, p_L] = \mathrm{i}\epsilon, \qquad [x_R, p_R] = \mathrm{i}\epsilon, \tag{7.73}$$

which can be used to determine the commutators of x and q:

$$[x, p] = [q, w] = \mathrm{i}\epsilon, \qquad [x, w] = [q, p] = 0. \tag{7.74}$$

This shows that (x, p) and (q, w) are pairs of conjugate variables. Interestingly, the winding number w commutes all other modes except q, but the latter disappears from the description. Hence, it can be interpreted as a number which labels different representations: if no other principle (like periodicity) forbids $w \neq 0$, then one can expect to have states with all possible w in the spectrum, each value of w forming a different sector. There are other interpretations from the point of view of T-duality and double field theory [3, 5, 8, 11].

The commutator of the modes with the Virasoro operators is

$$[L_m, \alpha_n] = -n\, \alpha_{m+n}. \tag{7.75}$$

as expected from (6.115). For $m = 0$, this reduces to

$$[L_0, \alpha_{-n}] = n\, \alpha_{-n}, \tag{7.76}$$

which shows that negative modes increase the energy. The commutator of the creation modes α_{-n} with the number operators is

$$[N_m, \alpha_{-n}] = \alpha_{-m} \delta_{m,n}. \tag{7.77}$$

7.1.6 Hilbert Space

The Hilbert space of the free scalar has the structure of a Fock space.

From (7.76), the momentum p commutes with the Hamiltonian L_0^+ such that it is a good quantum number to label the states:[7] this translates the fact that the action (7.1) does not depend on the conjugate variable x. As a consequence, there exists a family of vacua $|k\rangle$.

The vacua $|k\rangle$ are the states related to the vertex operators (7.33) through the state–operator correspondence:

$$|k\rangle := \lim_{z,\bar{z}\to 0} V_k(z,\bar{z}) |0\rangle = e^{i\epsilon kx} |0\rangle , \qquad (7.78)$$

where $|0\rangle$ is the $\mathrm{SL}(2,\mathbb{C})$ vacuum and x is the zero-mode of $X(z,\bar{z})$. That this identification is correct follows by applying the operator p:

$$p |k\rangle = k |k\rangle . \qquad (7.79)$$

The notation is consistent with the one of the $\mathrm{SL}(2,\mathbb{C})$ vacuum since $p |0\rangle = 0$.

The vacuum is annihilated by the action of the positive-frequency modes:

$$\forall n > 0 : \qquad \alpha_n |k\rangle = 0, \qquad (7.80)$$

which is equivalent to

$$N_n |k\rangle = 0. \qquad (7.81)$$

The different vacua are each ground state of a Fock space (they are all equivalent), but they are not ground states of the Hamiltonian since they have different energies:

$$L_0^+ |k\rangle = 2\epsilon \ell^2 k^2 |k\rangle , \qquad L_0^- |k\rangle = 0, \qquad (7.82)$$

using (7.70). The $\mathrm{SL}(2,\mathbb{C})$ vacuum is the lowest (highest) energy state if $\epsilon = 1$ ($\epsilon = -1$).

The Fock space $\mathcal{F}(k)$ built from the vacuum at momentum k is found by acting repetitively with the negative-frequency modes. A convenient basis, the oscillator basis, is given by the states:

$$\mathcal{F}(k) = \mathrm{Span}\left\{ |k; \{N_n\}\rangle \right\}, \qquad (7.83a)$$

$$|k; \{N_n\}\rangle := \prod_{n\geq 1} \frac{(\alpha_{-n})^{N_n}}{\sqrt{n^{N_n} N_n!}} |k\rangle , \qquad N_n \in \mathbb{N}^* \qquad (7.83b)$$

---

[7]To simplify the discussion, we do not consider winding but only vertex operators of the form (7.33).

(we do not distinguish the notations between the number operators and their eigenvalues). The full Hilbert space is given by

$$\mathcal{H} = \int_{\mathbb{R}} \mathrm{d}k \, \mathcal{F}(k).$$

(7.84)

Computation: Equation (7.78)
We provide a quick argument to justify the second form of (7.78). Take the limit of (7.57) with $w = 0$:

$$\lim_{z,\bar{z} \to 0} e^{i\epsilon k X(z,\bar{z})} |0\rangle = \lim_{z,\bar{z} \to 0} \exp i\epsilon k \left[x - i \frac{\ell^2}{2} p \ln |z|^2 \right.$$

$$\left. + i \sqrt{\frac{\ell^2}{2}} \sum_{n \neq 0} \frac{1}{n} \left(\alpha_n z^{-n} + \bar{\alpha}_n \bar{z}^{-n} \right) \right] |0\rangle$$

$$= \lim_{z,\bar{z} \to 0} \exp \left[i\epsilon kx - \epsilon k \sqrt{\frac{\ell^2}{2}} \sum_{n \neq 0} \frac{1}{n} \left(\alpha_n z^{-n} + \bar{\alpha}_n \bar{z}^{-n} \right) \right] |0\rangle \,.$$

The second term from the first line disappears because $p |0\rangle = 0$. For $\epsilon k > 0$, as $z, \bar{z} \to 0$, the terms with α_n and $\bar{\alpha}_n$ for $n < 0$ disappear since they are accompanied with a positive power of z^n and $\bar{z}^n$. The modes with $n > 0$ diverge but the minus sign makes the exponential to vanish. A more rigorous argument requires to normal order the exponential and then to use (7.80).

Computation: Equation (7.79)

$$p |k\rangle = \frac{1}{\ell^2} \frac{1}{2\pi i} \oint \left(\mathrm{d}z \, i \partial X(z) + \mathrm{d}\bar{z} \, i \bar{\partial} X(\bar{z}) \right) V_k(0,0) |0\rangle$$

$$= \frac{1}{\ell^2} \frac{1}{2\pi i} \oint \left(\frac{\mathrm{d}z}{z} \frac{\ell^2 k}{2} + \frac{\mathrm{d}\bar{z}}{\bar{z}} \frac{\ell^2 k}{2} \right) V_k(0,0) |0\rangle$$

$$= k \, V_k(0,0) |0\rangle$$

using (7.44).

Remark 7.4 (Fock Space and Verma Module Isomorphism) Note that, in the absence of the so-called null states, there is a one-to-one map between states in the α_{-n} oscillator basis and in the L_{-n} Virasoro basis. This translates an isomorphism between the Fock space and the Verma module of V_k. One hint for this relation is that applying α_{-n} and L_{-n} changes the weight (eigenvalue of L_0) by the same amount, and there are as many operators in both basis.

7.1.7 Euclidean and BPZ Conjugates

Since X is a real scalar field, it is self-adjoint (6.95) such that

$$x^\dagger = x \qquad p^\dagger = p, \qquad \alpha_n^\dagger = \alpha_{-n}. \tag{7.85}$$

This implies that the Virasoro operators (7.65) are Hermitian:

$$L_n^\dagger = L_{-n}, \tag{7.86}$$

as expected since $T(z)$ is self-adjoint for a free scalar field.

As a consequence of (7.85), the adjoint of the vacuum $|k\rangle$ follows from (7.78):

$$\langle k| = |k\rangle^\ddagger = \langle 0|e^{-i\epsilon k x}, \qquad \langle k|p = \langle k|k. \tag{7.87}$$

The BPZ conjugate (6.111) of the mode α_n is

$$\alpha_n^t = -(\pm 1)^n \alpha_{-n}, \tag{7.88}$$

where the sign depends on the choice of $I^\pm$ in (6.111). Using (7.53), this implies that the momentum operator gets a minus sign:[8]

$$p^t = -p, \qquad \langle -k| = |k\rangle^t. \tag{7.89}$$

The inner product between two vacua $|k\rangle$ and $|k'\rangle$ is normalized as:

$$\langle k|k'\rangle = 2\pi\,\delta(k - k') \tag{7.90}$$

such that the conjugate state (6.145) of the vacuum reads

$$\langle k^c| = \frac{1}{2\pi}\,\langle k|. \tag{7.91}$$

The Hermitian and BPZ conjugate states are related as:

$$|k\rangle^\ddagger = -\,|k\rangle^t, \tag{7.92}$$

which can be interpreted as a reality condition on $|k\rangle$.

---

[8]Be careful that $|k\rangle$ is not the state associated with the operator p through the state–operator correspondence. Instead, they are associated with V_k, see (7.78). This explains why $\langle k| \neq (|k\rangle)^t$ as in (6.136).

7.2   First-Order bc Ghost System

First-order systems describe two free fields called ghosts which have a first-order
action and whose conformal weights sum to 1. Commuting (resp. anti-commuting)
fields are often denoted by β and γ (resp. b and c) and correspondingly first-order
systems are also called $\beta\gamma$ or bc systems. We will introduce a sign $\epsilon = \pm 1$ to
denote the Grassmann parity of the fields and always write them as b and c. In
string theory, first-order systems describe the Faddeev–Popov ghosts associated with
reparametrizations and supersymmetries (Sects. 2.4 and 17.1).

7.2.1   Covariant Action

A first-order system is defined by two symmetric and traceless fields $b_{\mu_1\cdots\mu_\lambda}$ and
$c^{\mu_1\cdots\mu_{\lambda-1}}$ called ghosts. For fields of integer spins, the dynamics is governed by the
first-order action

$$S = \frac{1}{4\pi} \int d^2x \sqrt{g}\, g^{\mu\nu}\, b_{\mu\mu_1\cdots\mu_{\lambda-1}} \nabla_\nu c^{\mu_1\cdots\mu_{\lambda-1}} \tag{7.93}$$

after taking into account the symmetries of the field indices. Obviously, for $\lambda = 2$, one recovers the reparametrization ghost action (2.145). The action (7.93) is
invariant under Weyl transformations (the fields and covariant derivatives are inert)
such that it describes a CFT on flat space.

When the fields have half-integer spins (and often denoted as β and γ in this
case), they carry a spinor index. In this case, the action contains a Dirac matrix, and
the covariant derivative a spin connection.

The ghost action (7.93) is invariant under a global U(1) symmetry

$$b_{\mu_1\cdots\mu_n} \longrightarrow e^{-i\theta} b_{\mu_1\cdots\mu_n}, \qquad c^{\mu_1\cdots\mu_{n-1}} \longrightarrow e^{i\theta} c^{\mu_1\cdots\mu_{n-1}}. \tag{7.94}$$

7.2.2   Action on the Complex Plane

The simplest description of the system is on the complex plane. Due to the
conditions imposed on the fields, they have only two independent components for
all n, and the equations of motion imply that one is holomorphic, and the other
anti-holomorphic:

$$b(z) := b_{z\cdots z}(z), \qquad \bar{b}(\bar{z}) := b_{\bar{z}\cdots\bar{z}}(\bar{z}), \qquad c(z) := c^{z\cdots z}(\bar{z}), \qquad \bar{c}(\bar{z}) := c^{\bar{z}\cdots\bar{z}}(z). \tag{7.95}$$

In this language, the action becomes

$$S = \frac{1}{2\pi} \int d^2z \left(b\bar{\partial}c + \bar{b}\partial\bar{c}\right).$$

(7.96)

This action gives the correct equations of motion

$$\partial\bar{b} = 0, \qquad \bar{\partial}b = 0, \qquad \partial\bar{c} = 0, \qquad \bar{\partial}c = 0.$$

(7.97)

Since the fields split into holomorphic and anti-holomorphic sectors, it is convenient to study only the holomorphic sector as usual. This system is even simpler than the scalar field because the zero-modes do not couple both sectors.[9] All formulas for the anti-holomorphic sector are directly obtained from the holomorphic one by adding bars on quantities, except for conserved charges which have an index L or R and are both written explicitly.

The action describes a CFT, and the weight of the fields are given by

$$h(b) = \lambda, \qquad h(c) = 1 - \lambda, \qquad h(\bar{b}) = \lambda, \qquad h(\bar{c}) = 1 - \lambda,$$

(7.98)

where $\lambda = n$ if the fields are in a tensor representation, and $\lambda = n + 1/2$ if they are in a spinor-tensor representation. The holomorphic energy–momentum reads

$$T = -\lambda :b\partial c: + (1 - \lambda) :\partial b\,c:$$

(7.99a)

$$= -\lambda :\partial(bc): + :\partial b\,c:$$

(7.99b)

$$= (1 - \lambda) :\partial(bc): - :b\,\partial c:.$$

(7.99c)

Normal ordering is taken with respect to the $SL(2, \mathbb{C})$ vacuum (6.121).

Finally, both fields can be classically commuting or anti-commuting (see below for the quantum commutators):

$$b(z)c(w) = -\epsilon\, c(w)b(z), \quad b(z)b(w) = -\epsilon\, b(w)b(z), \quad c(z)c(w) = -\epsilon\, c(w)c(z),$$

(7.100)

where ϵ denotes the Grassmann parity

$$\epsilon = \begin{cases} +1 & \text{anti-commuting,} \\ -1 & \text{commuting.} \end{cases}$$

(7.101)

---

[9]For the scalar field, the coupling of both sectors happened because of the periodicity condition (7.62).

Sometimes, if $\epsilon = +1$, one denotes b and c, respectively, by β and γ. If b and c are ghosts arising from Faddeev–Popov gauge fixing, then $\epsilon = 1$ if λ is integer; and $\epsilon = -1$ if λ is half-integer ("wrong" spin–statistics assignment).

The U(1) global symmetry (7.94) reads infinitesimally

$$\delta b = -\mathrm{i}b, \qquad \delta c = \mathrm{i}c, \qquad \delta\bar{b} = -\mathrm{i}\bar{b}, \qquad \delta\bar{c} = \mathrm{i}\bar{c}. \tag{7.102}$$

It is generated by the conserved *ghost current* with components:

$$j(z) = -:b(z)c(z):, \qquad \bar{j}(\bar{z}) = -:\bar{b}(\bar{z})\bar{c}(\bar{z}): \tag{7.103}$$

and the associated charge is called the ghost number

$$N_{\mathrm{gh}} = N_{\mathrm{gh},L} + N_{\mathrm{gh},R}, \qquad N_{\mathrm{gh},L} = \oint \frac{\mathrm{d}z}{2\pi\mathrm{i}}\, j(z), \qquad N_{\mathrm{gh},R} = -\oint \frac{\mathrm{d}\bar{z}}{2\pi\mathrm{i}}\, \bar{j}(\bar{z}). \tag{7.104}$$

This charge counts the number of c ghosts minus the number of b ghosts, such that

$$N_{\mathrm{gh}}(c) = 1, \qquad N_{\mathrm{gh}}(b) = -1, \qquad N_{\mathrm{gh}}(\bar{c}) = 1, \qquad N_{\mathrm{gh}}(\bar{b}) = -1. \tag{7.105}$$

The propagator can be derived from the path integral

$$\int \mathrm{d}'b\,\mathrm{d}'c\, \frac{\delta}{\delta b(z)} \left[b(w)\mathrm{e}^{-S[b,c]} \right] = 0 \tag{7.106}$$

which gives the differential equation

$$\delta^{(2)}(z - w) + \frac{1}{2\pi}\, \langle b(w)\bar{\partial}c(z) \rangle = 0. \tag{7.107}$$

Using (B.2), the solution is easily found to be

$$\langle c(z)b(w) \rangle = \frac{1}{z - w}. \tag{7.108}$$

Remark 7.5 The propagator is constructed with the path integral. For convenience, the zero-modes are removed from the measure: reintroducing them, one finds that the propagator is computed not in the conformal vacuum (which has no operator insertion), but in a state with ghost insertions. This explains why the propagator (7.108) is not of the form (6.62b). However, this form is sufficient to extract the OPE as changing the vacuum does not introduce singular terms.

7.2.3 OPE

The OPEs between the b and c fields are found from the propagator (7.108):

$$c(z)b(w) \sim \frac{1}{z-w}, \qquad b(z)c(w) \sim \frac{\epsilon}{z-w}, \qquad (7.109a)$$

$$b(z)b(w) \sim 0, \qquad c(z)c(w) \sim 0. \qquad (7.109b)$$

The OPE of each ghost with T confirms the conformal weights in (7.98):

$$T(z)b(w) \sim \lambda \frac{b(w)}{(z-w)^2} + \frac{\partial b(w)}{z-w}, \qquad (7.110a)$$

$$T(z)c(w) \sim (1-\lambda) \frac{c(w)}{(z-w)^2} + \frac{\partial c(w)}{z-w}. \qquad (7.110b)$$

The OPE of T with itself is

$$T(z)T(w) \sim \frac{c_\lambda/2}{(z-w)^4} + \frac{2T(w)}{(z-w)^2} + \frac{\partial T(w)}{z-w}, \qquad (7.111)$$

where the central charge is

$$c_\lambda = 2\epsilon(-1 + 6\lambda - 6\lambda^2) = -2\epsilon\big(1 + 6\lambda(\lambda - 1)\big). \qquad (7.112)$$

Introducing the ghost charge:

$$q_\lambda = \epsilon(1 - 2\lambda), \qquad (7.113)$$

the central charge can also be written as

$$c_\lambda = \epsilon(1 - 3q_\lambda^2). \qquad (7.114)$$

This parameter will appear many times in this section and its meaning will become clearer as we proceed.

The OPE between the ghost current (7.103) and the b and c ghosts read

$$j(z)b(w) \sim -\frac{b(w)}{z-w}, \qquad (7.115a)$$

$$j(z)c(w) \sim \frac{c(w)}{z-w}. \qquad (7.115b)$$

The coefficients of the $(z - w)^{-1}$ terms correspond to the ghost number of the b and c fields (7.105). More generally, the ghost number $N_{\text{gh}}(\mathcal{O})$ of any operator $\mathcal{O}(z)$ is defined by

$$j(z)\mathcal{O}(w) \sim N_{\text{gh}}(\mathcal{O}) \frac{\mathcal{O}(w)}{z - w}. \tag{7.116}$$

The OPE for j with itself is

$$j(z)j(w) \sim \frac{\epsilon}{(z - w)^2}. \tag{7.117}$$

Finally, the OPE of the current with T reads

$$T(z)j(w) \sim \frac{q_\lambda}{(z - w)^3} + \frac{j(w)}{(z - w)^2} + \frac{\partial j(w)}{z - w}. \tag{7.118}$$

Due to the presence of the z^{-3} term, the current $j(z)$ is not a primary field if $q_\lambda \neq 0$, that is, if $\lambda \neq 1/2$. In that case, its transformation under changes of coordinates gets an anomalous contribution:

$$j(z) = \frac{dw}{dz} j'(w) + \frac{q_\lambda}{2} \frac{d}{dz} \ln \frac{dw}{dz} = \frac{dw}{dz} j'(w) + \frac{q_\lambda}{2} \frac{\partial_z^2 w}{\partial_z w}. \tag{7.119}$$

This implies in particular that the currents on the plane and on the cylinder ($w = \ln z$) are related by

$$j(z) = \frac{dw}{dz} \left(j^{\text{cyl}}(w) - \frac{q_\lambda}{2} \right), \tag{7.120}$$

which leads to the following relation between the ghost numbers on the plane and on the cylinder:

$$N_{\text{gh}} = N_{\text{gh}}^{\text{cyl}} - q_\lambda, \qquad N_{\text{gh},L} = N_{\text{gh},L}^{\text{cyl}} - \frac{q_\lambda}{2}, \qquad N_{\text{gh},R} = N_{\text{gh},R}^{\text{cyl}} - \frac{q_\lambda}{2}. \tag{7.121}$$

For this reason, it is important to make clear the space with respect to which is given the ghost number: if not explicitly stated, ghost numbers in this book are given on the plane.[10]

---

[10]Other references, especially old ones, give it on the cylinder. This can be easily recognized if some ghost numbers in the holomorphic sector are half-integers: for the reparametrization ghosts, q_λ is an integer such that the shift in (7.121) is a half-integer.

Due to this anomaly, one finds that the ghost number is not conserved on a curved space:

$$N^c - N^b = \frac{\epsilon\, q_\lambda}{2}\, \chi_g = (1 - 2\lambda)(g - 1), \qquad (7.122)$$

where χ_g is the Euler characteristics (2.4), N^b and N^c are the numbers of b and c operators. In string theory, where the only ghost insertions are zero-modes, this translates into a statement on the number of zero-modes to be inserted. Hence, this can be interpreted as a generalization of (2.72). For a proof, see, for example, [1, p. 397].

Computation: Equation (7.110a)

$$T(z)b(w) = \left(-\lambda :b(z)\partial c(z): + (1-\lambda):\partial b(z)\, c(z): \right) b(w)$$

$$\sim -\lambda :b(z)\partial \overline{c(z):\, b(w)} + (1-\lambda):\partial b(z)\, \overline{c(z):\, b(w)}$$

$$\sim -\lambda\, b(z)\partial_z \frac{1}{z-w} + (1-\lambda)\, \partial b(z)\, \frac{1}{z-w}$$

$$\sim \lambda \left(b(w) + \underline{(z-w)\partial b(w)} \right) \frac{1}{(z-w)^2} + (1-\lambda)\, \frac{\partial b(w)}{z-w}.$$

Computation: Equation (7.110b)

$$T(z)c(w) = \left(-\lambda :b(z)\partial c(z): + (1-\lambda):\partial b(z)c(z): \right) c(w)$$

$$\sim \epsilon\lambda :\partial c(z)\overline{b(z):\, c(w)} - \epsilon(1-\lambda):c(z)\partial \overline{b(z):\, c(w)}$$

$$\sim \lambda\, \frac{\partial c(z)}{z-w} - (1-\lambda)\, c(z)\, \partial_z \frac{1}{z-w}$$

$$\sim \lambda\, \frac{\partial c(w)}{z-w} + (1-\lambda)\left(c(w) + (z-w)\partial c(w) \right) \frac{1}{(z-w)^2}$$

$$\sim (1-\lambda)\, \frac{c(w)}{(z-w)^2} + \frac{\partial c(w)}{(z-w)^2}.$$

Computation: Equation (7.115a)

$$j(z)b(w) = -:b(z)c(z):\, b(w) \sim -:b(z)\overline{c(z):\, b(w)} \sim -\frac{b(z)}{z-w} \sim -\frac{b(w)}{z-w}.$$

Computation: Equation (7.115b)

$$j(z)c(w) = -:b(z)c(z):c(w) \sim \epsilon :c(z)\overline{b(z):}c(w) \sim \frac{c(z)}{z-w} \sim \frac{c(w)}{z-w}.$$

Computation: Equation (7.117)

$$j(z)j(w) = :b(z)c(z)::b(w)c(w):$$

$$\sim :\overline{b(z)c(z)}::\overline{b(w)c(w)}: + :b(z)\overline{c(z)}::\overline{b(w)}c(w): + :\overline{b(z)}c(z)::b(w)\overline{c(w)}:$$

$$\sim \frac{\epsilon}{(z-w)^2} + \frac{\epsilon :c(z)b(w):}{z-w} + \frac{:b(z)c(w):}{z-w} \sim \frac{\epsilon}{(z-w)^2}.$$

7.2.4 Mode Expansions

The b and c ghosts are expanded as

$$b(z) = \sum_{n \in \mathbb{Z}+\lambda+\nu} \frac{b_n}{z^{n+\lambda}}, \qquad c(z) = \sum_{n \in \mathbb{Z}+\lambda+\nu} \frac{c_n}{z^{n+1-\lambda}}, \tag{7.123}$$

where $\nu = 0, 1/2$ depends on ϵ and on the periodicity of the fields, see (6.102). The modes are extracted with the contour formulas

$$b_n = \oint \frac{dz}{2\pi i} z^{n+\lambda-1} b(z), \qquad c_n = \oint \frac{dz}{2\pi i} z^{n-\lambda} c(z). \tag{7.124}$$

Ghosts with $\lambda \in \mathbb{Z}$ have integer indices and $\nu = 0$ (we do not consider ghosts with twisted boundary conditions). On the other hand, ghosts with $\lambda \in \mathbb{Z}+1/2$ have integer indices and $\nu = 1/2$ in the R sector, and half-integer indices and $\nu = 0$ in the NS sector (see Sect. 6.4.4). The choices in the boundary conditions arise from the $\mathbb{Z}_2$ symmetry of the action:

$$b \longrightarrow -b, \qquad c \longrightarrow -c. \tag{7.125}$$

The number operators N_n^b and N_n^c are defined to count the numbers of excitations above the SL$(2, \mathbb{C})$ vacuum of b and c ghosts at level n:

$$N_n^b = :b_{-n}c_n:, \qquad N_n^c = \epsilon :c_{-n}b_n:. \tag{7.126}$$

The definitions follow from the commutators (7.134). Then, the level operators N^b and N^c are obtained by summing over n:

$$N^b = \sum_{n>0} n \, N_n^b, \qquad N^c = \sum_{n>0} n \, N_n^c.$$
(7.127)

The Virasoro operators are

$$L_m = \sum_n \left(n - (1-\lambda)m \right) : b_{m-n} c_n : = \sum_n (\lambda m - n) : b_n c_{m-n} :.$$
(7.128)

Of particular importance is the zero-mode

$$L_0 = - \sum_n n : b_n c_{-n} : = \sum_n n : b_{-n} c_n :.$$
(7.129)

We will give the expression of L_0 in terms of the level operators below, see (7.159). To do this, we will first need to change the normal ordering, which first requires to study the Hilbert space.

The modes of the ghost current are

$$j_m = - \sum_n : b_{m-n} c_n : = - \sum_n : b_n c_{m-n} :.$$
(7.130)

Note that the zero-mode of the current also equals the ghost number

$$N_{\mathrm{gh},L} = j_0 = - \sum_n : b_{-n} c_n :.$$
(7.131)

When both the holomorphic and anti-holomorphic sectors enter, it is convenient to introduce the combinations

$$b_n^\pm = b_n \pm \bar{b}_n, \qquad c_n^\pm = \frac{1}{2} (c_n \pm \bar{c}_n).$$
(7.132)

The normalization of $b_m^\pm$ is chosen to match one of $L_m^\pm$ (6.118). The normalization of $c_n^\pm$ is chosen such that (7.135) holds. Note the following useful identities:

$$b_n^- b_n^+ = 2 b_n \bar{b}_n, \qquad c_n^- c_n^+ = \frac{1}{2} c_n \bar{c}_n.$$
(7.133)

Computation: Equation (7.128)

$$T = -\lambda : b\partial c : + (1 - \lambda) : \partial bc :$$

$$= \sum_{m,n} \left(\lambda : b_m c_n : \frac{n+1-\lambda}{z^{m+\lambda} z^{m+2-\lambda}} - (1-\lambda) : b_m c_n : \frac{m+\lambda}{z^{m+\lambda+1} z^{m+1-\lambda}} \right)$$

$$= \sum_{m,n} \Big(\lambda (n+1-\lambda) - (1-\lambda)(m+\lambda) \Big) \frac{: b_m c_n :}{z^{m-n+2}}$$

$$= \sum_{m,n} \Big(\lambda (n+1-\lambda) - (1-\lambda)(m-n+\lambda) \Big) \frac{: b_{m-n} c_n :}{z^{m+2}}$$

$$= \sum_{m,n} (n - m + \lambda m) \frac{: b_{m-n} c_n :}{z^{m+2}} = \sum_m \frac{L_m}{z^{m+2}}.$$

The fourth line follows from shifting $m \to m - n$. The second equality in (7.128) follows by shifting $n \to m - n$.

Computation: Equation (7.130)

$$j = -:bc: = \sum_{m,n} \frac{: b_m c_n :}{z^{m+\lambda} z^{n+1-\lambda}} = \sum_{m,n} \frac{: b_{m-n} c_n :}{z^{m+1}} = \sum_m \frac{j_m}{z^{m+1}}.$$

7.2.5 Commutators

The (anti)commutators between the modes b_n and c_n read

$$[b_m, c_n]_\epsilon = \delta_{m+n,0}, \qquad [b_m, b_n]_\epsilon = 0, \qquad [c_m, c_n]_\epsilon = 0. \tag{7.134}$$

Therefore, the modes with $n < 0$ are creation operators and the modes with $n > 0$ are annihilation operators:

- a b ghost excitation at level $n > 0$ is created by b_{-n} and annihilated by c_n;
- a c ghost excitation at level $n > 0$ is created by c_{-n} and annihilated by b_n.

In terms of $b_m^\pm$ and $c_m^\pm$ (7.132), we have

$$[b_m^+, c_n^+]_\epsilon = \delta_{m+n}, \qquad [b_m^-, c_n^-]_\epsilon = \delta_{m+n}. \tag{7.135}$$

The commutators of the number operators with the modes are

$$[N_m^b, b_{-n}] = b_{-n} \delta_{m,n}, \qquad [N_m^c, c_{-n}] = c_{-n} \delta_{m,n}, \tag{7.136}$$

while those between the L_n and the ghost modes are

$$[L_m, b_n] = \big(m(\lambda - 1) - n\big)b_{m+n}, \qquad [L_m, c_n] = -(m\lambda + n)c_{m+n}, \qquad (7.137)$$

in agreement with (6.115). If $n \in \mathbb{Z}$, each ghost field has zero-modes b_0 and c_0 which commutes with L_0

$$[L_0, b_0] = 0, \qquad [L_0, c_0] = 0. \qquad (7.138)$$

The commutator of the current modes reads

$$[j_m, j_n] = m\,\delta_{m+n,0}. \qquad (7.139)$$

Then, the commutator with the Virasoro operators is

$$[L_m, j_n] = -n j_{m+n} + \frac{q\lambda}{2}\, m(m + 1)\delta_{m+n,0}. \qquad (7.140)$$

Finally, the commutators of the ghost number operator with the ghosts are

$$[N_{\mathrm{gh}}, b(w)] = -b(w), \qquad [N_{\mathrm{gh}}, c(w)] = c(w). \qquad (7.141)$$

Computation: Equation (7.134)

$$
\begin{aligned}
[b_m, c_n]_\epsilon &= \epsilon \oint_{C_0} \frac{\mathrm{d}w}{2\pi i}\, w^{-1} \oint_{C_w} \frac{\mathrm{d}z}{2\pi i}\, z^{-1}\, w^{n+\lambda} z^{m-\lambda+1} b(z)c(w) \\
&\sim \epsilon \oint_{C_0} \frac{\mathrm{d}w}{2\pi i}\, w^{-1} \oint_{C_w} \frac{\mathrm{d}z}{2\pi i}\, z^{-1}\, w^{n+\lambda} z^{m-\lambda+1} \frac{\epsilon}{z - w} \\
&= \oint_{C_0} \frac{\mathrm{d}w}{2\pi i}\, w^{m+n-1} = \delta_{m+n,0}.
\end{aligned}
$$

Computation: Equation (7.141)

$$[N_{\mathrm{gh}}, b(w)] = \oint \frac{\mathrm{d}z}{2\pi i}\, j(z)b(w) \sim -\oint \frac{\mathrm{d}z}{2\pi i} \frac{b(w)}{z - w} = -b(w).$$

The computation for c is similar.

7.2.6   Hilbert Space

The $SL(2, \mathbb{C})$ vacuum $|0\rangle$ (6.121) is defined by

$$\forall n > -\lambda: \quad b_n |0\rangle = 0, \qquad \forall n > \lambda - 1: \quad c_n |0\rangle = 0. \tag{7.142}$$

If $\lambda > 1$, there are positive modes which do not annihilate the vacuum.

To simplify the notation, we consider the case $\lambda \in \mathbb{Z}$, the half-integer case following by shifting the indices by $1/2$. Since the modes $\{c_1, \ldots, c_{\lambda-1}\}$ do not annihilate $|0\rangle$, one can create states

$$|n_1, \ldots, n_{\lambda-1}\rangle = c_1^{n_1} \cdots c_{\lambda-1}^{n_{\lambda-1}} |0\rangle \tag{7.143}$$

which have negative energies:

$$L_0 |n_1, \ldots, n_{\lambda-1}\rangle = -\left(\sum_{j=1}^{\lambda-1} j \, n_j\right) |n_1, \ldots, n_{\lambda-1}\rangle, \tag{7.144}$$

where (7.137) has been used. Moreover, this state is degenerate due to the existence of zero-modes since they commute with the Hamiltonian—see (7.138). As a consequence, it must be in a representation of the zero-mode algebra.

If the ghosts are commuting ($\epsilon = -1$), then it seems hard to make sense of the theory since one can find a state of arbitrarily negative energy since $n_i \in \mathbb{N}$. The zero-modes make the problem even worse. The appropriate interpretation of these states will be discussed in the context of the superstring theory for $\lambda = 3/2$ (superconformal ghosts).

In the rest of this section, we focus on the Grassmann odd case $\epsilon = 1$.

Energy Vacuum (Grassmann Odd)

Since $n_i = 0$ or $n_i = 1$ for anti-commuting ghosts ($\epsilon = 1$), there is a state of lowest energy. This is the *energy vacuum* (6.129). Since the zero-modes b_0 and c_0 commute with L_0, it is doubly degenerate. A convenient basis is

$$\{|\downarrow\rangle, |\uparrow\rangle\}, \tag{7.145}$$

where

$$|\downarrow\rangle := c_1 \cdots c_{\lambda-1} |0\rangle, \qquad |\uparrow\rangle := c_0 c_1 \cdots c_{\lambda-1} |0\rangle. \tag{7.146}$$

A general vacuum is a linear combination of the two basis vacua:

$$|\Omega\rangle = \omega_\downarrow |\downarrow\rangle + \omega_\uparrow |\uparrow\rangle, \qquad \omega_\downarrow, \omega_\uparrow \in \mathbb{C}. \tag{7.147}$$

The algebra of these vacua is the one of a two-state system:

$$b_0 \mid \uparrow\rangle = \mid \downarrow\rangle, \qquad c_0 \mid \downarrow\rangle = \mid \uparrow\rangle, \qquad b_0 \mid \downarrow\rangle = 0, \qquad c_0 \mid \uparrow\rangle = 0. \qquad (7.148)$$

Hence, for the vacuum $\mid \downarrow\rangle$ (resp. $\mid \uparrow\rangle$), b_0 (resp. c_0) acts as an annihilation operator, and conversely c_0 (resp. b_0) acts as a creation operator. Finally, both states are annihilated by all positive modes:

$$\forall n > 0 : \qquad b_n \mid \downarrow\rangle = b_n \mid \uparrow\rangle = 0, \qquad c_n \mid \downarrow\rangle = b_n \mid \downarrow\rangle = 0. \qquad (7.149)$$

Note that the SL(2, $\mathbb{C}$) vacuum can be recovered by acting with b_{-n} with $n < \lambda$:

$$|0\rangle = b_{1-\lambda} \cdots b_{-1} \mid \downarrow\rangle = b_{1-\lambda} \cdots b_{-1} b_0 \mid \uparrow\rangle . \qquad (7.150)$$

The zero-point energy (6.130) of these states is the conformal weight of the vacuum:

$$L_0 \mid \downarrow\rangle = a_\lambda \mid \downarrow\rangle , \qquad L_0 \mid \uparrow\rangle = a_\lambda \mid \uparrow\rangle , \qquad (7.151)$$

where a_λ can be written in various forms:

$$a_\lambda = -\sum_{n=1}^{\lambda-1} n = -\frac{\lambda(\lambda-1)}{2} = \frac{c_\lambda}{24} + \frac{2}{24}. \qquad (7.152)$$

Taking into account the anti-holomorphic sector leads to a four-fold degeneracy. The basis

$$\left\{ \mid \downarrow\downarrow\rangle , \mid \uparrow\downarrow\rangle , \mid \downarrow\uparrow\rangle , \mid \uparrow\uparrow\rangle \right\} \qquad (7.153)$$

is built as follows:

$$\mid \downarrow\downarrow\rangle := c_1 \bar{c}_1 \cdots c_{\lambda-1} \bar{c}_{\lambda-1} |0\rangle ,$$
$$\mid \uparrow\downarrow\rangle := c_0 \mid \downarrow\downarrow\rangle , \qquad \mid \downarrow\uparrow\rangle := \bar{c}_0 \mid \downarrow\downarrow\rangle , \qquad \mid \uparrow\uparrow\rangle := c_0 \bar{c}_0 \mid \downarrow\downarrow\rangle . \qquad (7.154)$$

The modes b_0 and $\bar{b}_0$ can be used to flip the arrows downward, leading to the following algebra:

$$c_0 \mid \downarrow\downarrow\rangle = \mid \uparrow\downarrow\rangle, \qquad \bar{c}_0 \mid \downarrow\downarrow\rangle = \mid \downarrow\uparrow\rangle, \qquad c_0 \mid \downarrow\uparrow\rangle = -\bar{c}_0 \mid \uparrow\downarrow\rangle = \mid \uparrow\uparrow\rangle,$$
$$b_0 \mid \uparrow\uparrow\rangle = \mid \downarrow\uparrow\rangle, \qquad \bar{b}_0 \mid \uparrow\uparrow\rangle = -\mid \uparrow\downarrow\rangle, \qquad b_0 \mid \uparrow\downarrow\rangle = \bar{b}_0 \mid \downarrow\uparrow\rangle = \mid \downarrow\downarrow\rangle .$$
$$(7.155a)$$

The vacua are annihilated by different combinations of the zero-modes:

$$b_0 \mid \downarrow\downarrow\rangle = \bar{b}_0 \mid \downarrow\downarrow\rangle = 0, \qquad c_0 \mid \uparrow\downarrow\rangle = \bar{b}_0 \mid \uparrow\downarrow\rangle = 0,$$
$$b_0 \mid \downarrow\uparrow\rangle = \bar{c}_0 \mid \downarrow\uparrow\rangle = 0, \qquad c_0 \mid \uparrow\uparrow\rangle = \bar{c}_0 \mid \uparrow\uparrow\rangle = 0. \tag{7.155b}$$

In these manipulations, one has to be careful to correctly anti-commute the modes with the ones hidden in the definitions of the vacua.

There is a second basis which is more natural when using the zero-modes $c_0^\pm$ and $b_0^\pm$ (7.132):

$$\left\{ \mid \downarrow\downarrow\rangle , \mid+\rangle , \mid-\rangle , \mid \uparrow\uparrow\rangle \right\}, \tag{7.156}$$

where the two vacua $\mid\pm\rangle$ are combinations of the $\mid \downarrow\uparrow\rangle$ and $\mid \uparrow\downarrow\rangle$ vacua:

$$\mid\pm\rangle = \mid \uparrow\downarrow\rangle \pm \mid \downarrow\uparrow\rangle . \tag{7.157}$$

The different vacua are naturally related by acting with $c_0^\pm$ and $b_0^\pm$ which act as raising and lowering operators:

$$c_0^\mp \mid \downarrow\downarrow\rangle = \frac{1}{2} \mid\pm\rangle , \qquad c_0^\mp \mid\pm\rangle = \pm \mid \uparrow\uparrow\rangle ,$$
$$b_0^\pm \mid\pm\rangle = \pm 2 \mid \downarrow\downarrow\rangle , \qquad b_0^\mp \mid \uparrow\uparrow\rangle = \pm \mid\pm\rangle . \tag{7.158a}$$

From the previous relations, it follows that the different vacua are annihilated by the zero-modes as follow:

$$b_0^+ \mid \downarrow\downarrow\rangle = b_0^- \mid \downarrow\downarrow\rangle = 0, \qquad c_0^- \mid-\rangle = b_0^+ \mid-\rangle = 0,$$
$$c_0^+ \mid+\rangle = b_0^- \mid+\rangle = 0 \qquad c_0^+ \mid \uparrow\uparrow\rangle = c_0^- \mid \uparrow\uparrow\rangle = 0. \tag{7.158b}$$

This also means that we have

$$c_0^- c_0^+ \mid \downarrow\downarrow\rangle = \frac{1}{2} \mid \uparrow\uparrow\rangle , \qquad b_0^+ b_0^- \mid \uparrow\uparrow\rangle = 2 \mid \downarrow\downarrow\rangle . \tag{7.158c}$$

Computation: Equation (7.158)

$$2 c_0^+ \mid\pm\rangle = (c_0 + \bar{c}_0) \mid \uparrow\downarrow\rangle \pm (c_0 + \bar{c}_0) \mid \downarrow\uparrow\rangle = \bar{c}_0 \mid \uparrow\downarrow\rangle \pm c_0 \mid \downarrow\uparrow\rangle = (-1 \pm 1) \mid \uparrow\uparrow\rangle$$
$$b_0^+ \mid\pm\rangle = (b_0 + \bar{b}_0) \mid \uparrow\downarrow\rangle \pm (b_0 + \bar{b}_0) \mid \downarrow\uparrow\rangle = b_0 \mid \uparrow\downarrow\rangle \pm \bar{b}_0 \mid \downarrow\uparrow\rangle = (1 \pm 1) \mid \downarrow\downarrow\rangle$$
$$2 c_0^\pm \mid \downarrow\downarrow\rangle = (c_0 \pm \bar{c}_0) \mid \downarrow\downarrow\rangle = c_0 \mid \downarrow\downarrow\rangle \pm \bar{c}_0 \mid \downarrow\downarrow\rangle = \mid \uparrow\downarrow\rangle \pm \mid \downarrow\uparrow\rangle = \mid\pm\rangle$$
$$b_0^\pm \mid \uparrow\uparrow\rangle = (b_0 \pm \bar{b}_0) \mid \uparrow\uparrow\rangle = b_0 \mid \uparrow\uparrow\rangle \pm \bar{b}_0 \mid \uparrow\uparrow\rangle = \mid \downarrow\uparrow\rangle \mp \mid \uparrow\downarrow\rangle = \mp \mid\mp\rangle$$

Energy Normal Ordering (Grassmann Odd)

We now turn towards the definition of the energy normal ordering (6.150). Ultimately, it will be found that $|\downarrow\rangle$ is the physical vacuum in string theory. For this reason, the energy normal ordering $\overset{*}{*}\cdots\overset{*}{*}$ is associated with the vacuum $|\downarrow\rangle$ in order to resolve the ambiguity of the zero-modes. In particular, b_0 is an annihilation operator in this case, while c_0 is a creation operator. In the rest of this section, we translate the normal ordering of expressions from the conformal vacuum to the energy vacuum.

The Virasoro operators L_n for $n \neq 0$ have no ordering problems since the modes which compose them commute. The expression of L_0 (7.129) in the energy ordering becomes

$$L_0 = \sum_n n \overset{*}{*} b_{-n} c_n \overset{*}{*} + a_\lambda = N^b + N^c + a_\lambda, \tag{7.159}$$

where a_λ is the zero-point energy (7.152) and N^b and N^c are the ghost mode numbers (7.127). The contribution of the non-zero modes is denoted by

$$\widehat{L}_0 = N^b + N^c. \tag{7.160}$$

The expression can be rewritten to encompass all modes:

$$L_m = \sum_n \left(n - (1 - \lambda)m\right) \overset{*}{*} b_m \, {}_n c_n \overset{*}{*} + a_\lambda \delta_{m,0} \tag{7.161}$$

Similarly, the expression of the ghost number is

$$N_{\text{gh},L} = j_0 = \sum_n \overset{*}{*} b_{-n} c_n \overset{*}{*} - \left(\frac{q_\lambda}{2} + \frac{1}{2}\right) \tag{7.162a}$$

$$= \sum_{n>0} \left(N_n^c - N_n^b\right) + \frac{1}{2}\left(N_0^c - N_0^b\right) - \frac{q_\lambda}{2}, \tag{7.162b}$$

and thus:

$$j_m = \sum_n \overset{*}{*} b_{m-n} c_n \overset{*}{*} - \left(\frac{q_\lambda}{2} + \frac{1}{2}\right) \delta_{m,0}. \tag{7.163}$$

It is useful to define the ghost number without ghost zero-modes:

$$\widehat{N}_{\text{gh},L} := \sum_{n>0} \left(N_n^c - N_n^b\right). \tag{7.164}$$

One can straightforwardly compute the ghost number of the vacua:

$$j_0 \mid \downarrow \rangle = (\lambda - 1) \mid \downarrow \rangle = \left(-\frac{q_\lambda}{2} - \frac{1}{2} \right) \mid \downarrow \rangle , \tag{7.165a}$$

$$j_0 \mid \uparrow \rangle = \lambda \mid \uparrow \rangle = \left(-\frac{q_\lambda}{2} + \frac{1}{2} \right) \mid \uparrow \rangle . \tag{7.165b}$$

This confirms that the SL(2, $\mathbb{C}$) vacuum has vanishing ghost number since $\mid \downarrow \rangle$ contains exactly $\lambda - 1$ ghosts:

$$j_0 \mid 0 \rangle = 0. \tag{7.166}$$

Using (7.121) allows to write the ghost numbers on the cylinder:

$$j_0^{\mathrm{cyl}} \mid \downarrow \rangle = -\frac{1}{2} \mid \downarrow \rangle , \qquad j_0^{\mathrm{cyl}} \mid \uparrow \rangle = \frac{1}{2} \mid \uparrow \rangle . \tag{7.167}$$

That both ghost numbers have same magnitude but opposite signs could be expected: since the ghost number changes as $N_{\mathrm{gh}} \to -N_{\mathrm{gh}}$ when $b \leftrightarrow c$, the mean value of the ghost number should be zero.

Remark 7.6 (Ghost Number Conventions) Since the ghost number is an additive quantum number, it is always possible to shift its definition by a constant. This can be used to set the ghost numbers of the vacua to some other values. For example, [1, p. 116] adds $q_\lambda/2$ to the ghost number in order to get $N_{\mathrm{gh}} = \pm 1/2$ on the plane (instead of the cylinder). We do not follow this convention in order to keep the symmetry between the vacuum ghost numbers on the cylinder.

Computation: Equation (7.159)
Start with (7.129) and use (6.157):

$$L_0 = -\sum_n n :b_n c_{-n}: = -\sum_{n \le -\lambda} n\, b_n c_{-n} + \epsilon \sum_{n > -\lambda} n\, c_{-n} b_n$$

$$= \sum_{n \ge \lambda} n\, b_{-n} c_n + \epsilon \sum_{n > -\lambda} n\, c_{-n} b_n$$

$$= \sum_{n \ge \lambda} n\, b_{-n} c_n + \epsilon \sum_{n > 0} n\, c_{-n} b_n + \epsilon \sum_{n=-\lambda+1}^{0} n\, c_{-n} b_n$$

$$= \sum_{n \ge \lambda} n\, b_{-n} c_n + \epsilon \sum_{n > 0} n\, c_{-n} b_n + \epsilon \sum_{n=0}^{\lambda-1} n\, b_{-n} c_n + a_\lambda$$

$$= \sum_{n>0} n\, b_{-n} c_n + \epsilon \sum_{n>0} n\, c_{-n} b_n + a_\lambda,$$

$$= \sum_{n} n_\star^\star\, b_{-n} c_{n\star}^\star + a_\lambda,$$

using that

$$\sum_{n=-\lambda+1}^{0} c_{-n} b_n = -\sum_{n=0}^{\lambda-1} n\, c_n b_{-n} = -\sum_{n=0}^{\lambda-1} n\,(-\epsilon\, b_{-n} c_n + 1) = \epsilon \sum_{n=0}^{\lambda-1} n\, b_{-n} c_n + a_\lambda.$$

The result also follows from (6.163).

Computation: Equation (7.162)

$$j_0 = -\sum_{n} :b_{-n} c_n: = -\sum_{n\geq\lambda} b_{-n} c_n + \epsilon \sum_{n>-\lambda} c_{-n} b_n$$

$$= -\sum_{n\geq\lambda} b_{-n} c_n + \epsilon \sum_{n>0} c_{-n} b_n + \epsilon \sum_{n=1}^{\lambda-1} c_n b_{-n} + \epsilon\, c_0 b_0$$

$$= -\sum_{n\geq\lambda} b_{-n} c_n + \epsilon \sum_{n>0} c_{-n} b_n - \sum_{n=1}^{\lambda-1} b_{-n} c_n + \epsilon(\lambda - 1) + \epsilon\, c_0 b_0$$

$$= -\sum_{n>0} b_{-n} c_n + \epsilon \sum_{n>0} c_{-n} b_n + \epsilon(\lambda - 1) + \epsilon\, c_0 b_0.$$

Finally, one can write

$$\epsilon(\lambda - 1) = -\frac{q_\lambda}{2} - \frac{\epsilon}{2}. \tag{7.168}$$

The result also follows from (6.163). The second expression is obtained by symmetrizing the last term such that

$$\epsilon\, c_0 b_0 + \epsilon(\lambda - 1) = \frac{\epsilon}{2}\, c_0 b_0 + \frac{1}{2}(-b_0 c_0 + \epsilon) + \epsilon(\lambda - 1)$$

$$= \frac{1}{2}\,(\epsilon\, c_0 b_0 - b_0 c_0) + \epsilon\left(\lambda - \frac{1}{2}\right).$$

Structure of the Hilbert Space (Grassmann Odd)

Since the zero-modes commute with the Hamiltonian and with all other negative-
and positive-frequency modes, the Hilbert space is decomposed in several subspaces, each associated to a zero-mode.[11]

Starting with the holomorphic sector only, the Hilbert space $\mathcal{H}_{\text{gh}}$ is

$$\mathcal{H}_{\text{gh}} = \mathcal{H}_{\text{gh},0} \oplus c_0 \mathcal{H}_{\text{gh},0}, \qquad \mathcal{H}_{\text{gh},0} := \mathcal{H}_{\text{gh}} \cap \ker b_0, \qquad (7.169)$$

which follows from the 2-state algebra (7.148). Obviously, one has $c_0 \mathcal{H}_{\text{gh},0} = \mathcal{H}_{\text{gh}} \cap$ $\ker c_0$. The oscillator basis of the Hilbert space $\mathcal{H}_{\text{gh},0}$ is generated by applying the negative-frequency modes and has the structure of a fermionic Fock space without zero-modes:

$$\mathcal{H}_{\text{gh},0} = \text{Span} \left\{ \left| \downarrow; \{N_n^b\}; \{N_n^c\} \right\rangle \right\}, \qquad (7.170a)$$

$$\left| \downarrow; \{N_n^b\}; \{N_n^c\} \right\rangle = \prod_{n \geq 1} (b_{-n})^{N_n^b} (c_{-n})^{N_n^c} \left| \downarrow \right\rangle, \qquad N_n^b, N_n^c \in \mathbb{N}^* \qquad (7.170b)$$

(again, number operators and their eigenvalues are not distinguished). This means that $\mathcal{H}_{\text{gh},0}$ can also be regarded as a Fock space built on the vacuum $| \downarrow \rangle$, for which c_0 and b_0 are, respectively, creation and annihilation operators. Conversely, c_0 and b_0 are, respectively, annihilation and creation operators for $c_0 \mathcal{H}_{\text{gh},0}$.

In particular, this means that any state can be written as the sum of two states

$$\psi = \psi_\downarrow + \psi_\uparrow, \qquad \psi_\downarrow \in \mathcal{H}_{\text{gh},0}, \qquad \psi_\uparrow \in c_0 \mathcal{H}_{\text{gh},0}, \qquad (7.171)$$

with $\psi_\downarrow$ and $\psi_\uparrow$ built, respectively, on top of the $| \downarrow \rangle$ and $| \uparrow \rangle$ vacua.

This pattern generalizes when considering both the holomorphic and anti-
holomorphic sectors. In that case, the Hilbert space is decomposed in four subspaces:[12]

$$\mathcal{H}_{\text{gh}} = \mathcal{H}_{\text{gh},0} \oplus c_0 \mathcal{H}_{\text{gh},0} \oplus \bar{c}_0 \mathcal{H}_{\text{gh},0} \oplus c_0 \bar{c}_0 \mathcal{H}_{\text{gh},0},$$

$$\mathcal{H}_{\text{gh},0} := \mathcal{H}_{\text{gh}} \cap \ker b_0 \cap \ker \bar{b}_0. \qquad (7.172)$$

---

[11] Due to the specific structure of the inner product defined below, these subspaces are not
orthonormal to each other.

[12] The reader should not get confused by the same symbol $\mathcal{H}_{\text{gh},0}$ as in the case of the holomorphic
sector.

Basis states of the Hilbert space $\mathcal{H}_{\mathrm{gh},0}$ are

$$\left| \downarrow\downarrow; \{N_n^b\}; \{N_n^c\}; \{\bar{N}_n^b\}; \{\bar{N}_n^c\} \right\rangle = \prod_{n \geq 1} (b_{-n})^{N_n^b} (\bar{b}_{-n})^{\bar{N}_n^b} (c_{-n})^{N_n^c} (\bar{c}_{-n})^{\bar{N}_n^c} \left| \downarrow\downarrow \right\rangle,$$

$$N_n^b, \bar{N}_n^b, N_n^c, \bar{N}_n^c \in \mathbb{N}^*.$$

$$(7.173)$$

A general state of $\mathcal{H}_{\mathrm{gh}}$ can be decomposed as

$$\psi = \psi_{\downarrow\downarrow} + \psi_{\uparrow\downarrow} + \psi_{\downarrow\uparrow} + \psi_{\uparrow\uparrow}, \qquad (7.174)$$

where each state is built by acting with negative-frequency modes on the corresponding vacuum.

In terms of the second basis (7.156), the Hilbert space admits a second decomposition:

$$\mathcal{H}_{\mathrm{gh}} = \mathcal{H}_{\mathrm{gh},0} \oplus c_0^+ \mathcal{H}_{\mathrm{gh},0} \oplus c_0^- \mathcal{H}_{\mathrm{gh},0} \oplus c_0^- c_0^+ \mathcal{H}_{\mathrm{gh},0},$$

$$\mathcal{H}_{\mathrm{gh},0} := \mathcal{H}_{\mathrm{gh}} \cap \ker b_0^- \cap \ker b_0^+. \qquad (7.175)$$

In view of applications to string theory, it is useful to introduce two more subspaces:

$$\mathcal{H}_{\mathrm{gh},\pm} := \mathcal{H}_{\mathrm{gh}} \cap \ker b_0^\pm = \mathcal{H}_{\mathrm{gh},0} \oplus c_0^\mp \mathcal{H}_{\mathrm{gh},0}, \qquad (7.176)$$

and the associated decomposition

$$\mathcal{H}_{\mathrm{gh}} = \mathcal{H}_{\mathrm{gh},\pm} \oplus c_0^\pm \mathcal{H}_{\mathrm{gh},\pm}. \qquad (7.177)$$

In off-shell closed string theory, the principal Hilbert space will be $\mathcal{H}_{\mathrm{gh}}^-$ due to the level-matching condition. In this case, $\mathcal{H}_{\mathrm{gh}}^-$ has the same structure as $\mathcal{H}_{\mathrm{gh}}$ in the pure holomorphic sector, and c_0^+ plays the same role as c_0. A state in $\mathcal{H}_{\mathrm{gh}}^-$ is built on top of the vacua $| \downarrow\downarrow \rangle$ and $|+\rangle$.

7.2.7   Euclidean and BPZ Conjugates

In order for the Virasoro operators to be Hermitian, the b_n and c_n must satisfy the following conditions:

$$b_n^\dagger = \epsilon b_{-n}, \qquad c_n^\dagger = c_{-n}. \qquad (7.178)$$

Hence, b_n is anti-Hermitian if $\epsilon = -1$. The BPZ conjugates of the modes are

$$b_n^t = (-1)^\lambda b_{-n}, \qquad c_n^t = (-1)^{1-\lambda} c_{-n}, \tag{7.179}$$

using $I^+(z)$ with (6.111).

In the rest of this section, we consider only the case $\epsilon = 1$ and $\lambda \in \mathbb{N}$. The adjoints of the vacuum read

$$| \downarrow \rangle^\ddagger = \langle 0 | c_{1-\lambda} \cdots c_{-1}, \qquad | \uparrow \rangle^\ddagger = \langle 0 | c_{1-\lambda} \cdots c_{-1} c_0. \tag{7.180}$$

The BPZ conjugates of the vacua are

$$\langle \downarrow | := | \downarrow \rangle^t = (-1)^{(1-\lambda)^2} \langle 0 | c_{-1} \cdots c_{1-\lambda},$$
$$\langle \uparrow | := | \uparrow \rangle^t = (-1)^{\lambda(1-\lambda)} \langle 0 | c_0 c_{-1} \cdots c_{1-\lambda}. \tag{7.181}$$

The signs are inconvenient but will disappear when considering both the left and right vacua together as in (7.154). We have the following relations:

$$\langle \downarrow | = (-1)^{a_\lambda + (1-\lambda)(2-\lambda)} | \downarrow \rangle^\ddagger, \qquad \langle \uparrow | = (-1)^{a_\lambda} | \uparrow \rangle^\ddagger, \tag{7.182}$$

where a_λ is the zero-point energy (7.152).

Computation: Equation (7.182)

To prove the relation, we can start from the BPZ conjugate $\langle \downarrow |$ and reorder the modes to bring them in the same order as the adjoint:

$$\langle \downarrow | = (-1)^{(1-\lambda)^2 + \frac{1}{2}(2-\lambda)(1-\lambda)} | \downarrow \rangle^\ddagger = (-1)^{-a_\lambda + (1-\lambda)(2-\lambda)} | \downarrow \rangle^\ddagger.$$

The reordering gives a factor (-1) to the power:

$$\sum_{i=1}^{\lambda-2} i = \frac{1}{2}(2-\lambda)(1-\lambda) = -a_\lambda + 1 - \lambda.$$

Similarly, for the second vacuum:

$$\langle \uparrow | = (-1)^{\lambda(1-\lambda) - \frac{1}{2}\lambda(1-\lambda)} | \uparrow \rangle^\ddagger = (-1)^{\frac{1}{2}\lambda(1-\lambda)} | \uparrow \rangle^\ddagger.$$

We can identify the power with (7.152).

Then, we have the following relations:

$$\langle \uparrow | b_0 = \langle \downarrow |, \qquad \langle \downarrow | c_0 = \langle \uparrow |, \qquad \langle \downarrow | b_0 = 0, \qquad \langle \uparrow | c_0 = 0. \tag{7.183}$$

Table 7.1 Summary of the first-order systems Remember that $h(b) = \lambda$ and $h(c) = 1 - \lambda$

	ϵ	λ	q_λ	c_λ	a_λ
b, c (diff.)	1	2	-3	-26	-1
β, γ (susy.)	-1	$3/2$	2	11	$3/8$
$\psi^\pm$	1	$1/2$	0	1	0
η, ξ	1	1	-1	-2	0

There is a subtlety in defining the inner product because the vacuum is degenerate. If we write the two vacua as vectors

$$| \downarrow \rangle = \begin{pmatrix} 0 \\ 1 \end{pmatrix}, \qquad | \uparrow \rangle = \begin{pmatrix} 1 \\ 0 \end{pmatrix}, \tag{7.184}$$

then the zero-modes have the following matrix representation:

$$b_0 = \begin{pmatrix} 0 & 0 \\ 1 & 0 \end{pmatrix}, \qquad c_0 = \begin{pmatrix} 0 & 1 \\ 0 & 0 \end{pmatrix}. \tag{7.185}$$

These matrices are not Hermitian as required by (7.178): since Hermiticity follows from the choice of an inner product, it means that the vacua cannot form an orthonormal basis. An appropriate choice for the inner products is[13]

$$\langle \downarrow | \downarrow \rangle = \langle \uparrow | \uparrow \rangle = 0,$$
$$\langle \uparrow | \downarrow \rangle = \langle \downarrow | c_0 | \downarrow \rangle = \langle 0 | c_{1-\lambda} \cdots c_{-1} c_0 c_1 \cdots c_{\lambda-1} | 0 \rangle = 1. \tag{7.186}$$

The effect of changing the definition of the inner product or to consider a nonorthonormal basis is represented by the insertion of c_0. The last condition implies that the conjugate state (6.145) to the SL(2, $\mathbb{C}$) vacuum is

$$\langle 0^c | = \langle 0 | c_{1-\lambda} \cdots c_{-1} c_0 c_1 \cdots c_{\lambda-1}, \qquad \langle \downarrow^c | = \langle \uparrow |. \tag{7.187}$$

7.2.8 Summary

In this section we summarize the values of the parameters for different theories of interest (Table 7.1). The (η, ξ) system will be introduced in Chap. 17 in the bosonization of the super-reparametrization (β, γ) ghosts. The $\psi^\pm$ system can be used to describe spin-$1/2$ fermions.

---

[13]To avoid confusions, let us note that the adjoint in (7.182) are defined only through the adjoint of the modes (6.110) but not with respect to the inner product given here, which would lead to exchanging $| \downarrow \rangle^\ddagger \sim \langle \uparrow |$ and $| \uparrow \rangle^\ddagger \sim \langle \downarrow |$.

7.3   Suggested Readings

- Free scalar: general references [10, sec. 4.1.3, 4.3, 4.6.2, 2, sec. 5.3.1, 6.3, 1, sec. 4.2, 9, 6], topological current and winding [4, 11, sec. 17.2–3].
- First-order system: general references [1, chap. 5, sec. 13.1, 6, sec. 4.15, 9, sec. 2.5], ghost vacua [7, sec. 15.3].

References

1. R. Blumenhagen, D. Lüst, S. Theisen, *Basic Concepts of String Theory*. English. 2013 edn. (Springer, Berlin, 2014)
2. P. Di Francesco, P. Mathieu, D. Senechal, *Conformal Field Theory*, 2nd edn. (Springer, Berlin, 1999)
3. O. Hohm, D. Lüst, B. Zwiebach, The spacetime of double field theory: review, remarks, and outlook. Fortschr. Phys. **61**(10), 926–966 (2013). https://doi.org/10.1002/prop.201300024. arXiv: 1309.2977
4. C.M. Hull, Doubled geometry and T-folds. J. High Energy Phys. **2007**(7), 080 (2007) https://doi.org/10.1088/1126-6708/2007/07/080. arXiv: hep-th/0605149
5. C. Hull, B. Zwiebach, Double field theory. J. High Energy Phys. **2009**(9), 099 (2009) https://doi.org/10.1088/1126-6708/2009/09/099. arXiv: 0904.4664
6. E. Kiritsis. *String Theory in a Nutshell*. English (Princeton University Press, Princeton, 2007)
7. I.D. Lawrie. *A Unified Grand Tour of Theoretical Physics*. English. 3rd edn. (CRC Press, Boca Raton, 2012)
8. E. Plauschinn, Non-geometric backgrounds in string theory (2018)
9. J. Polchinski, *String Theory: Volume 1, An Introduction to the Bosonic String* (Cambridge University Press, Cambridge, 2005)
10. D. Tong, Lectures on string theory (2009). arXiv: 0908.0333
11. B. Zwiebach, *A First Course in String Theory*. English. 2nd edn. (Cambridge University Press, Cambridge, 2009)

BRST Quantization

8

Abstract

The BRST quantization can be introduced either by following the standard QFT treatment (outlined in Sect. 3.2) or by translating it into the CFT language. One can then use all the CFT techniques to extract information on the spectrum, which makes this approach more powerful. Moreover, this also provides an elegant description of states and string fields. In this chapter, we set the stage of the BRST quantization using the CFT language, and we apply it to string theory. The main results of this chapter are a proof of the no-ghost theorem and a characterization of the BRST cohomology (physical states).

8.1 BRST for Reparametrization Invariance

The BRST symmetry we are interested in results from gauge fixing the reparametrization invariance. In this chapter, we focus on the holomorphic sector: since both sectors are independent, most results follow directly, except those concerning the zero-modes. We consider a generic matter CFT coupled to reparametrization ghosts

1. matter: central charge c_m, energy–momentum tensor T_m and Hilbert space $\mathcal{H}_m$;
2. reparametrization ghosts: bc ghost system (Sects. 2.3 and 7.2) with $\epsilon = +1$ and $\lambda = 2$, $c_{gh} = -26$, energy–momentum tensor T^{gh} and Hilbert space $\mathcal{H}_{gh}$.

The formulas for the reparametrization ghosts are summarized in Appendix B.3.5. For modes, the system (m, gh, b or c) is indicated as a superscript to not confuse it

© Springer Nature Switzerland AG 2021
H. Erbin, *String Field Theory*, Lecture Notes in Physics 980,
https://doi.org/10.1007/978-3-030-65321-7_8

with the mode index. The total central charge, energy–momentum tensor and Hilbert space are denoted by

$$c = c_m + c_{\text{gh}} = c_m - 26, \qquad T(z) = T^m(z) + T^{\text{gh}}(z), \qquad \mathcal{H} = \mathcal{H}_m \otimes \mathcal{H}_{\text{gh}}. \tag{8.1}$$

The goal is to find the physical states in the cohomology, that is, which are BRST closed

$$Q_B \, |\psi\rangle = 0 \tag{8.2}$$

but non-exact (Sect. 3.2): the latter statement can be understood as an equivalence between closed states under shift by exact states

$$|\psi\rangle \sim |\psi\rangle + Q_B \, |\Lambda\rangle \, . \tag{8.3}$$

We introduce the BRST current and study its CFT properties. Then, we give a computation of the BRST cohomology when the matter CFT contains at least two scalar fields.

8.2   BRST in the CFT Formalism

The BRST current can be found from (3.50) to be [12]

$$j_B(z) = \, :c(z) \left(T^m(z) + \frac{1}{2} T^{\text{gh}}(z) \right) : + \kappa \, \partial^2 c(z) \tag{8.4a}$$

$$= c(z) T^m(z) + \, :b(z)c(z)\partial c(z) : + \kappa \, \partial^2 c(z), \tag{8.4b}$$

and similarly for the anti-holomorphic sector. This can be derived from (3.53): the generator of infinitesimal changes of coordinates (given by the Lie derivative) is the energy–momentum tensor. The factor of $1/2$ comes from the expression (7.99) of the ghost energy–momentum tensor: the second term does not contribute, while the first has a factor of 2. Since the transformation of c in (3.53) has no factor, the $1/2$ is necessary to recover the correct normalization. Finally, one finds that the transformation of b is reproduced. The different computations can be checked using the OPEs given below. The last piece is a total derivative and does not contribute to the charge: for this reason, it cannot be derived from (3.53), and its coefficient will be determined below. Note that it is the only total derivative of dimension 1 and of ghost number 1.

The BRST charge is then obtained by the contour integral

$$Q_B = Q_{B,L} + Q_{B,R}, \qquad Q_{B,L} = \oint \frac{\mathrm{d}z}{2\pi\mathrm{i}} \, j_B(z), \qquad Q_{B,R} = \oint \frac{\mathrm{d}\bar{z}}{2\pi\mathrm{i}} \, \bar{j}_B(\bar{z}). \tag{8.5}$$

As usual, $Q_B \sim Q_{B,L}$ when considering only the holomorphic sectors such that we generally omit the index.

8.2.1   OPE

The OPE of the BRST current with T is

$$T(z)j_B(w) \sim \left(\frac{c_m}{2} - 4 - 6\kappa\right)\frac{c(w)}{(z-w)^4} + (3 - 2\kappa)\frac{\partial c(w)}{(z-w)^3}$$
$$+ \frac{j_B(w)}{(z-w)^2} + \frac{\partial j_B(w)}{z-w}. \tag{8.6}$$

Hence, the BRST current is a primary operator only if

$$c_m = 26, \qquad \kappa = \frac{3}{2}. \tag{8.7}$$

The BRST current must be primary; otherwise, the BRST symmetry is anomalous, which means that the theory is not consistent. This provides another derivation of the critical dimension. In this case, the OPE becomes

$$T(z)j_B(w) \sim \frac{j_B(w)}{(z-w)^2} + \frac{\partial j_B(w)}{z-w}. \tag{8.8}$$

Remark 8.1 (Critical Dimension in 2d Gravity) The value $c_m = 26$ (critical dimension) was obtained in Sect. 2.3 by requiring that the Liouville field decouples from the path integral. In $2d$ gravity, where this condition is not necessary (nor even desirable), the Liouville field is effectively part of the matter, such that $c_L + c_m = 26$. One can also study the BRST cohomology in this case.

The OPE of $j_B(z)$ with the ghosts is

$$j_B(z)b(w) \sim \frac{2\kappa}{(z-w)^3} + \frac{j(w)}{(z-w)^2} + \frac{T(w)}{z-w}, \tag{8.9a}$$

$$j_B(z)c(w) \sim \frac{:c(w)\partial c(w):}{z-w}. \tag{8.9b}$$

Similarly, the OPE with any matter weight h primary field ϕ is

$$j_B(z)\phi(w) \sim h\frac{c(w)\phi(w)}{(z-w)^2} + \frac{:h\,\partial c(w)\phi(w) + c(w)\partial\phi(w):}{z-w}, \tag{8.9c}$$

using that $c(w)^2 = 0$ to cancel one term.

The OPE with the ghost current is

$$j_B(z)j(w) \sim \frac{2\kappa + 1}{(z-w)^3} - \frac{2\partial c(w)}{(z-w)^2} - \frac{j_B(w)}{z-w}, \tag{8.10}$$

while the OPE with itself is (for $\kappa = 3/2$)

$$j_B(z)j_B(w) \sim -\frac{c_m - 18}{2} \frac{:c(w)\partial c(w):}{(z-w)^3} - \frac{c_m - 18}{4} \frac{:c(w)\partial^2 c(w):}{(z-w)^2}$$
$$- \frac{c_m - 26}{12} \frac{:c(w)\partial^3 c(w):}{z-w}. \tag{8.11}$$

There is no first-order pole if $c_m = 26$: as we will see shortly, this implies that the BRST charge is nilpotent.

8.2.2   Mode Expansions

The mode expansion of the BRST charge can be written equivalently as

$$Q_B = \sum_m :c_m \left(L_{-m}^m + \frac{1}{2} L_{-m}^{gh} \right): \tag{8.12a}$$

$$= \sum_m c_{-m} L_m^m + \frac{1}{2} \sum_{m,n} (n-m) :c_{-m} c_{-n} b_{m+n}: . \tag{8.12b}$$

In the energy ordering, this expression becomes

$$Q_B = \sum_m {}^\star_\star c_m \left(L_{-m}^m + \frac{1}{2} L_{-m}^{gh} \right){}^\star_\star - \frac{c_0}{2} \tag{8.13a}$$

$$= \sum_n c_m L_{-m}^m + \frac{1}{2} \sum_{m,n} (n-m) {}^\star_\star c_{-m} c_{-n} b_{m+n} {}^\star_\star - c_0, \tag{8.13b}$$

where the ordering constant is the same as in L_0^{gh} (as can be checked by comparing both sides of the anti-commutator). The simplest derivation of this term is to use the algebra and to ensure that it is consistent. The only ambiguity is in the second term, when one c does not commute with the b: this happens for $-n + (m+n) = 0$, such that the ordering ambiguity is proportional to c_0. Then, one finds that it is equal to $a_{gh} = -1$.

The BRST operator can be decomposed on the ghost zero-modes as

$$Q_B = c_0 L_0 - b_0 M + \widehat{Q}_B, \tag{8.14a}$$

where

$$\widehat{Q}_B = \sum_{m \neq 0} c_{-m} L_m^m - \frac{1}{2} \sum_{\substack{m,n \neq 0 \\ m+n \neq 0}} (m - n) \,{}^{\star}_{\star} c_{-m} c_{-n} b_{m+n} \,{}^{\star}_{\star} , \tag{8.14b}$$

$$M = \sum_{m \neq 0} m \, c_{-m} c_m . \tag{8.14c}$$

The interest of this decomposition is that L_0, M and $\widehat{Q}$ do not contain b_0 or c_0, which make it very useful to act on states decomposed according to the zero-modes (7.169). The nilpotency of the BRST operator implies the relations

$$[L_0, M] = [\widehat{Q}_B, M] = [\widehat{Q}_B, L_0] = 0, \qquad \widehat{Q}_B^2 = L_0 M . \tag{8.15}$$

Moreover, one has $N_{\mathrm{gh}}(\widehat{Q}_B) = 1$ and $N_{\mathrm{gh}}(M) = 2$.

8.2.3 Commutators

From the various OPEs, one can compute the (anti-)commutators of the BRST charge with the other operators. For the ghosts and a weight h primary field ϕ, one finds

$$\{Q_B, b(z)\} = T(z), \tag{8.16a}$$

$$\{Q_B, c(z)\} = c(z) \partial c(z), \tag{8.16b}$$

$$[Q_B, \phi(z)] = h \, \partial c(z) \phi(z) + c(z) \partial \phi(z). \tag{8.16c}$$

This reproduces correctly (3.53).

Two facts will be useful in string theory. First, (8.16c) is a total derivative for $h = 1$

$$[Q_B, \phi(z)] = \partial \big(c(z) \phi(z)\big). \tag{8.17}$$

Second, $c(z)\phi(z)$ is closed if $h = 1$

$$\{Q_B, c(z)\phi(z)\} = (1 - h) c(z) \partial c(z) \phi(z). \tag{8.18}$$

The commutator with the ghost current is

$$[Q_B, j(z)] = -j_B(z), \tag{8.19}$$

which confirms that the BRST charge increases the ghost number by 1

$$[N_{\text{gh}}, Q_B] = Q_B. \tag{8.20}$$

One finds that the BRST charge is nilpotent

$$\{Q_B, Q_B\} = 0 \tag{8.21}$$

and commutes with the energy–momentum tensor

$$[Q_B, T(z)] = 0 \tag{8.22}$$

only if the matter central charge corresponds to the critical dimension

$$c_m = 26. \tag{8.23}$$

The most important commutator for the modes is

$$L_n = \{Q_B, b_n\}. \tag{8.24}$$

Nilpotency of Q_B then implies that Q_B commutes with L_n

$$[Q_B, L_n] = 0. \tag{8.25}$$

8.3   BRST Cohomology: Two Flat Directions

The simplest case for studying the BRST cohomology is when the target spacetime has at least two non-compact flat directions represented by two free scalar fields (X^0, X^1) (Sect. 7.1). The remaining matter fields are arbitrary as long as the critical dimension $c_m = 26$ is reached. The reason for introducing two flat directions is that the cohomology is easily worked out by introducing light-cone (or complex) coordinates in target spacetime.

The field X^0 can be spacelike or timelike $\epsilon_0 = \pm 1$, while we consider X^1 to be always spacelike, $\epsilon_1 = 1$. The oscillators are denoted by α_m^0 and α_m^1, and the momenta of the Fock vacua by $k_\| = (k^0, k^1)$ such that

$$k_\|^2 = \epsilon_0 (k^0)^2 + (k^1)^2. \tag{8.26}$$

The rest of the matter sector, called the transverse sector $\perp$, is an arbitrary CFT with energy–momentum tensor $T^\perp$, central charge $c_\perp = 24$ and Hilbert space $\mathcal{H}_\perp$. The ghost together with the two scalar fields forms the longitudinal sector $\|$. The motivation for the names longitudinal and transverse will become clear later:

they will be identified with the light-cone and perpendicular directions in the target spacetime (and, correspondingly, with unphysical and physical states).

The Hilbert space of the theory is decomposed as

$$
\mathcal{H} := \mathcal{H}_{\parallel} \otimes \mathcal{H}_{\perp}, \qquad \mathcal{H}_{\parallel} := \int dk^0 \, \mathcal{F}_0(k^0) \otimes \int dk^1 \, \mathcal{F}_1(k^1) \otimes \mathcal{H}_{\mathrm{gh}}, \qquad (8.27)
$$

where $\mathcal{F}_0(k^0)$ and $\mathcal{F}_1(k^1)$ are the Fock spaces (7.83a) of the scalar fields X^0 and X^1, and $\mathcal{H}_{\mathrm{gh}}$ is the ghost Hilbert space (7.169). As a consequence, a generic state of $\mathcal{H}$ reads

$$
|\psi\rangle = |\psi_{\parallel}\rangle \otimes |\psi_{\perp}\rangle, \qquad (8.28)
$$

where $\psi_{\perp}$ is a generic state of the transverse matter CFT $\mathcal{H}_{\perp}$ and $\psi_{\parallel}$ is built by acting with oscillators on the Fock vacuum of $\mathcal{H}_{\parallel}$

$$
|\psi_{\parallel}\rangle = c_0^{N_0^c} \prod_{m>0} (\alpha^0_{-m})^{N_m^0} (\alpha^1_{-m})^{N_m^1} (b_{-m})^{N_m^b} (c_{-m})^{N_m^c} |k^0, k^1, \downarrow\rangle
$$

$$
|k^0, k^1, \downarrow\rangle := |k^0\rangle \otimes |k^1\rangle \otimes |\downarrow\rangle, \qquad N_m^0, N_m^1 \in \mathbb{N}, \qquad N_m^b, N_m^c = 0, 1. \qquad (8.29)
$$

Since the Virasoro modes commute with the ghost number, eigenstates of the Virasoro operators without zero-modes $\widehat{L}_0$, given by the sum of (7.68) and (7.160), can also be taken to be eigenstates of N_{gh}. It is also useful to define the Hilbert space of states lying in the kernel of b_0

$$
\mathcal{H}_0 = \mathcal{H} \cap \ker b_0 \qquad (8.30)
$$

such that

$$
\mathcal{H} = \mathcal{H}_0 \oplus c_0 \mathcal{H}_0. \qquad (8.31)
$$

The full L_0 operator reads

$$
L_0 = L_0^m + L_0^{\mathrm{gh}} = (L_0^m - 1) + N^b + N^c, \qquad (8.32)
$$

using (7.129) for L_0^{gh}. A more useful expression is obtained by separating the two sectors and by extracting the zero-modes using (7.67)

$$
L_0 = \left(L_0^{\perp} - m_{\parallel,L}^2 \ell^2 - 1\right) + \widehat{L}_0^{\parallel}, \qquad (8.33)
$$

where the longitudinal mass and total level operator are

$$m_{\parallel,L}^2 = -p_{\parallel,L}^2, \qquad \widehat{L}_0^{\parallel} = N^0 + N^1 + N^b + N^c \in \mathbb{N}. \qquad (8.34)$$

A state $|\psi\rangle$ is said to be on-shell if it is annihilated by L_0

$$\text{on-shell:} \qquad L_0 |\psi\rangle = 0. \qquad (8.35)$$

The absolute BRST cohomology $\mathcal{H}_{\text{abs}}(Q_B)$ defines the physical states (Sect. 3.2) and is given by the states $\psi \in \mathcal{H}$ that are Q_B-closed but not exact

$$\mathcal{H}_{\text{abs}}(Q_B) := \left\{ |\psi\rangle \in \mathcal{H} \mid Q_B |\psi\rangle = 0, \nexists |\chi\rangle \in \mathcal{H} \mid |\psi\rangle = Q_B |\chi\rangle \right\}. \qquad (8.36)$$

Since Q_B commutes with L_0, (8.25), the cohomology subspace is preserved under time evolution.

Before continuing, it is useful to outline the general strategy for studying the cohomology of a BRST operator Q in the CFT language. The idea is to find an operator Δ—called contracting homotopy operator—which, if it exists, trivializes the cohomology. Conversely, this implies that the cohomology is to be found within states that are annihilated by Δ or for which Δ is not defined. Then, it is possible to restrict Q on these subspaces: this is advantageous when the restriction of the BRST charge on these subspaces is a simpler. In fact, we will find that the reduced operator is itself a BRST operator, for which one can search for another contracting homotopy operator.[1]

Given a BRST operator Q, a contracting homotopy operator Δ for Q is an operator such that

$$\{Q, \Delta\} = 1. \qquad (8.37)$$

Interpreting Q as a derivative operator, Δ corresponds to the Green function or propagator. The existence of a well-defined Δ with empty kernel implies that the cohomology is empty because all closed states are exact. Indeed, consider a state $|\psi\rangle \in \mathcal{H}$, which is an eigenstate of Δ and closed $Q_B |\psi\rangle = 0$. Inserting (8.37) in front of the state gives

$$|\psi\rangle = \{Q_B, \Delta\} |\psi\rangle = Q_B(\Delta |\psi\rangle). \qquad (8.38)$$

If Δ is well-defined on $|\psi\rangle$ and $|\psi\rangle \notin \ker \Delta$, then $\Delta |\psi\rangle$ is another state in $\mathcal{H}$, which implies that $|\psi\rangle$ is exact. Hence, the BRST cohomology has to be found inside the subspaces $\ker \Delta$ or on which Δ is not defined.

---

[1] A similar strategy shows that there is no open string excitation for the open SFT in the tachyon vacuum.

8.3.1   Conditions on the States

In this subsection, we apply explicitly the strategy just discussed to get conditions on the states. A candidate contracting homotopy operator for Q_B is

$$\Delta := \frac{b_0}{L_0}, \tag{8.39}$$

thanks to (8.24)

$$L_0 = \{Q_B, b_0\}. \tag{8.40}$$

Indeed, suppose that $|\psi\rangle$ is an eigenstate of L_0 and that it is closed but not on-shell

$$Q_B |\psi\rangle = 0, \qquad L_0 |\psi\rangle \neq 0. \tag{8.41}$$

One can use (8.40) in order to write

$$|\psi\rangle = Q_B \left(\frac{b_0}{L_0} |\psi\rangle \right). \tag{8.42}$$

The operator inside the parenthesis is Δ defined above in (8.39). The formula (8.42) breaks down if ψ is in the kernel of L_0 since the inverse is not defined. This implies that a necessary condition for a L_0-eigenstate $|\psi\rangle$ to be in the BRST cohomology is to be on-shell (8.35). Considering explicitly the subset of states annihilated by b_0 is not needed at this stage since $\ker b_0 \subset \ker L_0$ for Q_B-closed states, according to (8.24). Hence, we conclude

$$\mathcal{H}_{\mathrm{abs}}(Q_B) \subset \ker L_0. \tag{8.43}$$

Note that this statement holds only at the level of vector spaces, i.e. when considering equivalence classes of states $|\psi\rangle \sim |\psi\rangle + Q |\Lambda\rangle$. This means that there exists a representative state of each equivalence class inside $\ker L_0$, but a generic state is not necessarily in $\ker L_0$. For example, consider a state $|\psi\rangle \in \ker L_0$ and closed. Then, $|\psi'\rangle = |\psi\rangle + Q_B |\Lambda\rangle$ with $|\Lambda\rangle \notin \ker L_0$ is still in $\mathcal{H}_{\mathrm{abs}}(Q_B)$ but $|\psi'\rangle \notin \ker L_0$ since $[L_0, Q_B] = 0$.

> **Computation: Equation (8.42)**
> For $L_0 |\psi\rangle \neq 0$, one has
>
> $$|\psi\rangle = \frac{L_0}{L_0} |\psi\rangle = \frac{1}{L_0} \{Q_B, b_0\} |\psi\rangle = \frac{1}{L_0} Q_B (b_0 |\psi\rangle),$$
>
> where the fact that $|\psi\rangle$ is closed has been used to cancel the second term of the anti-commutator. Note that L_0 commutes with both Q_B and b_0 such that it can be moved freely.

This shows that $\Delta = b_0/L_0$ given by (8.39) is not a contracting homotopy operator. A proper definition involves the projector P_0 on the kernel of L_0

$$|\psi\rangle \in \ker L_0 : \quad P_0 |\psi\rangle = |\psi\rangle, \qquad |\psi\rangle \in (\ker L_0)^\perp : \quad P_0 |\psi\rangle = 0. \qquad (8.44)$$

Then, the appropriate contracting homotopy operator reads $\Delta(1 - P_0)$, and (8.37) is changed to

$$\{Q_B, \Delta(1 - P_0)\} = (1 - P_0). \qquad (8.45)$$

This parallels completely the definition of the Green function in the presence of zero-modes, see (B.3). By abuse of language, we will also say that Δ is a contracting homotopy operator, remembering that this statement is correct only when multiplying with $(1 - P_0)$.

We will revisit these aspects later from the SFT perspective. In fact, we will find that Q_B is the kinetic operator of the gauge invariant theory, while Δ is the gauge fixed propagator in the Siegel gauge. This is expected from experience with standard gauge theories: the inverse of the kinetic operator (Green function) is not defined when the gauge invariance is not fixed.

The on-shell condition (8.35) is already a good starting point. In order to simplify the analysis further, one can restrict the question of computing the cohomology on the subspace

$$\mathcal{H}_0 := \mathcal{H} \cap \ker b_0 = \mathcal{H}_m \otimes \mathcal{H}_{\mathrm{gh},0}, \qquad (8.46)$$

where $\mathcal{H}_{\mathrm{gh},0} = \mathcal{H}_{\mathrm{gh}} \cap \ker b_0$ was defined in (7.2.6). This subspace contains all states $|\psi\rangle$ such that

$$|\psi\rangle \in \mathcal{H}_0 \quad \Longrightarrow \quad b_0 |\psi\rangle = 0. \qquad (8.47)$$

In this subspace, there is no exact state $|\psi\rangle$ with $L_0 |\psi\rangle \neq 0$ such that $b_0 |\psi\rangle = Q_B |\psi\rangle = 0$. Indeed, assuming these conditions, (8.42) leads to a contraction

$$b_0 |\psi\rangle = Q_B |\psi\rangle = 0, \quad L_0 |\psi\rangle \neq 0 \quad \Longrightarrow \quad |\psi\rangle = 0. \qquad (8.48)$$

Note that the converse statement is not true: there are on-shell states such that $b_0 |\psi\rangle \neq 0$. This also makes sense because the ghost Hilbert space can be decomposed with respect to the ghost zero-modes. The cohomology of Q_B in the subspace $\mathcal{H}_0$ is called the relative cohomology

$$\mathcal{H}_{\mathrm{rel}}(Q_B) := \mathcal{H}_0(Q_B) = \left\{ |\psi\rangle \in \mathcal{H}_0 \mid Q_B |\psi\rangle = 0, \nexists |\chi\rangle \in \mathcal{H} \mid |\psi\rangle = Q_B |\chi\rangle \right\}. \qquad (8.49)$$

The advantage of the subspace $b_0 = 0$ is to precisely pick the representative of $\mathcal{H}_{abs}$, which lies in $\ker L_0$. In particular, the operator L_0 is simple and has a direct physical interpretation as the worldsheet Hamiltonian. This condition is also meaningful in string theory because these states are also mass eigenstates, which have a nice spacetime interpretation, and it will later be interpreted in SFT as fixing the Siegel gauge. Moreover, it is implied by the choice of Δ in (8.39) as the contracting homotopy operator, which is particularly convenient to work with to derive the cohomology. However, there are other possible choices, which are interpreted as different gauge fixings.

After having built this cohomology, we can look for the full cohomology by relaxing the condition $b_0 = 0$. In view of the structure of the ghost Hilbert space (7.169), one can expect that $\mathcal{H}_{abs}(Q_B) = \mathcal{H}_{rel}(Q_B) \oplus c_0 \mathcal{H}_{rel}(Q_B)$, which is indeed the correct answer. But, we will see (building on Sect. 3.2.2) that, in fact, it is this cohomology that contains the physical states in string theory, instead of the absolute cohomology.

As a summary, we are looking for Q_B-closed non-exact states annihilated by b_0 and L_0

$$Q_B |\psi\rangle = 0, \qquad L_0 |\psi\rangle = 0, \qquad b_0 |\psi\rangle = 0. \tag{8.50}$$

8.3.2 Relative Cohomology

In (8.14a), the BRST operator was decomposed as

$$Q_B = c_0 L_0 - b_0 M + \widehat{Q}_B, \qquad \widehat{Q}_B^2 = L_0 M. \tag{8.51}$$

This shows that, on the subspace $L_0 = b_0 = 0$, $\widehat{Q}_B$ is nilpotent and equivalent to Q_B

$$|\psi\rangle \in \mathcal{H}_0 \cap \ker L_0 \quad \Longrightarrow \quad Q_B |\psi\rangle = \widehat{Q}_B |\psi\rangle, \qquad \widehat{Q}_B^2 |\psi\rangle = 0. \tag{8.52}$$

Hence, this implies that $\widehat{Q}_B$ is a proper BRST operator, and the relative cohomology of Q_B is isomorphic to the cohomology of $\widehat{Q}_B$

$$\mathcal{H}_0(Q_B) = \mathcal{H}_0(\widehat{Q}_B). \tag{8.53}$$

Next, we introduce light-cone coordinates in the target spacetime. While it does not allow to write Lorentz covariant expressions, it is helpful mathematically because it introduces a grading of the Hilbert space, for which powerful theorems exist (even if we will need only basic facts for our purpose).

Light-Cone Parametrization

The two scalar fields X^0 and X^1 are combined in a light-cone (if $\epsilon_0 = -1$) or complex (if $\epsilon_0 = 1$) fashion

$$X_L^{\pm} = \frac{1}{\sqrt{2}} \left(X_L^0 \pm \frac{i}{\sqrt{\epsilon_0}} X_L^1 \right). \tag{8.54}$$

The modes of $X^{\pm}$ are found by following (7.50):[2]

$$\alpha_n^{\pm} = \frac{1}{\sqrt{2}} \left(\alpha_n^0 \pm \frac{i}{\sqrt{\epsilon_0}} \alpha_n^1 \right), \qquad n \neq 0, \tag{8.55a}$$

$$x_L^{\pm} = \frac{1}{\sqrt{2}} \left(x_L^0 \pm \frac{i}{\sqrt{\epsilon_0}} x_L^1 \right), \qquad p_L^{\pm} = \frac{1}{\sqrt{2}} \left(p_L^0 \pm \frac{i}{\sqrt{\epsilon_0}} p_L^1 \right), \tag{8.55b}$$

The non-zero commutation relations are

$$\left[\alpha_m^+, \alpha_n^- \right] = \epsilon_0 \, m \, \delta_{m+n,0}, \qquad \left[x_L^{\pm}, p_L^{\mp} \right] = i\epsilon_0. \tag{8.56}$$

This implies that negative-frequency (creation) modes $\alpha_{-n}^{\pm}$ are canonically conjugate to positive-frequency (annihilation) modes $\alpha_n^{\mp}$. Note the similarity with the first-order system (7.134).

For later purposes, it is useful to note the following relations:

$$2\, p_L^+ p_L^- = \left(p_L^0 \right)^2 + \epsilon_0 \left(p_L^1 \right)^2 = \epsilon_0 \, p_{\parallel,L}^2, \tag{8.57a}$$

$$x^+ p^- + x^- p^+ = x^0 p^0 + \epsilon_0 \, x^1 p^1, \tag{8.57b}$$

$$\sum_n \alpha_n^+ \alpha_{m-n}^- = \frac{1}{2} \sum_n \left(\alpha_n^0 \alpha_{m-n}^0 + \epsilon_0 \, \alpha_n^1 \alpha_{m-n}^1 \right). \tag{8.57c}$$

In view of the commutators (8.56), the appropriate definitions of the light-cone number $N_n^{\pm}$ and level operators $N^{\pm}$ are

$$N_n^{\pm} = \frac{\epsilon_0}{n} \alpha_{-n}^{\pm} \alpha_n^{\mp}, \qquad N^{\pm} = \sum_{n>0} n \, N_n^{\pm}. \tag{8.58}$$

The insertion of ϵ_0 follows (7.63). Then, one finds the following relation:

$$N^+ + N^- = N^0 + N^1. \tag{8.59}$$

Using these definitions, the variables appearing in L_0 (8.33)

$$L_0 = \left(L_0^{\perp} - m_{\parallel,L}^2 \ell^2 - 1 \right) + \widehat{L}_0^{\parallel} \tag{8.60}$$

---

[2]For $\epsilon_0 = 1$, this convention matches the ones from [5] for $X^0 = X$ and $X^1 = \phi$. For $\epsilon = -1$, this convention matches [12].

can be rewritten as

$$m_{\parallel,L}^2 = -2\epsilon_0\, p_L^+ p_L^-, \qquad \widehat{L}_0^{\parallel} = N^+ + N^- + N^b + N^c. \tag{8.61}$$

The expression for the sum of the Virasoro operators (7.65) easily follows from (8.57):

$$L_m^0 + L_m^1 = \epsilon_0 \sum_n :\alpha_n^+ \alpha_{m-n}^- : = \epsilon_0 \sum_{n\neq 0,m} :\alpha_n^+ \alpha_{m-n}^- : + \epsilon_0\left(\alpha_0^- \alpha_m^+ + \alpha_m^+ \alpha_m^-\right).$$
$$\tag{8.62}$$

Computation: Equation (8.56)
For the modes $\alpha_m^{\pm}$, we have

$$[\alpha_m^+, \alpha_n^{\pm}] = \frac{1}{2}\left[\left(\alpha_m^0 + \frac{i}{\sqrt{\epsilon_0}}\alpha_m^1\right), \left(\alpha_n^0 \pm \frac{i}{\sqrt{\epsilon_0}}\alpha_n^1\right)\right]$$

$$= \frac{1}{2}\left([\alpha_m^0, \alpha_n^0] \mp \frac{1}{\epsilon_0}[\alpha_m^1, \alpha_n^1]\right) = \frac{\epsilon_0}{2}\, m\, \delta_{m+n,0}(1 \mp 1),$$

where we used (7.72). The other commutators follow similarly from (7.73), for example,

$$[x_L^-, p_L^{\pm}] = \frac{1}{2}\left[\left(x_L^0 - \frac{i}{\sqrt{\epsilon_0}}x_L^1\right), \left(p_L^0 \pm \frac{i}{\sqrt{\epsilon_0}}p_L^1\right)\right]$$

$$= \frac{1}{2}\left([x_L^0, p_L^0] \pm \epsilon_0[x_L^1, p_L^1]\right) = \frac{\epsilon_0}{2}(1 \pm 1).$$

Computation: Equation (8.57)
For the modes $\alpha_m^{\pm}$, we have

$$\sum_n \alpha_n^+ \alpha_{m-n}^- = \frac{1}{2}\sum_n\left(\alpha_n^0 + \frac{i}{\sqrt{\epsilon_0}}\alpha_n^1\right)\left(\alpha_{m-n}^0 - \frac{i}{\sqrt{\epsilon_0}}\alpha_{m-n}^1\right)$$

$$= \frac{1}{2}\sum_n\left(\alpha_n^0\alpha_{m-n}^0 + \epsilon_0\,\alpha_n^1\alpha_{m-n}^1 + \frac{i}{\sqrt{\epsilon_0}}(\alpha_{m-n}^0\alpha_n^1 - \alpha_n^0\alpha_{m-n}^1)\right).$$

The last two terms in parenthesis cancel as can be seen by shifting the sum $n \to m - n$ in one of the terms. Note that, for $m \neq 2n$, there is no cross-term only after summing over n.

The relations for the zero-modes follow simply by observing that expressions in both coordinates can be rewritten in terms of the 2-dimensional (spacetime) flat metric.

Computation: Equation (8.59)
Using (8.57), one finds

$$N^0 + N^1 = \sum_n n\left(N_n^0 + N_n^1\right) = \sum_n n\left(N_n^+ + N_n^-\right) = N^+ + N^-.$$

Reduced Cohomology
In terms of the light-cone variables, the reduced BRST operator $\widehat{Q}_B$ reads

$$\widehat{Q}_B = \sum_{m\neq 0} c_{-m}\left(L_m^\perp + \epsilon_0 \sum_n \alpha_n^+ \alpha_{m-n}^-\right) + \frac{1}{2}\sum_{m,n}(n-m):c_{-m}c_{-n}b_{m+n}:.$$

$$(8.63)$$

This operator can be further decomposed. Introducing the degree

$$\deg := N^+ - N^- + \widehat{N}^c - \widehat{N}^b \tag{8.64}$$

such that

$$\forall m \neq 0: \qquad \deg\left(\alpha_m^+\right) = \deg(c_m) = 1, \qquad \deg(\alpha_m^-) = \deg(b_m) = -1, \tag{8.65}$$

and $\deg = 0$ for the other variables, the operator $\widehat{Q}_B$ is decomposed as[3]

$$\widehat{Q}_B = Q_0 + Q_1 + Q_2, \qquad \deg(Q_j) = j, \tag{8.66a}$$

where

$$Q_1 = \sum_{m\neq 0} c_{-m}L_m^\perp + \sum_{\substack{m,n\neq 0 \\ m+n\neq 0}} {}^\star_\star c_{-m}\left(\epsilon_0\,\alpha_n^+ \alpha_{m-n}^- + \frac{1}{2}(m-n)\,c_{-m}b_{m+n}\right){}^\star_\star,$$

$$Q_0 = \sum_{n\neq 0}\alpha_0^+\,c_{-n}\alpha_n^-, \qquad Q_2 = \sum_{n\neq 0}\alpha_0^-\,c_{-n}\alpha_n^+. \tag{8.66b}$$

The nilpotency of $\widehat{Q}_B$ implies the following conditions on the Q_j:

$$Q_0^2 = Q_2^2 = 0, \qquad \{Q_0, Q_1\} = \{Q_1, Q_2\} = 0, \qquad Q_1^2 + \{Q_0, Q_2\} = 0. \tag{8.67}$$

---

[3]The general idea behind this decomposition is the notion of filtration, nicely explained in [1, sec. 3, 6].

Hence, Q_0 and Q_2 are both nilpotent and define a cohomology.

One can show that the cohomologies of $\widehat{Q}_B$ and Q_0 are isomorphic[4]

$$\mathcal{H}_0(\widehat{Q}_B) \simeq \mathcal{H}_0(Q_0) \tag{8.68}$$

under general conditions [5], in particular, if the cohomology is ghost-free (i.e. all states have $N_{\mathrm{gh}} = 1$).

The contracting homotopy operator for Q_0 is

$$\widehat{\Delta} := \frac{B}{\widehat{L}_0^{\|}}, \qquad B := \epsilon_0 \sum_{n \neq 0} \frac{1}{\alpha_0^+} \alpha_{-n}^+ b_n. \tag{8.69}$$

Indeed, it is straightforward to check that

$$\widehat{L}_0^{\|} = \{Q_0, B\} \quad \Longrightarrow \quad \{Q_0, \widehat{\Delta}\} = 1. \tag{8.70}$$

As a consequence, a necessary condition for a closed $\widehat{L}_0^{\|}$-eigenstate $|\psi\rangle$ to be in the cohomology of Q_0 is to be annihilated by $\widehat{L}_0^{\|}$

$$\widehat{L}_0^{\|} |\psi\rangle = 0, \quad \Longrightarrow \quad N^{\pm} |\psi\rangle = N^c |\psi\rangle = N^b |\psi\rangle = 0, \tag{8.71}$$

since $\widehat{L}_0^{\|}$ is a sum of positive integers. This means that the state ψ contains no-ghost or light-cone excitations $\alpha_{-n}^{\pm}$, b_{-n} and c_{-n}, and lies in the ground state of the Fock space $\mathcal{H}_{\|,0}$.

Then, we need to prove that this condition is sufficient: states with $\widehat{L}_0^{\|} = 0$ are closed. First, note that a state $|\psi\rangle \in \mathcal{H}_0$ with $\widehat{L}_0^{\|}$ has ghost number 1 since there are no-ghost excitations on top of the vacuum $|\downarrow\rangle$, which has $N_{\mathrm{gh}} = 1$. Second, $\widehat{L}_0$ and Q_0 commute, such that

$$0 = Q_0 \widehat{L}_0^{\|} |\psi\rangle = \widehat{L}_0^{\|} Q_0 |\psi\rangle. \tag{8.72}$$

Since Q_0 increases the ghost number by 1, one can invert $\widehat{L}_0^{\|} = N^b + N^c + \cdots$ in the last term since $\widehat{L}_0^{\|} \neq 0$ in this subspace. This gives

$$Q_0 |\psi\rangle = 0. \tag{8.73}$$

Hence, the condition $\widehat{L}_0^{\|} |\psi\rangle = 0$ is sufficient for $|\psi\rangle$ to be in the cohomology. This has to be contrasted with Sect. 8.3.1, where the condition $L_0^{\|} = 0$ is necessary but not sufficient.

---

[4]The role of Q_0 and Q_2 can be reversed by changing the sign in the definition of the degree and the role of $P_n^{\pm}$.

In this case, the on-shell condition (8.33) reduces to

$$L_0 = L_0^\perp - m_{\parallel,L}^2 \ell^2 - 1 = 0. \tag{8.74}$$

But, additional states can be found in ker B or in a subspace of $\mathcal{H}$ on which B is singular. We have ker $B = \ker \widehat{L}_0^\parallel$ such that nothing new can be found there. However, the operator B is not defined for states with vanishing momentum $\alpha_0^+ \propto p_L^+ = 0$. In fact, one must also have $\alpha_0^- \propto p_L^- = 0$ (otherwise, the contracting operator for Q_2 is well-defined and can be used instead). But, these states do not satisfy the on-shell condition (except for massless states with $L_0^\perp = 1$), as it will be clear later (see [13, sec. 2.2] for more details). For this reason, we assume that states have a generic non-zero momentum and that there is no pathology.

Full Relative Cohomology

This section aims to construct states in $\mathcal{H}_0(\widehat{Q}_B)$ from states in $\mathcal{H}(Q_0)$. We follow the construction from [5].

Given a state $|\psi_0\rangle \in \mathcal{H}_0(Q_0)$, the state $Q_1 |\psi_0\rangle$ is Q_0-closed since Q_0 and Q_1 anti-commute (8.67)

$$\{Q_0, Q_1\} |\psi_0\rangle = 0 \quad \Longrightarrow \quad Q_0(Q_1 |\psi_0\rangle) = 0. \tag{8.75}$$

Since $Q_1 |\psi_0\rangle$ is not in ker $\widehat{L}_0^\parallel$ (because Q_1 increases the ghost number by 1), the state $Q_1 |\psi_0\rangle$ is Q_0-exact and can be written as Q_0 of another state $|\psi_1\rangle$

$$Q_1 |\psi_0\rangle =: -Q_0 |\psi_1\rangle \quad \Longrightarrow \quad |\psi_1\rangle = -\frac{B}{\widehat{L}_0^\parallel} Q_1 |\psi_0\rangle. \tag{8.76}$$

Computation: Equation (8.76)

Start from the definition and insert (8.70) since $\widehat{L}_0$ is invertible

$$Q_1 |\psi_0\rangle = \left\{ Q_0, \frac{B}{\widehat{L}_0^\parallel} \right\} Q_1 |\psi_0\rangle = Q_0 \left(\frac{B}{\widehat{L}_0^\parallel} Q_1 |\psi_0\rangle \right).$$

The state $|\psi_1\rangle$ is identified with minus the state inside the parenthesis (up to a BRST exact state).

As for $|\psi_0\rangle$, apply $\{Q_0, Q_1\}$ on ψ_1

$$\{Q_0, Q_1\} |\psi_1\rangle = Q_0(Q_1 |\psi_1\rangle + Q_2 |\psi_0\rangle). \tag{8.77}$$

This implies that the combination in parenthesis is Q_0-closed and, for the same reason as above, it is exact

$$Q_1 |\psi_1\rangle + Q_2 |\psi_0\rangle = Q_0 |\psi_2\rangle, \qquad |\psi_2\rangle = -\frac{B}{\widehat{L}_0^{\parallel}} \left(Q_1 |\psi_1\rangle + Q_2 |\psi_0\rangle \right). \qquad (8.78)$$

Computation: Equation (8.77)

$$\{Q_0, Q_1\} |\psi_1\rangle = Q_0 Q_1 |\psi_1\rangle - Q_1^2 |\psi_0\rangle = Q_0 Q_1 |\psi_1\rangle + \{Q_0, Q_2\} |\psi_0\rangle.$$

The first equality follows from (8.76), and the second by using (8.67). The final result is obtained after using that $|\psi_0\rangle$ is Q_0-closed.

Iterating this procedure leads to a series of states

$$|\psi_{k+1}\rangle = -\frac{B}{\widehat{L}_0^{\parallel}} \left(Q_1 |\psi_k\rangle + Q_2 |\psi_{k-1}\rangle \right). \qquad (8.79)$$

We claim that a state in the relative cohomology $|\psi\rangle \in \mathcal{H}_0(\widehat{Q}_B)$ is built by summing all these states

$$|\psi\rangle = \sum_{k \in \mathbb{N}} |\psi_k\rangle. \qquad (8.80)$$

Indeed, it is easy to check that $|\psi\rangle$ is $\widehat{Q}_B$-closed

$$\widehat{Q}_B |\psi\rangle = 0. \qquad (8.81)$$

We leave aside the proof that ψ is not exact (see [5]). Note that ψ and ψ_0 have the same ghost numbers

$$N_{\mathrm{gh}}(\psi) = N_{\mathrm{gh}}(\psi_0) = 1 \qquad (8.82)$$

since $N_{\mathrm{gh}}(BQ_j) = 0$.

In fact, since ψ_0 does not contain longitudinal modes, it is annihilated by Q_1 and Q_2 (these operators contain either a ghost creation operator together with a light-cone annihilation operator, or the reverse)

$$Q_1 |\psi_0\rangle = Q_2 |\psi_0\rangle = 0. \qquad (8.83)$$

As a consequence, one has $\psi_k = 0$ for $k \geq 1$ and $\psi = \psi_0$.

Computation: Equation (8.81)

$$\widehat{Q}_B \,|\psi\rangle = \sum_{k \subset \mathbb{N}} \widehat{Q}_B \,|\psi_k\rangle$$

$$= Q_0 \,|\psi_0\rangle + \underbrace{Q_1 \,|\psi_0\rangle + Q_0 \,|\psi_1\rangle}_{=0} + \underbrace{Q_2 \,|\psi_0\rangle + Q_1 \,|\psi_1\rangle + Q_0 \,|\psi_2\rangle}_{=0} + \cdots$$

$$= 0.$$

8.3.3 Absolute Cohomology, States and No-Ghost Theorem

The absolute cohomology is constructed from the relative cohomology

$$\mathcal{H}_{\text{abs}}(Q_B) = \mathcal{H}_{\text{rel}}(Q_B) \oplus c_0 \,\mathcal{H}_{\text{rel}}(Q_B). \tag{8.84}$$

The interested reader is referred to [5] for the proof. A simple motivation is that the Hilbert space is decomposed in terms of the ghost zero-modes as in (7.169). Since the zero-modes commute with $\widehat{Q}_0$, linear combination of states in $\mathcal{H}_{\text{rel}}(Q_B)$ and $c_0\mathcal{H}_{\text{rel}}(Q_B)$ is expected to be in the cohomology. Obviously, one has to work out the other terms of Q_B and prove that there are no other states.

It looks like there is a doubling of the physical states, one built on $|\downarrow\rangle$ and one on $|\uparrow\rangle$. The remedy is to impose the condition $b_0 = 0$ on the states (see also Sect. 3.2.2 and [13, sec. 2.2] for more details). As already pointed out, states in $\mathcal{H}_{\text{abs}}$ form equivalence class under $|\psi\rangle \sim |\psi\rangle + Q_B \,|\Lambda\rangle$, and it is necessary to select a single representative. This is what the condition $b_0 = 0$ achieves. Obviously, it is always possible to add BRST exact states to write another representative (e.g. to restore the Lorentz covariance).

The last step is to discuss the no-ghost theorem: the latter states that there is no negative-norm states in the BRST cohomology of string theory. This follows straightforwardly from the condition $\widehat{L}_0 = 0$: it implies that there are no-ghost and no light-cone excitations. The ghosts and the time direction (if X^0 is timelike) are responsible for negative-norm states. Hence, the cohomology has no negative-norm states if the transverse CFT is unitary (which implies that all states in $\mathcal{H}_\perp$ have a positive-definite inner product).

Physical states $|\psi\rangle \in \mathcal{H}_{\text{rel}}(Q_B)$ are thus of the form

$$|\psi\rangle = |k^0, k^1, \downarrow\rangle \otimes |\psi_\perp\rangle, \qquad |\psi_\perp\rangle \in \mathcal{H}_\perp, \tag{8.85a}$$

$$\left(L_0^\perp - m_{\parallel,L}^2 \ell^2 - 1\right) |\psi\rangle = 0, \qquad p_{L,\parallel}^2 = -m_{\parallel,L}^2 \ell^2. \tag{8.85b}$$

This form can be made covariant: taking a state of the form $|\psi\rangle \otimes |\downarrow\rangle$ with $|\psi\rangle \in \mathcal{H}_m$, acting with Q_B, implies the equivalence with the old covariant quantization

$$\left(L_0^m - 1\right)|\psi\rangle = 0, \qquad \forall n > 0: \quad L_n^m |\psi\rangle = 0. \tag{8.86}$$

This means that ψ must be a weight 1 primary field of the matter CFT.

Remark 8.2 (Open String) The results of this section provide, in fact, the cohomology for the open string after taking $p_L = p$ (instead of $p_L = p/2$ for the closed string).

8.3.4 Cohomology for Holomorphic and Anti-holomorphic Sectors

It remains to generalize the computation of the cohomology when considering both the holomorphic and anti-holomorphic sectors.

In this case, the BRST operator is

$$Q_B = c_0 L_0 - b_0 M + \widehat{Q}_B + \bar{c}_0 \bar{L}_0 - \bar{b}_0 \bar{M} + \widehat{\bar{Q}}_B. \tag{8.87}$$

It is useful to rewrite this expression in terms of $L_0^\pm$, $b_0^\pm$ and $c_0^\pm$

$$Q_B = c_0^+ L_0^+ - b_0^+ M^+ + c_0^- L_0^- - b_0^- M^- + \widehat{Q}_B^+, \tag{8.88}$$

where

$$L_0^+ = \left(L_0^{\perp+} - \frac{m_\parallel^2 \ell^2}{2} - 2\right) + \widehat{L}_0^{\parallel+}, \qquad L_0^- = L_0^{\perp-} + \widehat{L}_0^{\parallel-} \tag{8.89}$$

and

$$M^\pm := \frac{1}{2}(M \pm \bar{M}). \tag{8.90}$$

Because of the relations $L_0^\pm = \{Q_B, b_0^\pm\}$, we find that states in the cohomology must be on-shell $L_0^+ = 0$ and must satisfy the level-matching condition $L_0^- = 0$[5]

$$L_0^+ |\psi\rangle = L_0^- |\psi\rangle = 0. \tag{8.91}$$

Again, it is possible to reduce the cohomology by imposing conditions on the zero-modes such that the above conditions are automatically satisfied (see

---

[5]In the current case, the propagator is less easily identified. We will come back on its definition later.

also Sect. 3.2.2). Imposing first the condition $b_0^- = 0$ defines the semi-relative cohomology. The relative cohomology is found by imposing $b_0^\pm = 0$ and in fact corresponds to the physical space (see [13, sec. 2.3] for more details). The rest of the derivation follows straightforwardly because the two sectors commute: we find that the cohomology is ghost-free and has no light-cone excitations

$$\widehat{L}_0^{\parallel \pm} = N^0 \pm \bar{N}^0 + N^1 \pm \bar{N}^1 + N^b \pm \bar{N}^b + N^c \pm \bar{N}^c = 0. \tag{8.92}$$

In general, it is simpler to work with a covariant expression and to impose the necessary conditions. Taking a state $|\psi\rangle \otimes |\downarrow\downarrow\rangle$ with $|\psi\rangle \in \mathcal{H}_m$, we find that ψ is a weight $(1, 1)$ primary field of the matter CFT

$$(L_0^m + \bar{L}_0^m - 2)|\psi\rangle = 0, \qquad (L_0^m - \bar{L}_0^m)|\psi\rangle = 0,$$
$$\forall n > 0: \quad L_n^m |\psi\rangle = \bar{L}_n^m |\psi\rangle = 0. \tag{8.93}$$

An important point is that the usual mass-shell condition $k^2 = -m^2$ is provided by the first condition only. This also shows that states in the cohomology naturally appear with $c\bar{c}$ insertion since

$$|\downarrow\downarrow\rangle = c(0)\bar{c}(0)|0\rangle = c_1 \bar{c}_1 |0\rangle. \tag{8.94}$$

This hints at rewriting of scattering amplitudes in terms of unintegrated states (3.29) only.

A state is said to be of level $(\ell, \bar{\ell})$ and denoted as $\psi_{\ell,\bar{\ell}}$ if it satisfies

$$\widehat{L}_0 |\psi_{\ell,\bar{\ell}}\rangle = \ell |\psi_{\ell,\bar{\ell}}\rangle, \qquad \widehat{\bar{L}}_0 |\psi_{\ell,\bar{\ell}}\rangle = \bar{\ell} |\psi_{\ell,\bar{\ell}}\rangle. \tag{8.95}$$

Example 8.1: Closed String Tachyon

As an example, let us construct the state $\psi_{0,0}$ with level zero for a spacetime with D non-compact dimensions. In this case, the transverse CFT contains $D - 2$ free scalars that combine with X^0 and X^1 into D scalars X^μ. The Fock space is built on the vacuum $|k\rangle$, and we define the mass such that on-shell condition reduces to the standard QFT expression

$$k^2 = -m^2, \qquad m^2 := \frac{2}{\ell^2}(N + \bar{N} - 2), \tag{8.96}$$

where N and $\bar{N}$ are the matter level operators. The state in the remaining transverse CFT (without the $D - 2$ scalars) is the SL(2, $\mathbb{C}$) vacuum with $L_0^\perp = \bar{L}_0^\perp = 0$ (this is the state with the lowest energy for a unitary CFT). In this case, the on-shell condition reads

$$m^2 \ell^2 = -4 < 0. \tag{8.97}$$

Since the mass is negative, this state is a tachyon. The vertex operator associated to the state reads

$$\mathcal{V}(k, z, \bar{z}) = c(z)\bar{c}(\bar{z})e^{ik \cdot X(z,\bar{z})}. \tag{8.98}$$

◄

8.4 Summary

In this chapter, we have described the BRST quantization from the CFT point of view. We have first considered only the holomorphic sector (equivalently, the open string). We proved that the cohomology does not contain negative-norm states, and we provided an explicit way to construct the states. Finally, we glued together both sectors and characterized the BRST cohomology of the closed string.

What is the next step? We could move to computations of on-shell string amplitudes, but this falls outside the scope of this book. We can also start to consider string field theory. Indeed, the BRST equation $Q_B |\psi\rangle = 0$ and the equivalence $|\psi\rangle \sim |\psi\rangle + Q_B |\Lambda\rangle$ completely characterize the states. In QFT, states are solutions of the linearized equations of motion; hence, the BRST equation can provide a starting point for building the action. This is the topic of Chap. 10.

8.5 Suggested Readings

- The general method to construct the absolute cohomology follows [5, 12]. Other works and reviews include [2, 4, 7–11].
- String states are discussed in [3, sec. 3.3, , 12, sec. 4.1].

References

1. M. Asano, M. Natsuume, The no-ghost theorem for string theory in curved backgrounds with a flat timelike direction. Nuclear Phys. B **588**(1–2), 453–470 (2000). https://doi.org/10.1016/S0550-3213(00)00495-8. arXiv: hep-th/0005002
2. A. Bilal, Remarks on the BRST-cohomology for $c_M > 1$ matter coupled to Liouville gravity. Phys. Lett. B **282**(3–4), 309–313 (1992). https://doi.org/10.1016/0370-2693(92)90644-J. arXiv: hep-th/9202035
3. R. Blumenhagen, D. Lüst, S. Theisen, *Basic Concepts of String Theory*. English. 2013 edition (Springer, Berlin, 2014)
4. P. Bouwknegt, J. McCarthy, K. Pilch, BRST analysis of physical states for 2D (super) gravity coupled to (super) conformal matter, in *New Symmetry Principles in Quantum Field Theory* (Springer, Berlin, 1992), pp. 413–422. https://doi.org/10.1007/978-1-4615-3472-3_17. arXiv: hep-th/9110031

5. P. Bouwknegt, J. McCarthy, K. Pilch, BRST analysis of physical states for 2D gravity coupled to $c \leq 1$ matter. Commun. Math. Phys. **145**(3), 541–560 (1992). https://doi.org/10.1007/BF02099397
6. T.Y. Chow, You could have invented spectral sequences. en. Notices AMS **53**(1), 5 (2006)
7. J. Distler, P. Nelson, New discrete states of strings near a black hole. Nuclear Phys. B **374**(1), 123–155 (1992). https://doi.org/10.1016/0550-3213(92)90479-U
8. K. Itoh, BRST quantization of Polyakov's two-dimensional gravity. Nuclear Phys. B **342**(2), 449–470 (1990). https://doi.org/10.1016/0550-3213(90)90198-M
9. K. Itoh, N. Ohta, Spectrum of two-dimensional (super)gravity. Progr. Theoret. Phys. Suppl. **110**, 97–116 (1992). https://doi.org/10.1143/PTPS.110.97. arXiv: hep-th/9201034
10. S. Mukhi, Extra States in C < 1 String Theory (1991). arXiv: hep-th/9111013
11. N. Ohta, Discrete States in Two-Dimensional Quantum Gravity (1992). arXiv: hep-th/9206012
12. J. Polchinski, *String Theory: Volume 1, An Introduction to the Bosonic String* (Cambridge University Press, Cambridge, 2005)
13. C.B. Thorn, String field theory. Phys. Rep. **175**(1–2), 1–101 (1989). https://doi.org/10.1016/0370-1573(89)90015-X

Part II

String Field Theory

String Field

9

Abstract

In this chapter, we introduce general concepts about the string field. The goal is to give an idea of which type of object it is and the different possibilities for describing it. We will see that the string field is a functional and, for this reason, it is more convenient to work with the associated ket field, which can itself be represented in momentum space. We focus on what to expect from a free field, taking inspiration from the worldsheet theory. The interpretation becomes more difficult when taking into account the interactions.

9.1   Field Functional

A string field, after quantization, is an operator that creates or destroys a string at a given time. Since a string is a 1-dimensional extended object, the string field Ψ must depend on the spatial positions of each point of the string denoted collectively as X^μ. Hence, the string field is a functional $\Psi[X^\mu]$. The fact that it is a functional rather than a function makes the construction of a field theory much more challenging: it asks for revisiting all concepts we know in point-particle QFT without any prior experience with a simple model.[1]

It is important that the dependence is only on the shape and not on the parametrization. However, it is simpler to first work with a specific parametrization $X(\sigma)$ and make sure that nothing depends on it at the end (equivalent to imposing the invariance under reparametrization of the worldsheet). This leads to work with a functional $\Psi[X(\sigma)]$ of fields on the worldsheet (at fixed time). To proceed, one

---

[1]The problem is not in working with the wordline formalism and writing a BRST field theory, but really to take into account the spatial extension of the objects. In fact, generalizing further to functionals of extended ($p > 1$)-dimensional objects—branes—shows that SFT is the simplest of such field theories.

© Springer Nature Switzerland AG 2021
H. Erbin, *String Field Theory*, Lecture Notes in Physics 980,
https://doi.org/10.1007/978-3-030-65321-7_9

should first determine the degrees of freedom of the string and then find the interactions. The simplest way to achieve the first step is to perform a second-quantization of the string wave functional: the string field is written as a linear combination of first-quantized states with spacetime wave functionals as coefficients.[2] This provides a free Hamiltonian; trying to add interactions perturbatively does not work well.

It is not possible to go very far with this approach, and one is lead to choose a specific gauge, breaking the manifest invariance under reparametrizations. The simplest is the light-cone gauge since one works only with the physical degrees of freedom of the string. While this approach is interesting to gain some intuitions and show that, in principle, it is possible to build a string field theory, it requires making various assumptions and ends up with problems (especially for superstrings).[3]

Since worldsheet reparametrization invariance is just a kind of gauge symmetry—maybe less familiar than the non-Abelian gauge symmetries in Yang–Mills, but still a gauge symmetry—one may surmise that it should be possible to gauge fix this symmetry and introduce a BRST symmetry in its place. This is the programme of the BRST (or covariant) string field theory in which the string field depends not only on the worldsheet (at fixed time) but also on the ghosts: $\Psi[X(\sigma), c(\sigma)]$. There is no dependence on the b ghost because the latter is the conjugate momentum of the c ghost: in the operator language, $b(\sigma) \sim \frac{\delta}{\delta c(\sigma)}$.

The BRST formalism has the major advantage to allow to move easily from $D = 26$ dimensions—described by X^μ scalars ($\mu = 0, \ldots, 25$)—to a (possibly curved) D-dimensional spacetime and a string with some internal structure—described by a more general CFT, in which D scalars X^μ represent the non-compact dimensions and the remaining system with central charge $26 - D$ describes the compactification and structure. It is sufficient to consider the string field as a general functional of all the worldsheet fields. For simplicity, we will continue to write X in the functional dependence, keeping the other matter fields implicit.

It is complicated to find an explicit expression for the string field as a functional of $X(\sigma)$ and $c(\sigma)$. In fact, the field written in this way is in the *position representation* and, as usual in quantum mechanics, one can choose to work with the ket representation ket Ψ:

$$\Psi[X(\sigma), c(\sigma)] := \langle X(\sigma), c(\sigma)|\Psi\rangle . \tag{9.1}$$

It is often more convenient to work with $|\Psi\rangle$ (which we will also denote simply as Ψ, not distinguishing between states and operators). This field ket will be used throughout the book to represent the string field.

---

[2]The description of the first-quantized states depends on the CFT used to describe the theory. This explains the lack of manifest background independence of SFT. Unfortunately, no better approach has been found until now.

[3]While this approach has been mostly abandoned, recent results show that it can still be used when defined with a proper regularization [1–5].

Writing a field theory in terms of $|\Psi\rangle$ may not be intuitive since in point-particle QFT, one is used to work with the position or momentum representation. In fact, there is a very simple way to recover a formulation in terms of spacetime point-particle fields, which can be used almost whenever there is a doubt about what is going on. Indeed, as is well-known from standard worldsheet string theory, the string states behave like a collection of particles. This is because the modes of the CFT fields (like α_n^μ) carry spacetime indices (Lorentz, group representation...) such that the states themselves carry indices. Indeed, these quantum numbers classify eigenstates of the operators L_0 and $\bar{L}_0$. On the other hand, positions and shapes are not eigenstates of any simple CFT operator.

9.2 Field Expansion

It follows that the second-quantized string field can be written as a linear combination of first-quantized off-shell states $|\phi_\alpha(k)\rangle = \mathcal{V}_\alpha(k; 0, 0)|0\rangle$ (which form a basis of the CFT Hilbert space $\mathcal{H}$):

$$|\Psi\rangle = \sum_\alpha \int \frac{d^D k}{(2\pi)^D} \, \psi_\alpha(k) \, |\phi_\alpha(k)\rangle, \qquad (9.2)$$

where k is the D-dimensional momentum of the string (conjugated to the position of the centre-of-mass) and α is a collection of discrete quantum numbers (Lorentz indices, group representation...). When inserting this expansion inside the action, we find that it reduces to a standard field theory with an infinite number of particles described by the spacetime fields $\psi_\alpha(k)$ (momentum representation). The fields can also be written in the position representation by Fourier transforming only the momentum k to the centre-of-mass x:

$$\psi_\alpha(x) = \int \frac{d^D k}{(2\pi)^D} \, e^{ik \cdot x} \psi_\alpha(k). \qquad (9.3)$$

However, we will see that it is often not convenient because the action is non-local in position space (including, for example, exponentials of derivatives).

The physical intuition is that the string is a non-local object in spacetime. It can be expressed in momentum space through a Fourier transformation: variables dual to non-compact (resp. compact) dimensions are continuous (discrete). As a consequence, the momentum is continuous since the centre-of-mass moves in the non-compact spacetime, while the string itself has a finite extension and the associated modes are discrete but still not bounded (and similarly for compact dimensions). This indicates that the spectrum is the collection of a set of continuous and discrete modes. Hence, the non-locality of the string (due to the spatial extension) is traded for an infinite number of modes that behave like standard particles. In this description, the non-locality arises (1) in the infinite number of fields, (2) in the coupling between the modes and (3) as a complicated momentum dependence of the action.

When we are not interested in the spacetime properties, we will write a generic basis of the Hilbert space $\mathcal{H}$ as $\{\phi_r\}$:

$$|\Psi\rangle = \sum_r \psi_r |\phi_r\rangle. \tag{9.4}$$

The sum over r includes discrete and continuous labels.

Example 9.1: Scalar Field

In order to illustrate the notations for a point-particle, consider a scalar field $\phi(x)$. It can be expanded in Fourier modes as

$$\phi(x) = \int \frac{d^D k}{(2\pi)^D} \phi(k) e^{ik\cdot x}. \tag{9.5}$$

The corresponding ket $|\phi\rangle$ is found by expanding on a basis $\{|k\rangle\}$:

$$|\phi\rangle = \int \frac{d^D k}{(2\pi)^D} \phi(k) |k\rangle, \qquad \phi(k) = \langle k|\phi\rangle. \tag{9.6}$$

Similarly, the position space field is defined from the basis $\{|x\rangle\}$ such that

$$\phi(x) = \langle x|\phi\rangle = \int \frac{d^D k}{(2\pi)^D} \langle x|k\rangle \langle k|\phi\rangle, \qquad \langle x|k\rangle = e^{ik\cdot x}. \tag{9.7}$$

◀

9.3   Summary

In this chapter, we introduced general ideas about what a string field is. We now need to write an action. In general, one proceeds in two steps:

1. build the kinetic term (free theory):
   (a) equations of motion → physical states
   (b) equivalence relation → gauge symmetry
2. add interactions and deform the gauge transformation.

We consider the first point in the next chapter, but we will have to introduce more machinery in order to discuss interactions.

9.4 Suggested Readings

- General discussions of the string field and the ideas of string field theory can be found in [8, sec. 4, 10].
- Light-cone SFT is reviewed in [6, chap. 6, 7, chap. 9, 9].

References

1. Y. Baba, N. Ishibashi, K. Murakami, Light-cone gauge string field theory in noncritical dimensions. J. High Energy Phys. **2009**(12), 010–010 (2009). https://doi.org/10.1088/1126-6708/2009/12/010. arXiv: 0909.4675
2. N. Ishibashi, Multiloop Amplitudes of Light-Cone Gauge String Field Theory for Type II Superstrings' (2018). arXiv: 1810.03801
3. N. Ishibashi, K. Murakami, Multiloop amplitudes of light-cone gauge bosonic string field theory in noncritical dimensions. J. High Energy Phys. **2013**(9), 53 (2013). https://doi.org/10.1007/JHEP09(2013)053. arXiv: 1307.6001
4. N. Ishibashi, K. Murakami, Worldsheet theory of light-cone gauge noncritical strings on higher genus Riemann surfaces. J. High Energy Phys. **2016**(6), 87 (2016). https://doi.org/10.1007/JHEP06(2016)087. arXiv: 1603.08337
5. N. Ishibashi, K. Murakami, Multiloop amplitudes of light-cone gauge super-string field theory: odd spin structure contributions. J. High Energy Phys. **2018**(3), 63 (2018). https://doi.org/10.1007/JHEP03(2018)063. arXiv: 1712.09049
6. M. Kaku, *Introduction to Superstrings and M-Theory* (Springer, Berlin, 1999)
7. M. Kaku, *Strings, Conformal Fields, and M-Theory* (Springer, Berlin, 1999)
8. J. Polchinski, *What Is String Theory?* (1994). arXiv: hep-th/9411028
9. C.B. Thorn, String field theory. Phys. Rep. **175**(1–2), 1–101 (1989). https://doi.org/10.1016/0370-1573(89)90015-X
10. B. Zwiebach, Closed String Field Theory: An Introduction (1993). arXiv: hep-th/9305026

Free BRST String Field Theory

10

Abstract

In this chapter, we construct the BRST (or covariant) free bosonic string field theories. It is useful to first ignore the interactions in order to introduce some general tools and structures in a simpler setting. Moreover, the free SFT is easily constructed and does not require as much input as the interactions. In this chapter, we discuss mostly the open string, keeping the closed string for the last section. We start by describing the classical theory: equations of motion, action, gauge invariance and gauge fixing. Then, we perform the path integral quantization and compute the action in terms of spacetime fields for the first two levels (tachyon and gauge field).

10.1 Classical Action for the Open String

While the rest of this book discusses the closed string only, the ideas of this chapter are simpler to introduce for the open string. For this reason, we make an exception and start this chapter with the open string. There are also a few subtleties which can be more easily explained once the general structure is understood. Everything needed for the open string for this chapter can be found in Chap. 8; in fact, describing the open string (at this level) is equivalent to considering only the holomorphic sector of the CFT and setting $p_L = p$ (instead of $p/2$). We consider a generic matter CFT in addition to the ghost system, and we denote as $\mathcal{H}$ the space of states. The open and closed string fields are denoted, respectively, by Φ and Ψ, such that it is clear which theory is studied.

An action can be either constructed from first principles or derived from the equations of motion. Since the fundamental structure of string field theory is not (really) known, one needs to rely on the second approach. But do we already know the (free) equations of motion for the string field? The answer is yes. But, before

© Springer Nature Switzerland AG 2021
H. Erbin, *String Field Theory*, Lecture Notes in Physics 980,
https://doi.org/10.1007/978-3-030-65321-7_10

showing how these can be found from the worldsheet formalism, we will study the case of the point-particle to fix ideas and notations.

10.1.1 Warm-Up: Point-Particle

The free (or linearized) equation of motion for a scalar particle reads

$$(-\Delta + m^2)\phi(x) = 0. \tag{10.1}$$

Solutions to this equation provide one-particle state of the free theory: a convenient basis is $\{e^{ikx}\}$, where each state satisfies the on-shell condition

$$k^2 = -m^2. \tag{10.2}$$

The field $\phi(x)$ is decomposed on the basis as

$$\phi(x) = \int dk\, \phi(k)e^{ikx}, \tag{10.3}$$

where $\phi(k)$ are the coefficients of the expansion. Since the field is off-shell, the condition $k^2 = -m^2$ is not imposed. Following Chap. 9, the field can also be represented as a ket:

$$\phi(x) = \langle x|\phi\rangle, \qquad \phi(k) = \langle k|\phi\rangle, \tag{10.4}$$

or, conversely,

$$|\phi\rangle = \int dx\, \phi(x)\,|x\rangle = \int dk\, \phi(k)\,|k\rangle. \tag{10.5}$$

Writing the kinetic operator as a kernel:

$$K(x, x') := \langle x|K\,|x'\rangle = \delta(x - x')\left(-\Delta_x + m^2\right), \tag{10.6}$$

the equations of motion read

$$\int dx'\, K(x, x')\phi(x') = 0 \quad \Longleftrightarrow \quad K\,|\phi\rangle = 0. \tag{10.7}$$

An action can easily be found from the equation of motion by multiplying with $\phi(x)$ and integrating:

$$S = \frac{1}{2}\int dx\, \phi(x)\left(-\Delta + m^2\right)\phi(x) = \frac{1}{2}\int dx dx'\, \phi(x)K(x, x')\phi(x'). \tag{10.8}$$

It is straightforward to write the action in terms of the ket:

$$S = \frac{1}{2} \langle \phi | K | \phi \rangle. \tag{10.9}$$

There is one hidden assumption in the previous lines: the definition of a scalar product. A natural inner product is provided in the usual quantum mechanics by associating a bra to a ket. Similarly, integration provides another definition of the inner product when working with functions. We will find that the definition of the inner product requires more care in closed SFT. To summarize, and to write the kinetic term of the action, one needs the linearized equation of motion and an appropriate inner product on the space of states.

10.1.2 Open String Action

The worldsheet equation that yields precisely all the string physical states $|\psi\rangle$ is the BRST condition:

$$Q_B |\psi\rangle = 0. \tag{10.10}$$

Considering the open string field Φ to be a linear combination of all possible one-string states $|\psi\rangle$

$$\Phi \in \mathcal{H}, \tag{10.11}$$

the equation of motion is

$$Q_B |\Phi\rangle = 0. \tag{10.12}$$

Moving away from the physical state condition, the string field Φ is off-shell and is expanded on a general basis $\{\phi_r\}$ of $\mathcal{H}$. This presents a first difficulty because the worldsheet approach—and the description of amplitudes—looks ill-defined for off-shell states: extending the usual formalism will be the topic of Chap. 11. However, this is not necessary for the free theory, and we can directly proceed.

Next, we need to find an inner product $\langle \cdot, \cdot \rangle$ on the Hilbert space $\mathcal{H}$. A natural candidate is the BPZ inner product since it is not degenerate

$$\langle A, B \rangle := \langle A | B \rangle, \tag{10.13}$$

where $\langle A | = |A\rangle^t$ is the BPZ conjugate (6.98) of $|A\rangle$, using I^-. This leads to the action:

$$S = \frac{1}{2} \langle \Phi, Q_B \Phi \rangle = \frac{1}{2} \langle \Phi | Q_B | \Phi \rangle. \tag{10.14}$$

Due to the definition of the BPZ product, the action is equivalent to a 2-point correlation function on the disk.

The inner product satisfies the following identities:

$$\langle A, B \rangle = (-1)^{|A||B|} \langle B, A \rangle, \qquad \langle Q_B A, B \rangle = -(-1)^{|A|} \langle A, Q_B B \rangle, \qquad (10.15)$$

where $|A|$ denotes the Grassmann parity of the operator A.

A first consistency check is to verify that the ghost number of the string can be defined such that the action is not vanishing. Indeed, the ghost number anomaly on the disk implies that the total ghost number must be $N_{gh} = 3$. Since physical states have $N_{gh} = 1$, it is reasonable to take the string field to satisfy the same condition, even off-shell:

$$N_{gh}(\Phi) = 1. \qquad (10.16)$$

This condition means that there is no ghost at the classical level beyond the one of the energy vacuum $| \downarrow \rangle$, which has $N_{gh} = 1$. Moreover, the BRST charge has $N_{gh}(Q_B) = 1$, such that the action has ghost number 3.

One needs to find the Grassmann parity of the string field. Using the properties of the BPZ inner product, the string field should be Grassmann odd

$$|\Phi| = 1 \qquad (10.17)$$

for the action to be even. This is in agreement with the fact that the string field has ghost number 1 and that the ghosts are Grassmann odd. One must impose a reality condition on the string field (a complex field would behave like two real fields and have too many states). The appropriate reality condition identifies the Euclidean and BPZ conjugates:

$$|\Phi\rangle^{\ddagger} = |\Phi\rangle^{t}. \qquad (10.18)$$

That this relation is correct will be checked a posteriori for the tachyon field in Sect. 10.4.

Computation: Equation (10.17)

$$\langle \Phi, Q_B \Phi \rangle = (-1)^{|\Phi|(|Q_B \Phi|)} \langle Q_B \Phi, \Phi \rangle = (-1)^{|\Phi|(1+|\Phi|)} \langle Q_B \Phi, \Phi \rangle$$

$$= \langle Q_B \Phi, \Phi \rangle = -(-1)^{|\Phi|} \langle \Phi, Q_B \Phi \rangle,$$

where we used both properties in (10.15), together with the fact that $|\Phi|(1 + |\Phi|)$ is necessarily even. In order for the bracket to be non-zero, one must have $|\Phi| = 1$.

Since the Hilbert space splits as $\mathcal{H} = \mathcal{H}_0 \oplus c_0 \mathcal{H}_0$ with $\mathcal{H}_0 = \mathcal{H} \cap \ker b_0$, see (8.31), it is natural to split the field as (this is discussed further in Sect. 10.2)

$$|\Phi\rangle = |\Phi_\downarrow\rangle + c_0 \, |\widetilde{\Phi}_\downarrow\rangle, \tag{10.19}$$

where

$$\Phi_\downarrow, \widetilde{\Phi}_\downarrow \in \mathcal{H}_0 \quad \Longrightarrow \quad b_0 \, |\Phi_\downarrow\rangle = b_0 \, |\widetilde{\Phi}_\downarrow\rangle = 0. \tag{10.20}$$

The ghost number of each component is

$$N_{\mathrm{gh}}(\Phi_\downarrow) = 1, \qquad N_{\mathrm{gh}}(\widetilde{\Phi}_\downarrow) = 0. \tag{10.21}$$

Remembering the decomposition (8.14a) of the BRST operator

$$Q_B = c_0 L_0 - b_0 M + \widehat{Q}_B, \tag{10.22}$$

inserting the decomposition (10.19) in the action (10.14) gives

$$S = \frac{1}{2} \langle \Phi_\downarrow | c_0 L_0 \, |\Phi_\downarrow\rangle + \frac{1}{2} \langle \widetilde{\Phi}_\downarrow | c_0 M \, |\widetilde{\Phi}_\downarrow\rangle + \langle \widetilde{\Phi}_\downarrow | c_0 \widehat{Q}_B \, |\Phi_\downarrow\rangle. \tag{10.23}$$

The equations of motion are obtained by varying the different fields:

$$0 = -M \, |\widetilde{\Phi}_\downarrow\rangle + \widehat{Q}_B \, |\Phi_\downarrow\rangle, \qquad 0 = c_0 L_0 \, |\Phi_\downarrow\rangle + c_0 \widehat{Q}_B \, |\widetilde{\Phi}_\downarrow\rangle. \tag{10.24}$$

Computation: Equation (10.23)
Let us introduce the projector $\Pi_s = b_0 c_0$ on the space $\mathcal{H}_0 = \mathcal{H} \cap \ker b_0$ and the orthogonal projector $\bar{\Pi}_s = c_0 b_0$ such that

$$|\Phi\rangle = |\Phi_\downarrow\rangle + |\Phi_\uparrow\rangle, \qquad |\Phi_\downarrow\rangle = \Pi_s \, |\Phi\rangle, \qquad |\Phi_\uparrow\rangle = \bar{\Pi}_s \, |\Phi\rangle. \tag{10.25}$$

We then have

$$\Pi_s Q_B \, |\Phi\rangle = -b_0 M \, |\Phi_\uparrow\rangle + \widehat{Q}_B \, |\Phi_\downarrow\rangle, \qquad \bar{\Pi}_s Q_B \, |\Phi\rangle = c_0 L_0 \, |\Phi_\downarrow\rangle + \widehat{Q}_B \, |\Phi_\uparrow\rangle, \tag{10.26}$$

using

$$[\Pi_s, \widehat{Q}_B] = [\Pi_s, M] = [\Pi_s, L_0] = 0. \tag{10.27}$$

Then, we need the fact that $\Pi_s^\dagger = \bar\Pi_s$ to compute the action:

$$
S = \frac{1}{2} \langle \Phi, Q_B \Phi \rangle
$$

$$
= \frac{1}{2} \langle \Pi_s \Phi + \bar\Pi_s \Phi, Q_B \Phi \rangle
$$

$$
= \frac{1}{2} \langle \Pi_s \Phi, \bar\Pi_s Q_B \Phi \rangle + \frac{1}{2} \langle \bar\Pi_s \Phi, \Pi_s Q_B \Phi \rangle
$$

$$
= \frac{1}{2} \langle \Phi_\downarrow, c_0 L_0 \Phi_\downarrow + \widehat{Q}_B \Phi_\uparrow \rangle + \frac{1}{2} \langle \Phi_\uparrow, -b_0 M \Phi_\uparrow + \widehat{Q}_B \Phi_\downarrow \rangle
$$

$$
= \frac{1}{2} \langle \Phi_\downarrow, c_0 L_0 \Phi_\downarrow \rangle + \frac{1}{2} \langle \Phi_\downarrow, \widehat{Q}_B \Phi_\uparrow \rangle - \frac{1}{2} \langle \Phi_\uparrow, b_0 M \Phi_\uparrow \rangle + \frac{1}{2} \langle \Phi_\uparrow, \widehat{Q}_B \Phi_\downarrow \rangle.
$$

The result follows by setting $|\Phi_\uparrow\rangle = c_0 |\widetilde\Phi\rangle$, using (10.15) and that the BPZ conjugate of c_0 is $-c_0$.

10.1.3 Gauge Invariance

In writing the action, only the condition that the states are BRST closed has been used. One needs to interpret the condition that the states are not BRST exact, or phrased differently that the two states differing by a BRST exact state are equivalent:

$$
|\phi\rangle \sim |\psi\rangle + Q_B |\lambda\rangle. \tag{10.28}
$$

Uplifting this condition to the string field, the most direct interpretation is that it corresponds to a gauge invariance:

$$
|\Phi\rangle \longrightarrow |\Phi'\rangle = |\Phi\rangle + \delta_\Lambda |\Phi\rangle, \qquad \delta_\Lambda |\Phi\rangle = Q_B |\Lambda\rangle \qquad N_{\mathrm{gh}}(\Lambda) = 0. \tag{10.29}
$$

In order for the ghost numbers to match, the gauge parameter has vanishing ghost number. The action (10.14) is obviously invariant since the BRST charge is nilpotent.

10.1.4 Siegel Gauge

In writing the action (10.14), the condition $b_0 |\psi\rangle = 0$ has *not* been imposed on the string field. In Sect. 3.2.2, this condition was found by restricting the BRST cohomology, projecting out states built on the ghost vacuum $|\uparrow\rangle$, as required by the behaviour of the on-shell scattering amplitudes. In Chap. 8, we obtained it by finding that the absolute cohomology contains twice more states as necessary. This was also understood as a way to work with a specific representative of the BRST cohomology. Since the field is off-shell and the action computes off-shell Green functions, these arguments cannot be used, which explains why we did not use this condition earlier.

On the other hand, the condition

$$b_0 |\Phi\rangle = 0 \tag{10.30}$$

can be interpreted as a gauge fixing condition, called *Siegel gauge*. It can be reached from any field through a gauge transformation (10.29) with

$$|\Lambda\rangle = -\Delta |\Phi\rangle, \qquad \Delta = \frac{b_0}{L_0}, \tag{10.31}$$

where Δ was defined in (8.39) and will be identified with the propagator. Note that $b_0 = 0$ does not imply $L_0 = 0$ since the string field is not BRST closed.

This gauge choice is well-defined and completely fixes the gauge symmetry off-shell, meaning that no solution of the equation of motion is pure gauge after the gauge fixing. This is shown as follows: assume that $|\psi\rangle = Q_B |\chi\rangle$ is an off-shell pure gauge state with $L_0 \neq 0$, and then, because it is also annihilated by b_0, one finds

$$0 = \{Q_B, b_0\} |\psi\rangle = L_0 |\psi\rangle, \tag{10.32}$$

which yields a contradiction.

The gauge fixing condition breaks down for $L_0 = 0$, but this does not pose any problem when working with Feynman diagrams since they are not physical by themselves (nor are the off-shell and on-shell Green functions). Only the sum giving the scattering amplitudes (truncated on-shell Green functions) is physical; in this case, the singularity $L_0 = 0$ corresponds to the on-shell condition, and it is well-known how such infrared divergences for intermediate states are removed (through the LSZ prescription, mass renormalization and tadpole cancellation).

Computation: Equation (10.31)
Performing a gauge transformation gives

$$b_0 |\Phi'\rangle = b_0 |\Phi\rangle + b_0 Q_B |\Lambda\rangle = 0. \tag{10.33}$$

Then, one writes

$$b_0 |\Phi\rangle = b_0 [Q_B, \Delta] |\Phi\rangle = b_0 Q_B \Delta |\Phi\rangle, \tag{10.34}$$

using the relation (8.37), the expression (8.39) and the fact that $b_0^2 = 0$. Plugging this back in the first equation gives

$$b_0 Q_B (\Delta |\Phi\rangle + |\Lambda\rangle) = 0. \tag{10.35}$$

The factor of b_0 can be removed by multiplying with c_0, and the parentheses should vanish (since it is not identically closed), which means that (10.31) holds up to a BRST exact state.

Example 10.1: Gauge Fixing and Singularity

In Maxwell's theory, the gauge transformation

$$A'_\mu = A_\mu + \partial_\mu \lambda \tag{10.36}$$

is used to impose the Lorentz condition

$$\partial^\mu A'_\mu = 0 \implies \Delta \lambda = -\partial^\mu A_\mu. \tag{10.37}$$

In momentum space, the parameter reads

$$\lambda = -\frac{k^\mu}{k^2} A_\mu. \tag{10.38}$$

It is singular when k is on-shell, $k^2 = 0$. However, this does not prevent from computing Feynman diagrams. ◄

To understand the effect of the gauge fixing on the string field components, decompose the field as (10.19) $|\Phi\rangle = |\Phi_\downarrow\rangle + c_0 |\widetilde{\Phi}_\downarrow\rangle$. Then, imposing the condition (10.30) yields

$$|\widetilde{\Phi}_\downarrow\rangle = 0 \implies |\Phi\rangle = |\Phi_\downarrow\rangle. \tag{10.39}$$

This has the expected effect of dividing by two the number of states and shows that they are not physical.

Plugging this condition in the action (10.23) leads to gauge fixed action

$$S = \frac{1}{2} \langle\Phi|c_0 L_0 |\Phi\rangle, \tag{10.40}$$

for which the equation of motion is

$$L_0 |\Phi\rangle = 0. \tag{10.41}$$

But, note that this equation contains much less information than the original (10.12): as $|\widetilde{\Phi}_\downarrow\rangle$ is truncated from (10.40), a part of the equations of motion is lost. The missing equation can be found by setting $|\widetilde{\Phi}\rangle = 0$ in (10.24) and must be imposed on top of the action:

$$\widehat{Q}_B |\Phi\rangle = 0. \tag{10.42}$$

It is called *out-of-Siegel gauge constraint* and is equivalent to the Gauss constraint in electromagnetism: the equations of motion for pure gauge states contain also the physical fields, and thus, when one fixes a gauge, these relations are lost and must be imposed on the side of the action. This procedure mimics what happens in the old covariant theory, where the Virasoro constraints are imposed after choosing the flat gauge (if Φ contains no ghost on top of $|\downarrow\rangle$, and then $\widehat{Q}_B = 0$ implies $L_n = 0$, see Sect. 8.3.3). Moreover, the states that do not satisfy the condition $b_0 = 0$ do not propagate: this restricts the external states to be considered in amplitudes.

Remark 10.1 Another way to derive (10.40) is to insert $\{b_0, c_0\} = 1$ in the action:

$$
\begin{aligned}
S &= \frac{1}{2} \langle\Phi| Q_B \{c_0, b_0\} |\Phi\rangle = \frac{1}{2} \langle\Phi| Q_B b_0 c_0 |\Phi\rangle \\
&= \frac{1}{2} \langle\Phi| \{b_0, Q_B\} c_0 |\Phi\rangle - \frac{1}{2} \langle\Phi| b_0 Q_B c_0 |\Phi\rangle \\
&= \frac{1}{2} \langle\Phi| c_0 L_0 |\Phi\rangle .
\end{aligned}
$$

The drawback of this computation is that it does not show directly how the constraints (10.42) arise.

Remark 10.2 (Generalized Gauge Fixing) It is possible to generalize the Siegel gauge, in the same way that the Feynman gauge generalizes the Lorentz gauge. This has been studied in [1,2].

In this section, we have motivated different properties and adopted some normalizations. The simplest way to check that they are consistent is to derive the action in terms of the spacetime fields and check that it has the expected properties from standard QFT. This will be the topic of Sect. 10.4.

10.2 Open String Field Expansion, Parity and Ghost Number

A basis for the off-shell Hilbert space $\mathcal{H}$ is denoted by $\{\phi_r\}$, where the ghost numbers and parity of the states are written as

$$n_r := N_{\mathrm{gh}}(\phi_r), \qquad |\phi_r| = n_r \mod 2. \tag{10.43}$$

The corresponding basis of dual (or conjugate) states $\{\phi_r^c\}$ is defined by (6.145):

$$\langle \phi_r^c | \phi_s \rangle = \delta_{rs}. \tag{10.44}$$

The basis states can be decomposed according to the ghost zero-modes

$$|\phi_r\rangle = |\phi_{\downarrow,r}\rangle + |\phi_{\uparrow,r}\rangle, \qquad b_0 \, |\phi_{\downarrow,r}\rangle = c_0 \, |\phi_{\uparrow,r}\rangle = 0. \tag{10.45}$$

Finally, each state $\psi_\uparrow \in c_0 \mathcal{H}$ can be associated to a state $\widetilde{\psi}$:

$$|\psi_\uparrow\rangle = c_0 \, |\widetilde{\psi}_\downarrow\rangle, \qquad b_0 \, |\widetilde{\psi}_\downarrow\rangle = 0, \qquad N_{\text{gh}}(\psi_\uparrow) = N_{\text{gh}}\big(\widetilde{\psi}_\downarrow\big) + 1. \tag{10.46}$$

More details can be found in Sect. 11.2.

Any field Φ can be expanded as

$$|\Phi\rangle = \sum_r \psi_r \, |\phi_r\rangle, \tag{10.47}$$

where the ψ_r are spacetime fields (remembering that r denotes collectively the continuous and discrete quantum numbers).[1]

Obviously, the coefficients do not carry a ghost number since they are not worldsheet operators. However, they can be Grassmann even or odd such that each term of the sum has the same parity, so that the field has a definite parity:

$$\forall r: \qquad |\Phi| = |\psi_r| \, |\phi_r|. \tag{10.48}$$

If the field is Grassmann odd (resp. even), then the coefficients ψ_r and the basis states must have opposite (resp. identical) parities, such that $|\Phi| = 1$.

Since the parity results from worldsheet ghosts and there would be Grassmann odd states even in a purely bosonic theory, it suggests that the parity of the coefficients ψ_r is also related to a *spacetime ghost number* G defined as

$$G(\psi_r) = 1 - n_r. \tag{10.49}$$

The normalization is chosen such that the component of a classical string field ($N_{\text{gh}} = 1$) is classical spacetime fields with $G = 0$ (no ghost). We will see later that this definition makes sense.

---

[1] The notation is slightly ambiguous: from (10.45), it looks like both components of ϕ_r have the same coefficient ψ_r. But, in fact, one sums over all linearly independent states: in terms of the components of ϕ_r, different basis can be considered; for example, $\{\phi_{\downarrow,r}, \phi_{\uparrow,r}\}$, or $\{\phi_{\downarrow,r} \pm \phi_{\uparrow,r}\}$. A more precise expression can be found in (10.54) and (10.56).

A quantum string field Φ generally contains components Φ_n of all worldsheet ghost numbers n:

$$\Phi = \sum_{n \in \mathbb{Z}} \Phi_n, \qquad N_{\mathrm{gh}}(\Phi_n) = n. \tag{10.50}$$

The projections on the positive and negative (cylinder) ghost numbers are denoted by $\Phi_\pm$:

$$\Phi = \Phi_+ + \Phi_-, \qquad \Phi_+ = \sum_{n>1} \Phi_n, \qquad \Phi_- = \sum_{n \leq 1} \Phi_n. \tag{10.51}$$

The shift in the indices is explained by the relation (B.56) between the cylinder and plane ghost numbers.

For a field Φ_n of fixed ghost number, coefficients of the expansion vanish whenever the ghost number of the basis state does not match the one of the field:

$$\forall n_r \neq n : \quad \psi_r = 0. \tag{10.52}$$

Another possibility to define the field Φ_n is to insert a delta function:

$$|\Phi_n\rangle = \delta(N_{\mathrm{gh}} - n)|\Psi\rangle = \sum_r \delta(n_r - n)\,\psi_r\,|\phi_r\rangle. \tag{10.53}$$

According to (10.45), a string field Φ can also be separated in terms of the ghost zero-modes:

$$|\Phi\rangle = |\Phi_\downarrow\rangle + |\Phi_\uparrow\rangle = |\Phi_\downarrow\rangle + c_0\,|\widetilde{\Phi}_\downarrow\rangle, \tag{10.54a}$$

$$|\Phi_\uparrow\rangle = c_0\,|\widetilde{\Phi}_\downarrow\rangle, \qquad |\widetilde{\Phi}_\downarrow\rangle = b_0\,|\Phi_\uparrow\rangle, \tag{10.54b}$$

where the components satisfy the constraints

$$b_0\,|\Phi_\downarrow\rangle = 0, \qquad c_0\,|\Phi_\uparrow\rangle = 0, \qquad b_0\,|\widetilde{\Phi}_\downarrow\rangle = 0. \tag{10.55}$$

The fields $|\Phi_\downarrow\rangle$ and $|\Phi_\uparrow\rangle$ (or $|\widetilde{\Phi}_\downarrow\rangle$) are called the down and top components, and they can be expanded as

$$|\Phi_\downarrow\rangle = \sum_r \psi_{\downarrow,r}\,|\phi_{\downarrow,r}\rangle, \qquad |\Phi_\uparrow\rangle = \sum_r \psi_{\uparrow,r}\,|\phi_{\uparrow,r}\rangle. \tag{10.56}$$

10.3   Path Integral Quantization

The string field theory can be quantized with a path integral:

$$Z = \int d\Phi_{cl} \, e^{-S[\Phi_{cl}]} = \int d\Phi_{cl} \, e^{-\frac{1}{2}\langle\Phi_{cl}|Q_B|\Phi_{cl}\rangle}. \tag{10.57}$$

An index has been added to the field to emphasize that it is the classical field (no spacetime ghosts). The simplest way to define the measure is to use the expansion (9.4) such that

$$Z = \int \prod_s d\psi_s \, e^{-S[\{\psi_r\}]}. \tag{10.58}$$

10.3.1   Tentative Faddeev–Popov Gauge Fixing

The action can be gauge fixed using the Faddeev–Popov formalism. The gauge fixing condition is

$$F(\Phi_{cl}) := b_0 \, |\Phi_{cl}\rangle = 0. \tag{10.59}$$

Its variation under a gauge transformation (10.29) reads

$$\delta F = b_0 Q_B \, |\Lambda_{cl}\rangle, \tag{10.60}$$

which implies that the Faddeev–Popov determinant is

$$\det \frac{\delta F}{\delta \Lambda_{cl}} = \det b_0 Q_B. \tag{10.61}$$

This determinant is rewritten as a path integral by introducing a ghost C and an anti-ghost B' string fields (the prime on B' will become clear below):

$$\det b_0 Q_B = \int dB' dC \, e^{-S_{FP}}, \qquad S_{FP} = -\langle B'|b_0 Q_B |C\rangle. \tag{10.62}$$

The ghost numbers are attributed by selecting the same ghost number for the C ghost and for the gauge parameter and then requiring that the Faddeev–Popov action is non-vanishing:

$$N_{gh}(B') = 3, \qquad N_{gh}(C) = 0. \tag{10.63}$$

The ghosts can be expanded as

$$|B'\rangle = \delta(N_{\text{gh}} - 3) \sum_r b'_r \, |\phi_r\rangle, \qquad |C\rangle = \delta(N_{\text{gh}}) \sum_r c_r \, |\phi_r\rangle, \tag{10.64}$$

where the coefficients b_r and c_r are Grassmann odd in order for the determinant formula to make sense:

$$|b_r| = |c_r| = 1. \tag{10.65}$$

Then, since the basis states appearing in B' and C are, respectively, odd and even, this implies

$$|B'| = 0, \qquad |C| = 1. \tag{10.66}$$

However, there is a redundancy in the gauge fixing because the Faddeev–Popov action is itself invariant under two independent transformations:

$$\delta \, |C\rangle = Q_B \, |\Lambda_{-1}\rangle, \qquad N_{\text{gh}}(\Lambda_{-1}) = -1, \tag{10.67a}$$

$$\delta \, |B'\rangle = b_0 \, |\Lambda'\rangle, \qquad N_{\text{gh}}(\Lambda') = 4. \tag{10.67b}$$

This residual invariance arises because not all $|\Lambda_{\text{cl}}\rangle$ generate a gauge transformation. Indeed, if

$$|\Lambda\rangle = |\Lambda_0\rangle + Q_B \, |\Lambda_{-1}\rangle, \tag{10.68}$$

the field transforms as

$$|\Phi'_{\text{cl}}\rangle \longrightarrow |\Phi_{\text{cl}}\rangle + Q_B \, |\Lambda_0\rangle, \tag{10.69}$$

and there is no trace left of $|\Lambda_{-1}\rangle$, so it should not be counted.

The second invariance (10.67b) is not problematic because b_0 is an algebraic operator (the Faddeev–Popov action associated to the determinant has no dynamics). The decompositions of the gauge parameter Λ' and the B' field into components (10.54) read

$$|B'\rangle = |B'_\downarrow\rangle + c_0 \, |B\rangle, \qquad |B\rangle := |\widetilde{B}'_\downarrow\rangle, \tag{10.70a}$$

$$|\Lambda'\rangle = |\Lambda'_\downarrow\rangle + c_0 \, |\widetilde{\Lambda}'_\downarrow\rangle. \tag{10.70b}$$

The gauge transformations act on the components as

$$\delta \, |B'_\downarrow\rangle = |\widetilde{\Lambda}'_\downarrow\rangle, \qquad \delta \, |B\rangle = 0. \tag{10.71}$$

This shows that B is gauge invariant and $B'_\downarrow$ can be completely removed by the gauge transformation. This makes sense because $B'_\downarrow$ does not appear in the action (10.62). The gauge transformation (10.67b) can be used to fix the gauge:

$$|F'\rangle = c_0 |B'\rangle = 0 \quad \Longrightarrow \quad |B'_\downarrow\rangle = 0. \tag{10.72}$$

This fixes completely the gauge invariance since the field B is restricted to satisfy $b_0 |B\rangle = 0$, and the component form (10.71) of the gauge transformation shows that no transformation is allowed. Moreover, there is no need to introduce a Faddeev–Popov determinant for this gauge fixing because the corresponding ghosts would not couple to the other fields (and this would continue to hold even in the presence of interactions, see Remark 10.4). Indeed, from the absence of derivatives in the gauge transformation, one finds that the determinant is constant, and thus a ghost representation is not necessary:

$$\det \frac{\delta F'}{\delta \Lambda'} = \det c_0 b_0 = \det c_0 \det b_0 = \frac{1}{2} \det\{b_0, c_0\} = \frac{1}{2}. \tag{10.73}$$

Then, redefining the measure, the partition function and action reduce to

$$\Delta_{\text{FP}} = \int dB \, dC \, e^{-S_{\text{FP}}[B,C]}, \qquad S_{\text{FP}} = \langle B|Q_B|C\rangle. \tag{10.74}$$

Note that the field B satisfies

$$b_0 |B\rangle = 0, \qquad N_{\text{gh}}(B) = 2, \qquad |B| = 1. \tag{10.75}$$

Since both fields are Grassmann odd, the action can be rewritten in a symmetric way:

$$S_{\text{FP}} = \frac{1}{2}\Big(\langle B|Q_B|C\rangle + \langle C|Q_B|B\rangle \Big). \tag{10.76}$$

Remark 10.3 (Ghost and Anti-ghost Definitions) The definition of the anti-ghost B and ghost C is appropriate because the worldsheet and spacetime ghost numbers are related by a minus sign (and a shift of one unit). In the BV formalism, we will see that the fields contain the matter and ghost fields, while the antifields contain the anti-ghosts. These two sets are, respectively, defined with $N_{\text{gh}} \leq 1$ and $N_{\text{gh}} > 1$.

The constraint $b_0 |B\rangle = 0$ can be lifted by adding a top component:

$$|B\rangle = |B_\downarrow\rangle + c_0 |\widetilde{B}_\downarrow\rangle \tag{10.77}$$

together with the gauge invariance

$$\delta \, |B\rangle = Q_B \, |\Lambda_1\rangle. \tag{10.78}$$

Note the difference with (10.70): while $B = \widetilde{B}'_\downarrow$ was the top component of the B' field, here, it is defined to be the down component, such that $|B_\downarrow\rangle = |\widetilde{B}'_\downarrow\rangle$. However, for the moment, we keep B to satisfy $b_0 \, |B\rangle = 0$.

Remark 10.4 (Decoupling of the Ghosts) Since the theory is free, the Faddeev–Popov action (10.74) could be ignored and absorbed in the normalization because it does not couple to the field. On the other hand, when interactions are included, the gauge transformation is modified and the ghosts couple to the matter fields. But this is true only for the C transformation (10.67a), not for (10.67b). Then it means that ghosts introduced for gauge fixing (10.67b) will never couple to the matter and other ghosts.

The invariance (10.67a) is a gauge invariance for C and must be treated in the same way as (10.29). Then, following the Faddeev–Popov procedure, one is lead to introduce new ghosts for the ghosts. But, the same structure appears again. This leads to a residual gauge invariance, which has the same form. This process continues recursively, and one finds an infinite tower of ghosts.

10.3.2 Tower of Ghosts

In order to simplify the notations, all the fields are denoted by Φ_n, where n gives the ghost number:

- $\Phi_1 := \Phi_{cl}$ is the original physical field;
- $\Phi_0 := C$ and, more generally, Φ_n with $n < 1$ are ghosts;
- $\Phi_2 := B$ and, more generally, Φ_n with $n > 1$ are anti-ghosts.

The recipe is that each pair of ghost fields (Φ_{n+2}, Φ_{-n}) is associated to a gauge parameter Λ_{-n-1} with $n \geq 0$. It is then natural to gather all the fields in a single field

$$|\Phi\rangle = \sum_n |\Phi_n\rangle \tag{10.79}$$

satisfying the gauge fixing constraint:

$$b_0 \, |\Phi\rangle = 0 \quad \Longrightarrow \quad b_0 \, |\Phi_n\rangle = 0. \tag{10.80}$$

For $n \leq 1$, these constraints are gauge fixing conditions for the invariance $\delta \, |\Phi_n\rangle = Q_B \Lambda_n$. For $n > 1$, they arise by considering only the top component of the B field.

Finally, the gauge fixing condition can be incorporated inside the action by using a Lagrange multiplier β, which is an auxiliary string field containing also components of all ghost numbers:

$$|\beta\rangle = \sum_{n\in\mathbb{Z}} |\beta_n\rangle. \tag{10.81}$$

The path integral then reads

$$Z = \int d\Phi d\beta \, e^{-S[\Phi,\beta]}, \tag{10.82}$$

where

$$S[\Phi,\beta] = \frac{1}{2}\langle\Phi|Q_B|\Phi\rangle + \langle\beta|b_0|\Phi\rangle \tag{10.83a}$$

$$= \sum_{n\in\mathbb{Z}}\left(\frac{1}{2}\langle\Phi_{2-n}|Q_B|\Phi_n\rangle + \langle\beta_{4-n}|b_0|\Phi_n\rangle\right). \tag{10.83b}$$

The first term of the action has the same form as the classical action (10.14) but now includes fields at every ghost number. The complete BV analysis is relegated to the interacting theory.

Removing the auxiliary field $\beta = 0$, one finds that the action is invariant under the extended gauge transformation

$$\delta|\Phi\rangle = Q_B|\Lambda\rangle, \tag{10.84}$$

where the gauge parameter has also components of all ghost numbers:

$$|\Lambda\rangle = \sum_{n\in\mathbb{Z}} |\Lambda_n\rangle. \tag{10.85}$$

10.4 Spacetime Action

In order to make the string field action more concrete, and as emphasized in Chap. 9, it is useful to expand the string field in spacetime fields and write the action for the lowest modes. This also helps to check that the normalization chosen until here correctly reproduces the standard QFT normalizations. For simplicity, we focus on the open bosonic string in $D = 26$.

We build the string field from the vacuum $|k, \downarrow\rangle$ (Chap. 8) by acting with the ghost positive-frequency modes b_{-n} and c_{-n} and the zero-mode c_0 and scalar oscillators $i\alpha^\mu_{-n}$.

Up to level $\ell = 1$, the classical open string field can be expanded as

$$|\Phi\rangle = \frac{1}{\sqrt{\alpha'}} \int \frac{\mathrm{d}^D k}{(2\pi)^D} \left(T(k) + A_\mu(k)\alpha^\mu_{-1} + i\sqrt{\frac{\alpha'}{2}} B(k)b_{-1}c_0 + \cdots \right) |k, \downarrow\rangle \tag{10.86}$$

before gauge fixing. The spacetime fields are $T(k)$, $A_\mu(k)$ and $B(k)$, and their roles will be interpreted below. The first two terms are part of the $|\Phi_\downarrow\rangle$ component, while the last term is part of the $|\Phi_\uparrow\rangle$ component. All terms are correctly Grassmann even, and they have vanishing spacetime ghost numbers. The normalizations are chosen in order to retrieve the canonical normalization in QFT. The factor of i in front of B is needed for the field B to be real (as can be seen below, this leads to the expected factor ik_μ that maps to ∂_μ in position space).

Equation (10.12) leads to the following equations of motion of the spacetime fields:

$$\left(\alpha' k^2 - 1\right)T(k) = 0, \qquad k^2 A_\mu(k) + ik_\mu B(k) = 0,$$
$$k^\mu A_\mu(k) + iB(k) = 0. \tag{10.87}$$

Moreover, plugging the last equation into the second one gives

$$k^2 A_\mu(k) - k_\mu k \cdot A(k) = 0. \tag{10.88}$$

After Fourier transformation, the equations in position space read

$$\left(\alpha' \Delta + 1\right) T = 0, \qquad B = \partial^\mu A_\mu, \qquad \Delta A_\mu = \partial_\mu B. \tag{10.89}$$

This shows that $T(k)$ is a tachyon with mass $m^2 = -1/\alpha'$ and $A_\mu(k)$ is a massless gauge field. The field $B(k)$ is the Nakanishi–Lautrup auxiliary field, and it is completely fixed once A_μ is known since its equation has no derivative. Siegel gauge imposes $B = 0$, which shows that it generalizes the Feynman gauge to the string field.

Computation: Equation (10.87)
Keeping only the levels 0 and 1 terms in the string field, it is sufficient to truncate the BRST operator as

$$Q_B = c_0 L_0 - b_0 M + \widehat{Q}_B,$$
$$M \sim 2c_{-1}c_1, \qquad \widehat{Q}_B \sim c_1 L^m_{-1} + c_{-1}L^m_1, \tag{10.90}$$
$$L^m_1 \sim \alpha_0 \cdot \alpha_1, \qquad L^m_{-1} \sim \alpha_0 \cdot \alpha_{-1}.$$

Acting on the string field gives

$$
Q_B \, |\Phi\rangle = \frac{1}{\sqrt{\alpha'}} \int \frac{\mathrm{d}^D k}{(2\pi)^D} \Big(T(k) c_0 L_0 \, |k, \downarrow\rangle + A_\mu(k)
$$

$$
\times \Big(c_0 L_0 + \eta_{\nu\rho} c_{-1} \alpha_1^\nu \alpha_0^\rho \Big) \alpha_{-1}^\mu \, |k, \downarrow\rangle
$$

$$
+ \mathrm{i}\sqrt{\frac{\alpha'}{2}} \, B(k) \Big(-2 b_0 c_{-1} c_1 + \eta_{\nu\rho} c_1 \alpha_{-1}^\nu \alpha_0^\rho \Big) b_{-1} c_0 \, |k, \downarrow\rangle \Big)
$$

$$
= \frac{1}{\sqrt{\alpha'}} \int \frac{\mathrm{d}^D k}{(2\pi)^D} \Big(T(k) (\alpha' k^2 - 1) c_0 \, |k, \downarrow\rangle
$$

$$
+ A_\mu(k) \Big(\alpha' k^2 c_0 \alpha_{-1}^\mu + \sqrt{2\alpha'} \eta_{\nu\rho} \eta^{\mu\nu} k^\rho \, c_{-1} \Big) |k, \downarrow\rangle
$$

$$
+ \mathrm{i}\sqrt{\frac{\alpha'}{2}} \, B(k) \Big(2 c_{-1} + \sqrt{2\alpha'} \eta_{\nu\rho} k^\rho \alpha_{-1}^\nu c_0 \Big) |k, \downarrow\rangle \Big)
$$

$$
= \frac{1}{\sqrt{\alpha'}} \int \frac{\mathrm{d}^D k}{(2\pi)^D} \Big(T(k) (\alpha' k^2 - 1) c_0 \, |k, \downarrow\rangle
$$

$$
+ \alpha' \Big(A_\mu(k) k^2 + \mathrm{i} k^\mu B(k) \Big) c_0 \alpha_{-1}^\mu \, |k, \downarrow\rangle
$$

$$
+ \sqrt{2\alpha'} \Big(k^\mu A_\mu(k) + \mathrm{i} B(k) \Big) c_{-1} \, |k, \downarrow\rangle \Big).
$$

One needs to be careful when anti-commuting the ghosts, and we used that $p_L = k$ and $\alpha_0 = \sqrt{2\alpha'} k$ for the open string. It remains to require that the coefficient of each state vanishes.

In order to confirm that A_μ is indeed a gauge field, we must study the gauge transformation. The gauge parameter is expanded at the first level:

$$
|\Lambda\rangle = \frac{\mathrm{i}}{\sqrt{2\alpha'}} \int \frac{\mathrm{d}^D k}{(2\pi)^D} \big(\lambda(k) \, b_{-1} \, |k, \downarrow\rangle + \cdots \big). \tag{10.91}
$$

Note that $b_{-1} | \downarrow\rangle$ is the SL(2, $\mathbb{C}$) ghost vacuum. Since

$$
Q_B \, |\Lambda\rangle = \frac{\mathrm{i}}{\sqrt{\alpha'}} \int \frac{\mathrm{d}^D k}{(2\pi)^D} \, \lambda(k) \left(-\sqrt{\frac{\alpha'}{2}} \, k^2 \, b_{-1} c_0 + k_\mu \alpha_{-1}^\mu \right) |k, \downarrow\rangle , \tag{10.92}
$$

matching the coefficients in (10.29) gives

$$
\delta A_\mu = -\mathrm{i} k_\mu \lambda, \qquad \delta B = k^2 \lambda. \tag{10.93}
$$

This is the appropriate transformation for a U(1) gauge field.

Finally, one can derive the action; for simplicity, we work in the Siegel gauge. We consider only the tachyon component:

$$|T\rangle = \int \frac{d^D k}{(2\pi)^D} T(k) c_1 |k, 0\rangle, \qquad (10.94)$$

with $c_1 |0\rangle = |\downarrow\rangle$. The BPZ conjugate and the Hermitian conjugate are, respectively,

$$\langle T| = \int \frac{d^D k}{(2\pi)^D} T(k) \langle -k, 0|c_{-1}, \qquad (10.95a)$$

$$\langle T^{\ddagger}| = \int \frac{d^D k}{(2\pi)^D} T(k)^* \langle k, 0|c_{-1}, \qquad (10.95b)$$

since $c_1^t = c_{-1}$ when using the operator I^- in (6.111). Imposing equality of both leads to the reality condition

$$T(k)^* = T(-k), \qquad (10.96)$$

which agrees with the fact that the tachyon is real (the integration measure changes as $d^D k \to -d^D k$, but the contour is reversed).

Then, the action reads

$$S[T] = \frac{1}{2} \int \frac{d^D k}{(2\pi)^D} T(-k) \left(k^2 - \frac{1}{\alpha'}\right) T(k). \qquad (10.97)$$

This shows that the action is canonically normalized as it should for a real scalar field. Similarly, one can compute the action for the gauge field:

$$S[A] = \frac{1}{2} \int \frac{d^D k}{(2\pi)^D} A_\mu(-k) k^2 A^\mu(k). \qquad (10.98)$$

The correct normalization of the tachyon (real scalar field of negative mass) gives a justification a posteriori for the normalization of the action (10.14). Typically, string field actions are normalized in this way, by requiring that the first physical spacetime fields have the correct normalization. Note how this implies the correct normalization for all the other physical fields. Generalizing this computation for higher levels, one always finds the kinetic term to be

$$\frac{L_0^+}{2} = \frac{1}{2}(k^2 + m^2), \qquad (10.99)$$

which is the canonical normalization.

Computation: Equation (10.97)

$$\langle T | c_0 L_0 | T \rangle = \frac{1}{\alpha'} \int \frac{d^D k}{(2\pi)^D} \frac{d^D k'}{(2\pi)^D} \, T(k) T(k') \, \langle -k', 0 | c_{-1} c_0 L_0 c_1 | k, 0 \rangle$$

$$= \frac{1}{\alpha'} \int \frac{d^D k}{(2\pi)^D} \frac{d^D k'}{(2\pi)^D} \, T(k) T(k') (\alpha' k^2 - 1) \, \langle -k', 0 | c_{-1} c_0 c_1 | k, 0 \rangle$$

$$= \frac{1}{\alpha'} \int \frac{d^D k}{(2\pi)^D} \, d^D k' \, T(k) T(k') (\alpha' k^2 - 1) \delta^{(D)}(k + k'),$$

where we used $\langle 0 | c_{-1} c_0 c_1 | 0 \rangle = 1$ and $\langle k' | k \rangle = (2\pi)^D \delta^{(D)}(k + k')$.

10.5   Closed String

The derivation of the BRST free action for the closed string is very similar. The starting point is the equation of motion

$$Q_B | \Psi \rangle = 0 \tag{10.100}$$

for the closed string field $| \Psi \rangle$. The difference with (10.12) is that the BRST charge Q_B now includes both the left- and right-moving sectors. In the case of the open string, the field Φ was free of any constraint: we will see shortly that this is not the case for the closed string.

The next step is to find an inner product $\langle \cdot, \cdot \rangle$ to write the action:

$$S = \frac{1}{2} \langle \Psi, Q_B \Psi \rangle. \tag{10.101}$$

Following the open string, it seems logical to give the string field Ψ the same ghost number as the states in the cohomology:

$$N_{gh}(\Psi) = 2. \tag{10.102}$$

In this case, the ghost number of the arguments of $\langle \cdot, \cdot \rangle$ in (10.101) is $N_{gh} = 5$. The ghost number anomaly requires the total ghost number to be 6, that is,

$$N_{gh}(\langle \cdot, \cdot \rangle) = 1. \tag{10.103}$$

There is no other choice because $N_{gh}(\Psi)$ must be integer. The simplest solution is to insert one c zero-mode c_0 or $\bar{c}_0$, or a linear combination. The BRST operator Q_B contains both $L_0^{\pm}$ (see the decomposition (8.88)): the natural expectation (and by analogy with the open string) is that the gauge fixed equation of motion (to be discussed below) should be equivalent to the on-shell equation $L_0^+ = 0$ (see also Sect. 8.3.4). This is possible only if the insertion is c_0^-. With this insertion, $\langle \cdot, \cdot \rangle$ can be formed from the BPZ product:

$$\langle A, B \rangle = \langle A | c_0^- | B \rangle . \tag{10.104}$$

Then, the action reads

$$S = \frac{1}{2} \langle \Psi | c_0^- Q_B | \Psi \rangle . \tag{10.105}$$

However, the presence of c_0^- has a drastic effect because it annihilates part of the string field. Decomposing the Hilbert space as in (7.175)

$$\mathcal{H} = \mathcal{H}^- \oplus c_0^- \mathcal{H}^-, \qquad \mathcal{H}^- := \mathcal{H} \cap \ker b_0^-, \tag{10.106}$$

the string field reads

$$|\Psi\rangle = |\Psi_-\rangle + c_0^- |\widetilde{\Psi}_-\rangle, \qquad \Psi_-, \widetilde{\Psi}_- \in \mathcal{H}^-, \tag{10.107}$$

such that

$$c_0^- |\Psi\rangle = c_0^- |\Psi_-\rangle. \tag{10.108}$$

The problem in such cases is that the kinetic term may become non-invertible. This motivates to project out the component $\widetilde{\Psi}_-$ by imposing the following constraint on the string field:

$$b_0^- |\Psi\rangle = 0. \tag{10.109}$$

The constraint (10.109) is stronger than the constraint $L_0^- = 0$ for states in the cohomology (Sect. 8.3.1), so there is no information lost on-shell by imposing it. For this reason, we will also impose the level-matching condition:

$$L_0^- |\Psi\rangle = 0, \tag{10.110}$$

such that

$$\Psi \subset \mathcal{H}^- \cap \ker L_0^- . \tag{10.111}$$

This will later be motivated by studying the propagator and the off-shell scattering amplitudes. To avoid introducing more notations, we will not use a new symbol for this space and keep implicit that $\Psi \in \ker L_0^-$.

The necessity of this condition can be understood differently. We had found that it is necessary to ensure that the closed string parametrization is invariant under translations along the string (Sect. 3.2.2). Since there is no BRST symmetry associated to this symmetry, one needs to keep the constraint.[2] This suggests that one may enlarge further the gauge symmetry and interpret (10.109) as a gauge fixing condition. This would be quite desirable: one could argue that a fundamental field should be completely described by the Lagrangian (if such a description exists) and that it should not be necessary to supplement it with constraints imposed by hand. While this can be achieved at the free level, this idea runs into problems in the presence of interactions (Sect. 13.3.1) and the interpretation is not clear.[3]

The action (10.105) is gauge invariant under

$$|\Psi\rangle \longrightarrow |\Psi'\rangle = |\Psi\rangle + \delta_\Lambda |\Psi\rangle\,, \qquad \delta_\Lambda |\Psi\rangle = Q_B |\Lambda\rangle\,, \tag{10.112}$$

where the gauge parameter has ghost number 1 and also lives in $\mathcal{H}^- \cap \ker L_0^-$:

$$N_{\mathrm{gh}}(\Lambda) = 1\,, \qquad L_0^- |\Lambda\rangle = 0\,, \qquad b_0^- |\Lambda\rangle = 0\,. \tag{10.113}$$

As for the open string, the gauge invariance (10.112) can be gauge fixed in the Siegel gauge:

$$b_0^+ |\Psi\rangle = 0\,. \tag{10.114}$$

Then, the action reduces to

$$S = \frac{1}{2}\, \langle\Psi| c_0^- c_0^+ L_0^+ |\Psi\rangle = \frac{1}{4}\, \langle\Psi| c_0 \bar{c}_0 L_0^+ |\Psi\rangle\,. \tag{10.115}$$

The equation of motion is to the on-shell condition as expected:

$$L_0^+ |\Psi\rangle = 0\,. \tag{10.116}$$

Additional constraints must be imposed to ensure that only the physical degrees of freedom propagate.

---

[2] Yet another reason can be found in Sect. 3.2.2 (see also Sect. 8.3.4): to motivate the need of the b_0^+ condition, we could take the on-shell limit from off-shell states because L_0^+ is continuous. However, the L_0^- operator is discrete, and there is no such limit we can consider [13]. So we must always impose this condition, both off- and on-shell.

[3] A recent proposal can be found in [8].

> **Computation: Equation (10.115)**
>
> $$c_0^- Q_B = (c_0 - \bar{c}_0)(c_0 L_0 + \bar{c}_0 \bar{L}_0) = c_0 \bar{c}_0 (L_0 + \bar{L}_0).$$

10.6 Summary

In this chapter, we have shown how the BRST conditions defining the cohomology can be interpreted as an equation of motion for a string field together with a gauge invariance. We found a subtlety for the closed string due to the ghost number anomaly and because of the level-matching condition. Then, we studied several basic properties in order to prove that the free action has the expected properties.

The next step is to add the interactions to the action, but we do not know first principles to write them. For this reason, we need to take a detour and consider off-shell amplitudes. By introducing a factorization of the amplitudes, it is possible to rewrite them as Feynman diagrams, where fundamental interactions are connected by propagators (which we will find to match the one in the Siegel gauge). This can be used to extract the interacting terms of the action.

10.7 Suggested Readings

- The free BRST string field theory is discussed in detail in [13] (see also [4, chap. 7, 5, chap. 9, 10, chap. 11]). Shorter discussions can be found in [1, 9, sec. 4, 12, 14, 15].
- Spacetime fields and actions are discussed in [9, 11, sec. 4].
- Gauge fixing [1, 2, 3, 6, sec. 2.1, 7, 11, sec. 6.5, 7.2, 7.4].
- General properties of string field (reality, parity, etc.) [1, 16].

References

1. M. Asano, M. Kato, New covariant gauges in string field theory. Progr. Theoret. Phys. **117**(3), 569–587 (2007). https://doi.org/10.1143/PTP.117.569. arXiv: hep-th/0611189
2. M. Asano, M. Kato, General linear gauges and amplitudes in open string field theory. Nuclear Phys. B **807**(1–2), 348–372 (2009). https://doi.org/10.1016/j.nuclphysb.2008.09.004. arXiv: 0807.5010
3. M. Bochicchio, String field theory in the siegel gauge. Phys. Lett. B **188**(3), 330–334 (1987). https://doi.org/10.1016/0370-2693(87)91391-8
4. M. Kaku, *Introduction to Superstrings and M-Theory* (Springer, Berlin, 1999)
5. M. Kaku, *Strings, Conformal Fields, and M-Theory* (Springer, Berlin, 1999)
6. M. Kroyter, Y. Okawa, M. Schnabl, S. Torii, B. Zwiebach, Open superstring field theory I: gauge fixing, ghost structure, and propagator. J. High Energy Phys. **2012**(3), 30 (2012). https://doi.org/10.1007/JHEP03(2012)030. arXiv: 1201.1761

7. J.M.F. Labastida, M. Pernici, BRST quantization in the siegel gauge. Phys. Lett. **194**(4), 511–517 (1987). https://doi.org/10.1016/0370-2693(87)90226-7
8. Y. Okawa, *Closed String Field Theory without the Level-Matching Condition* (2018). http://www.hri.res.in/~strings/okawa.pdf
9. J. Polchinski, What Is String Theory? (1994). arXiv: hep-th/9411028
10. W. Siegel, *Introduction to String Field Theory* (World Scientific, Singapore, 2001)
11. W. Taylor, B. Zwiebach, D-branes, tachyons, and string field theory, in *Strings, Branes and Extra Dimensions* (World Scientific, Singapore, 2004), pp. 641–760. https://doi.org/10.1142/9789812702821_0012. arXiv: hep-th/0311017
12. C.B. Thorn, Perturbation theory for quantized string fields. Nuclear Phys. B **287**(Suppl. C), 61–92 (1987). https://doi.org/10.1016/0550-3213(87)90096-4
13. C.B. Thorn, String field theory. Phys. Rep. **175**(1–2), 1–101 (1989). https://doi.org/10.1016/0370-1573(89)90015-X
14. E. Witten, Non-commutative geometry and string field theory. Nuclear Phys. B **268**(2), 253–294 (1986). https://doi.org/10.1016/0550-3213(86)90155-0
15. B. Zwiebach, Closed String Field Theory: An Introduction (1993). arXiv: hep-th/9305026
16. B. Zwiebach, Closed string field theory: quantum action and the BV master equation. Nuclear Phys. B **390**(1), 33–152 (1993). https://doi.org/10.1016/0550-3213(93)90388-6. arXiv: hep-th/9206084

Introduction to Off-Shell String Theory

11

Abstract

In this chapter, we introduce a framework to describe off-shell amplitudes in string theory. We first start by motivating various concepts—in particular, local coordinates and factorization—by focusing on the 3- and 4-point amplitudes. We then prepare the stage for a general description of off-shell amplitudes. We focus again on the closed bosonic string only.

11.1 Motivations

11.1.1 3-Point Function

The tree-level 3-point amplitude of 3 weight h_i vertex operators[1] $\mathcal{V}_i$ is given by

$$A_{0,3} = \left\langle \prod_{i=1}^{3} \mathcal{V}_i(z_i) \right\rangle_{S^2} \propto (z_1 - z_2)^{h_3 - h_1 - h_2} \times \text{perms} \times \text{c.c.} \qquad (11.1)$$

There is no integration since $\dim \mathcal{M}_{0,3} = 0$.

The amplitude is independent of the z_i only if the matter state is on-shell, $h_i = 0$, for example, if $\mathcal{V}_i = c\bar{c}V_i$ with $h(V_i) = 1$. Indeed, if $h_i \neq 0$, then $A_{0,3}$ is not

---

[1] The quantum number (k, j) of the vertex operator is mostly irrelevant for the discussion of the current and next chapters, and they are omitted. We will distinguish them by a number and reintroduce the momentum k when necessary. We also omit the overall normalization of the amplitudes.

© Springer Nature Switzerland AG 2021
H. Erbin, *String Field Theory*, Lecture Notes in Physics 980,
https://doi.org/10.1007/978-3-030-65321-7_11

invariant under conformal transformations (6.38):

$$z \longrightarrow f_g(z) = \frac{az+b}{cz+d} \in \mathrm{SL}(2, \mathbb{C}) \tag{11.2}$$

(it transforms covariantly). This is a consequence of the punctures: the presence of the latter modifies locally the metric, since they act as sources of negative curvature. When performing a conformal transformation, the metric around the punctures changes in a different way as away from them. This implies that the final result depends on the metric chosen around the punctures. This looks puzzling because the original path integral derivation (Chap. 3) indicates that the 3-point amplitude should not depend on the locations of the operators because its moduli space is empty (hence, all choices of z_i should be equivalent).

The solution is to introduce local coordinates w_i with a flat metric $|dw_i|^2$ around each puncture conventionally located at $w_i = 0$. The local coordinates are defined by the maps:

$$z = f_i(w_i), \qquad z_i = f_i(0). \tag{11.3}$$

This is also useful to characterize in a simpler way the dependence of off-shell amplitudes rather than using the metric around the punctures (computations may be more difficult with a general metric).

The expression of a local operator in the local coordinate system is found by applying the corresponding change of coordinates (6.48):

$$f \circ \mathscr{V}(w) = f'(w)^h \overline{f'(w)}^{\bar{h}} \, \mathscr{V}\big(f(w)\big). \tag{11.4}$$

The amplitude reads then

$$A_{0,3} = \left\langle \prod_{i=1}^{3} f_i \circ \mathscr{V}_i(0) \right\rangle_{S^2} = \left(\prod_{i=1}^{3} f_i'(0)^{h_i} \overline{f_i'(0)}^{\bar{h}_i} \right) \left\langle \prod_{i=1}^{3} \mathscr{V}_i\big(f_i(0)\big) \right\rangle_{S^2} \tag{11.5a}$$

$$\propto \left(\prod_{i=1}^{3} f_i'(0)^{h_i} \overline{f_i'(0)}^{\bar{h}_i} \right) \big(f_1(0) - f_2(0)\big)^{h_3 - h_1 - h_2} \times \text{perms} \times \text{c.c.} \tag{11.5b}$$

The amplitude depends on the local coordinate choice f_i, but not on the metric around the punctures. It is also invariant under $\mathrm{SL}(2, \mathbb{C})$; the transformation (11.2) written in terms of the local coordinates is

$$f_i \longrightarrow \frac{af_i + b}{cf_i + d} \tag{11.6}$$

from which we get

$$f_i' \longrightarrow \frac{f_i'}{(cf_i + d)^2}, \qquad f_i - f_j \longrightarrow \frac{f_i - f_j}{(cf_i + d)(cf_j + d)}. \qquad (11.7)$$

All together, this implies the invariance of the 3-point amplitude since the factors in the denominator cancel. When the states are on-shell $h_i = 0$, the dependence in the local coordinate cancels, showing that the latter is non-physical.

One can ask how Feynman graphs can be constructed in string theory. By definition, the amplitude is the sum of Feynman graphs contributing at that order in the loop expansion and for the given number of external legs. The Feynman graphs are themselves built from a set of Feynman rules. These correspond to the data of the fundamental interactions together with the definition of a propagator. Since a tree-level cubic interaction is the interaction of the lowest order, it makes sense to promote it to a fundamental cubic vertex[2] $\mathcal{V}_{0,3}$:

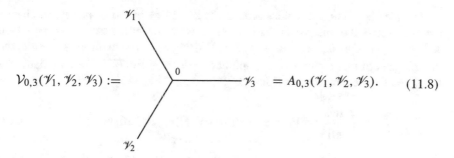

$$\mathcal{V}_{0,3}(\mathcal{V}_1, \mathcal{V}_2, \mathcal{V}_3) := \qquad\qquad\qquad = A_{0,3}(\mathcal{V}_1, \mathcal{V}_2, \mathcal{V}_3). \qquad (11.8)$$

The index 0 reminds that it is a tree-level interaction.

11.1.2 4-Point Function

The tree-level 4-point amplitude is expressed as

$$A_{0,4} = \int d^2 z_4 \left\langle \prod_{i=1}^{3} c\bar{c} V_i(z_i)\, V_4(z_4) \right\rangle_{S^2}. \qquad (11.9)$$

The conformal weights are denoted by $h(V_i) = h_i$. For on-shell states, $h_i = 1$; while there is no dependence on the positions z_1, z_2 and z_3, there are divergences for

$$z_4 \longrightarrow z_1, z_2, z_3, \qquad (11.10)$$

---

[2]The notation will become clear later and should not be confused with the vertex operators.

corresponding to collisions of punctures in the integration process. Moreover, the expression does not look symmetric: it would be more satisfactory if all the insertions were accompanied by ghost insertions and if all the puncture locations were treated on an equal footing.

Example 11.1: Tachyons

Given tachyon states $V_i = e^{ik_i \cdot X}$, the amplitude reads

$$A_{0,4} \propto \prod_{\substack{i,j=1 \\ i<j}}^{3} |z_i - z_j|^{2+k_i \cdot k_j} \int d^2 z_4 \prod_{i=1}^{3} |z_4 - z_i|^{k_i \cdot k_4}. \tag{11.11}$$

The integral diverges for $z_4 \to z_i$ if $k_i \cdot k_4 \leq 0$. This can happen for physical values of the momenta k_i. ◄

The idea is to cut out regions around z_1, z_2 and z_3 in the z_4-plane and change the interpretation of these contributions. First, we consider the case $z_4 \to z_3$, which corresponds to cutting a region around z_3. Writing $z_4 = q y_4$ with $y_4 \in \mathbb{C}$ fixed, the contribution of this region to the amplitude is denoted by $\mathcal{F}_{0,4}^{(s)}$. For simplicity, we take $z_3 = 0$. The contribution reads

$$\mathcal{F}_{0,4}^{(s)} = \int \frac{d^2 q}{|q|^2} \langle c\bar{c} V_1(z_1) c\bar{c} V_2(z_2) c\bar{c} V_3(0) |q y_4|^2 V_4(q y_4) \rangle. \tag{11.12}$$

The implicit radial ordering pushes V_3 to the left of V_4, and using the OPE between the b and c ghosts gives

$$\mathcal{F}_{0,4}^{(s)} = -\int \frac{d^2 q}{|q|^2} \Big\langle c\bar{c} V_1(z_1) c\bar{c} V_2(z_2) \oint_{|w|=|q|^{1/2}} dw \, w \, b(w)$$
$$\times \oint_{|w|=|q|^{1/2}} d\bar{w} \, \bar{w} \, \overline{b(w)} \, c\bar{c} V_4(q y_4) c\bar{c} V_3(0) \Big\rangle. \tag{11.13}$$

The sign arises by anti-commuting c and $\bar{b}$. The integration variable q can be removed from the argument of V_4 using the L_0 and $\bar{L}_0$ operators:

$$\mathcal{F}_{0,4}^{(s)} = -\int \frac{d^2 q}{|q|^2} \Big\langle c\bar{c} V_1(z_1) c\bar{c} V_2(z_2) \oint dw \, w \, b(w)$$
$$\times \oint d\bar{w} \, \bar{w} \, \overline{b(w)} \, q^{L_0} \bar{q}^{\bar{L}_0} c\bar{c} V_4(y_4) c\bar{c} V_3(0) \Big\rangle. \tag{11.14}$$

This expression is more satisfactory because all vertex operators are accompanied with c ghost insertions and none of the arguments are integrated over. But, in fact, even better can be achieved.

Inserting two complete sets of states $\{\phi_r\}$ (see Sect. 11.2) inside this expression gives (restoring a generic z_3-dependence):

$$\mathcal{F}_{0,4}^{(s)} = \langle c\bar{c}V_1(z_1)c\bar{c}V_2(z_2)\phi_r(0)\rangle \, \langle c\bar{c}V_3(z_3)c\bar{c}V_4(y_4)\phi_s(0)\rangle$$

$$\times \int \frac{\mathrm{d}^2 q}{|q|^2} \left\langle \phi_r^c q^{L_0}\bar{q}^{\bar{L}_0}b_0\bar{b}_0\phi_s^c \right\rangle, \tag{11.15}$$

where the sum over r and s is implicit. The conjugate states ϕ_r^c are defined by $\langle \phi_r^c|\phi_s\rangle = \delta_{rs}$. The first two terms are cubic interactions (11.8), and the last term connects both. It is then tempting to identify the latter with a propagator Δ

$$\Delta(\phi_r^c, \phi_s^c) := \langle \phi_r^c| \, \Delta \, |\phi_s^c\rangle := -\int \frac{\mathrm{d}^2 q}{|q|^2} \left\langle \phi_r^c q^{L_0}\bar{q}^{\bar{L}_0}b_0\bar{b}_0\phi_s^c \right\rangle, \tag{11.16}$$

such that

$$\mathcal{F}_{0,4}^{(s)} = \mathcal{V}_{0,3}\big(c\bar{c}V_1(z_1), c\bar{c}V_2(z_2), \phi_r(0)\big) \times \Delta(\phi_r^c, \phi_s^c)$$

$$\times \mathcal{V}_{0,3}\big(c\bar{c}V_3(z_3), c\bar{c}V_4(y_4), \phi_s(0)\big)$$

$$\tag{11.17}$$

To make this more precise, change the coordinates as

$$q = \mathrm{e}^{-s+i\theta}, \qquad s \in \mathbb{R}_+, \qquad \theta \in [0, 2\pi), \tag{11.18}$$

such that the integral becomes

$$\int \frac{\mathrm{d}^2 q}{|q|^2} q^{L_0}\bar{q}^{\bar{L}_0} = 2 \int_0^\infty \mathrm{d}s \int_0^{2\pi} \mathrm{d}\theta \, \mathrm{e}^{-s(L_0+\bar{L}_0)}\mathrm{e}^{i\theta(L_0-\bar{L}_0)} = \frac{2}{L_0 + \bar{L}_0} \delta_{L_0, \bar{L}_0}. \tag{11.19}$$

This shows that the propagator can be rewritten as

$$\Delta = -\frac{2b_0\bar{b}_0}{L_0 + \bar{L}_0}\,\delta_{L_0,\bar{L}_0} = \frac{b_0^+}{L_0^+}\,b_0^-\delta_{L_0^-,0},\qquad(11.20)$$

where $L_0^\pm = L_0 \pm \bar{L}_0$ and $b_0^\pm = b_0 \pm \bar{b}_0$. The sign is added by anticipating the normalization to be derived later. Its properties will be studied in detail in Sect. 14.2.2.

Taking the basis states $\phi_r := \phi_\alpha(k)$ to be eigenstates of L_0 and $\bar{L}_0$

$$L_0\,|\phi_\alpha(k)\rangle = \bar{L}_0\,|\phi_\alpha(k)\rangle = \frac{\alpha'}{4}\,(k^2 + m_\alpha^2)\,|\phi_\alpha(k)\rangle\qquad(11.21)$$

allows to rewrite the last term of $\mathcal{F}_{0,4}^{(s)}$ as

$$\Delta_{\alpha\beta}(k) = \int\frac{\mathrm{d}^2q}{|q|^2}\,\left\langle\phi_\alpha^c(k)q^{L_0}\bar{q}^{\bar{L}_0}b_0\bar{b}_0\phi_\beta^c(-k)\right\rangle = \frac{M_{\alpha\beta}(k)}{k^2 + m_\alpha^2}.\qquad(11.22)$$

The finite-dimensional matrix $M_{\alpha\beta}$ gives the overlap of states of identical masses:

$$M_{\alpha\beta}(k) := \frac{2}{\alpha'}\,\langle\phi_\alpha^c(k)|\,b_0^+b_0^-\,|\phi_\beta^c(-k)\rangle.\qquad(11.23)$$

The propagator depends only on one momentum because $\langle k|k'\rangle \sim \delta^{(D)}(k - k')$. This is exactly the standard propagator one finds in QFT, and this justifies the above claim. The contribution $\mathcal{F}_{0,4}^{(s)}$ to the amplitude can be seen as a s-channel Feynman graph obtained by gluing two cubic fundamental vertices with a propagator. We will see later the interpretation in terms of Riemann surfaces.

The same procedure can be followed by considering $z_4 \sim z_2$ and $z_4 \sim z_1$. This leads to contributions $\mathcal{F}_{0,4}^{(t)}$ and $\mathcal{F}_{0,4}^{(u)}$ corresponding to t- and u-channel Feynman graphs:

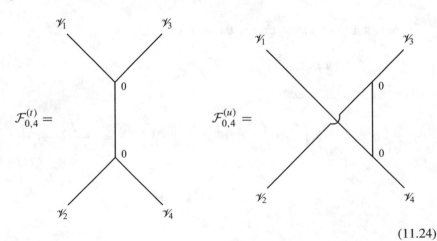

$$(11.24)$$

In general, the sum of the three contributions $\mathcal{F}_{0,4}^{(s,t,u)}$ does not reproduce the full amplitude $A_{0,4}$. Said differently, the regions cut in the z_4-plane do not cover it completely. It is then natural to interpret the remaining part as a fundamental tree-level quartic interaction denoted by

$$\mathcal{V}_{0,4} = \qquad \text{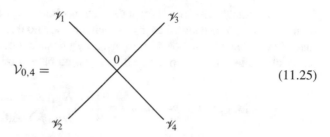} \qquad (11.25)$$

such that

$$A_{0,4} = \mathcal{F}_{0,4}^{(s)} + \mathcal{F}_{0,4}^{(t)} + \mathcal{F}_{0,4}^{(u)} + \mathcal{V}_{0,4}. \qquad (11.26)$$

Up to (11.14), it was sufficient to consider on-shell states, but the insertion of the complete basis requires to consider also off-shell states since on-shell states do not form a basis of the Hilbert space. As discussed for the 3-point functions, it is necessary to introduce local coordinates to describe off-shell states properly.

With the 3- and 4-point functions, we motivated the use of off-shell states and introduced the two important ideas of local coordinates and amplitude factorization. We also indicated that amplitudes can be written in a more symmetric way (see also the discussion at the end of Sect. 3.1.2). In the rest of this chapter, we give additional ideas on off-shell string theory.

Remark 11.1 (Riemann Surface Interpretation) The interpretation of the insertion of a propagator in terms of Riemann surface consists in gluing two of them thanks to the plumbing fixture procedure (Sect. 12.3).

11.2 Off-Shell States

An off-shell state is a generic state of the CFT Hilbert space

$$\mathcal{H} = \mathcal{H}_m \otimes \mathcal{H}_{\text{gh}} \qquad (11.27)$$

without any constraint. A basis for the off-shell states is denoted by

$$\mathcal{H} = \text{Span}\{|\phi_r\rangle\}. \qquad (11.28)$$

Since there is no constraint on the states, the ghost number of ϕ_r is arbitrary and denoted as

$$n_r := N_{\text{gh}}(\phi_r) \in \mathbb{Z} \tag{11.29}$$

(the ghost number is restricted for states in the cohomology of Q_B). The Grassmann parity of a state ϕ_r is denoted as $|\phi_r|$. When there are no fermions in the matter sector (usually the case for the bosonic string), only ghosts are odd. Then, the Grassmann parity of a state is odd (resp. even) if its ghost number is odd (resp. even):

$$|\phi_r| := N_{\text{gh}}(\phi_r) \mod 2 = \begin{cases} 0 & N_{\text{gh}}(\phi_r) \text{ even,} \\ 1 & N_{\text{gh}}(\phi_r) \text{ odd.} \end{cases} \tag{11.30}$$

The dual basis $\{|\phi_r^c\rangle\}$ is defined from the BPZ inner product:

$$\langle \phi_r^c | \phi_s \rangle = \delta_{rs}. \tag{11.31}$$

Denoting the ghost numbers of the dual states by

$$n_r^c := N_{\text{gh}}(\phi_r^c), \tag{11.32}$$

the product is non-vanishing if

$$n_r^c + n_r = 6, \tag{11.33}$$

due to the ghost number anomaly on the sphere. This condition cannot be satisfied if the dual state ϕ_r^c is simply taken to be the BPZ conjugate ϕ_r^t since the BPZ conjugation does not change the ghost number. This implies that

$$\langle \phi_r | \phi_s \rangle = 0 \tag{11.34}$$

from the ghost anomaly for every state, except for the closed string states with $N_{\text{gh}} = 3$ (in fact, the inner product of these states is also zero as can be seen after investigation). One can show that

$$\langle \phi_r | \phi_s^c \rangle = (-1)^{|\phi_r|} \delta_{rs}. \tag{11.35}$$

Hence, the resolution of the identity can be written in the two equivalent ways

$$1 = \sum_r |\phi_r\rangle \langle \phi_r^c| = \sum_r (-1)^{|\phi_r|} |\phi_r^c\rangle \langle \phi_r|. \tag{11.36}$$

Following (7.176), the Hilbert space can be decomposed as

$$\mathcal{H} = \mathcal{H}_\pm \oplus c_0^\pm \mathcal{H}_\pm, \tag{11.37}$$

where

$$\mathcal{H}_\pm := \mathcal{H} \cap \ker b_0^\pm = \mathcal{H}_0 \oplus c_0^\mp \mathcal{H}_0, \qquad \mathcal{H}_0 := \mathcal{H} \cap \ker b_0^- \cap \ker b_0^+. \tag{11.38}$$

In fact, we will find that a consistent description of the off-shell amplitudes for the closed string requires imposing some conditions on the states even at the off-shell level. The off-shell states will have to satisfy the level-matching condition and to be annihilated by b_0^-:

$$L_0^- |\phi\rangle = 0, \qquad b_0^- |\phi\rangle = 0. \tag{11.39}$$

This implies that the off-shell states will be elements of $\mathcal{H}^- \cap \ker L_0^-$. This will appear as consistency conditions on the geometry of the moduli space and by studying the propagator. In general, we shall work with $\mathcal{H}$ and indicate when necessary the restriction to $\mathcal{H}^-$ (keeping the condition $\ker L_0^-$ implicit to avoid new notations).

The Hilbert space $\mathcal{H}$ can be separated according to the ghost zero-modes

$$\mathcal{H} \sim \mathcal{H}_{\downarrow\downarrow} \oplus \mathcal{H}_{\downarrow\uparrow} \oplus \mathcal{H}_{\uparrow\downarrow} \oplus \mathcal{H}_{\uparrow\uparrow}, \tag{11.40}$$

with the following definitions:

$$\mathcal{H}_{\downarrow\uparrow} \sim \mathcal{H}_0, \qquad \mathcal{H}_{\downarrow\uparrow} \sim \bar{c}_0 \mathcal{H}_{\downarrow\downarrow} \qquad \mathcal{H}_{\uparrow\downarrow} \sim c_0 \mathcal{H}_{\downarrow\downarrow} \qquad \mathcal{H}_{\uparrow\uparrow} \sim c_0 \bar{c}_0 \mathcal{H}_{\downarrow\downarrow}. \tag{11.41}$$

Accordingly, every basis state can be split as

$$\phi_r = \phi_{\downarrow\downarrow,r} + \phi_{\downarrow\uparrow,r} + \phi_{\uparrow\downarrow,r} + \phi_{\uparrow\uparrow,r} \tag{11.42}$$

such that

$$b_0 |\phi_{\downarrow\downarrow,r}\rangle = \bar{b}_0 |\phi_{\downarrow\downarrow,r}\rangle = 0, \qquad b_0 |\phi_{\downarrow\uparrow,r}\rangle = \bar{c}_0 |\phi_{\downarrow\uparrow,r}\rangle = 0,$$
$$c_0 |\phi_{\uparrow\downarrow,r}\rangle = \bar{b}_0 |\phi_{\uparrow\downarrow,r}\rangle = 0, \qquad c_0 |\phi_{\uparrow\uparrow,r}\rangle = \bar{c}_0 |\phi_{\uparrow\uparrow,r}\rangle = 0. \tag{11.43}$$

Moreover, the basis can be indexed such that

$$|\phi_{\downarrow\uparrow,r}\rangle = \bar{c}_0 |\phi_{\downarrow\downarrow,r}\rangle \qquad |\phi_{\uparrow\downarrow,r}\rangle = c_0 |\phi_{\downarrow\downarrow,r}\rangle \qquad |\phi_{\uparrow\uparrow,r}\rangle = c_0 \bar{c}_0 |\phi_{\downarrow\downarrow,r}\rangle. \tag{11.44}$$

A dual state ϕ_r^c is also expanded:

$$\phi_r^c = \phi_{\downarrow\downarrow,r}^c + \phi_{\downarrow\uparrow,r}^c + \phi_{\uparrow\downarrow,r}^c + \phi_{\uparrow\uparrow,r}^c, \tag{11.45}$$

and the components satisfy

$$\langle\phi_{\downarrow\downarrow,r}^c| c_0 = \langle\phi_{\downarrow\downarrow,r}^c| \bar{c}_0 = 0, \qquad \langle\phi_{\downarrow\uparrow,r}^c| c_0 = \langle\phi_{\downarrow\uparrow,r}^c| \bar{b}_0 = 0,$$
$$\langle\phi_{\uparrow\downarrow,r}^c| b_0 = \langle\phi_{\uparrow\downarrow,r}^c| \bar{c}_0 = 0, \qquad \langle\phi_{\uparrow\uparrow,r}^c| b_0 = \langle\phi_{\uparrow\uparrow,r}^c| \bar{b}_0 = 0. \tag{11.46}$$

The indexing of the basis is chosen such that

$$\langle\phi_{\downarrow\uparrow,r}^c| = \langle\phi_{\downarrow\downarrow,r}^c| \bar{b}_0 \qquad \langle\phi_{\uparrow\downarrow,r}^c| = \langle\phi_{\downarrow\downarrow,r}^c| b_0 \qquad \langle\phi_{\uparrow\uparrow,r}^c| = \langle\phi_{\downarrow\downarrow,r}^c| \bar{b}_0 b_0, \tag{11.47}$$

such that

$$\langle\phi_{x,r}^c|\phi_{y,s}\rangle = \delta_{xy}\delta_{rs}, \tag{11.48}$$

where $x, y = \downarrow\downarrow, \uparrow\downarrow, \downarrow\uparrow, \uparrow\uparrow$.

Consider the Hilbert space $\mathcal{H}^-$; then, a basis state must satisfy

$$b_0^- |\phi_r\rangle = 0 \quad \implies \quad b_0 |\phi_r\rangle = \bar{b}_0 |\phi_r\rangle. \tag{11.49}$$

The expansion (11.42) gives the relation

$$\phi_{\uparrow\downarrow,r} + \phi_{\uparrow\uparrow,r} = \phi_{\downarrow\downarrow,r} + \phi_{\downarrow\uparrow,r} \tag{11.50}$$

such that

$$\phi_r = 2(\phi_{\downarrow\downarrow,r} + \phi_{\downarrow\uparrow,r}). \tag{11.51}$$

For convenience, the factor of 2 can be omitted (which amounts to rescaling the basis states).

11.3   Off-Shell Amplitudes

In this section, we provide a guideline of what we need to look for in order to write an off-shell amplitude. The geometrical tools will be described in the next chapter and the construction of off-shell amplitudes in the following one.

11.3.1   Amplitudes from the Marked Moduli Space

In Chap. 3, the scattering amplitudes were written as an integral over the moduli space $\mathcal{M}_g$ of the Riemann surface Σ_g. As a consequence, the moduli of $\mathcal{M}_g$ and the positions of the vertex operators are not treated on an equal footing. Moreover, the insertions of operators are not symmetric since some are integrated, and the others have factors of c. These problems can be solved by reinterpreting the scattering amplitudes in a more geometrical way.

The key is to consider the punctures where vertex operators are inserted as part of the geometry and not as external data added on top of the Riemann surface Σ_g.

Then, the worldsheet with the external states is described as a punctured (or marked) Riemann surface $\Sigma_{g,n}$, which is a Riemann surface Σ_g with n punctures (marked points) z_i. The Euler number of such a surface was given in (3.4):

$$\chi_{g,n} := \chi(\Sigma_{g,n}) = 2 - 2g - n. \tag{11.52}$$

This makes sense since punctures can be interpreted as disks (boundaries). Note that the punctures are labelled and thus distinguishable.

Since the marked points are distinguished, marked Riemann surfaces with identical g and n but with punctures located at different points are seen as different (this statement requires some care for $g = 0$ and $g = 1$ due to the presence of CKV). The corresponding moduli space is denoted by $\mathcal{M}_{g,n}$, and it can be viewed as a fibre bundle with $\mathcal{M}_g$ as the base and the puncture positions as the fibre. The dimension of $\mathcal{M}_{g,n}$ is

$$\mathsf{M}_{g,n} := \dim_{\mathbb{R}} \mathcal{M}_{g,n} = 6g - 6 + 2n, \qquad \text{for} \quad \begin{cases} g \geq 2, \\ g = 1, n \geq 1, \\ g = 0, n \geq 3. \end{cases} \tag{11.53}$$

These cases are equivalent to $\chi_{g,n} < 0$, that is, when the surfaces have a negative curvature. The corresponding coordinates are denoted by t_λ, $\lambda = 1, \ldots, \mathsf{M}_{g,n}$. Comparing (11.53) with (2.51) and (2.93), this corresponds to the situation where $\Sigma_{g,n}$ has no CKV left unfixed.

Example 11.2: 4-Punctured Sphere $\Sigma_{0,4}$

The positions of the punctures are denoted by z_i with $i = 1, \ldots, 4$. Since there are three CKVs, the positions of three punctures (say z_1, z_2 and z_3) can be fixed, leaving only one position that characterizes $\Sigma_{0,4}$. Hence, the moduli space has dimension $\mathsf{M}_{0,4} = 2$, and $\mathcal{M}_{0,4}$ is parametrized by $\{z_4\}$. ◄

Example 11.3: 2-Punctured Torus $\Sigma_{1,2}$

The positions of the punctures are denoted by z_i with $i = 1, 2$. One puncture can be fixed using the single CKV of the surface, which leaves one position. Together with the moduli parameter τ of the torus, this gives $M_{1,2} = 4$, and the coordinates of $\mathcal{M}_{1,2}$ are $\{z_2, \tau\}$. ◄

The g-loop n-point scattering amplitude with external states $\{\mathcal{V}_i\}$ can be written as an integral over $\mathcal{M}_{g,n}$ of some $M_{g,n}$-form $\omega_{M_{g,n}}^{(g,n)}$:

$$A_{g,n}(\mathcal{V}_1, \ldots, \mathcal{V}_n) = \int_{\mathcal{M}_{g,n}} \omega_{M_{g,n}}^{g,n}(\mathcal{V}_1, \ldots, \mathcal{V}_n). \tag{11.54}$$

The integration over $M_{g,n}$ has the correct dimension to reproduce the formulas from Sect. 3.1.

While it is possible to derive this amplitude from the path integral (see the comments at the end of Sect. 3.1.2), we will make only use of the properties of CFT on Riemann surfaces in the next chapter. This provides an alternative point of view on the computation of scattering amplitudes and how to derive the formulas, which can be helpful when the manipulation of the path integral is more complicated (for example, with the superstring).

The expression of the form $\omega_{M_{g,n}}^{g,n}$ must (1) provide a measure on the moduli space and (2) extract a function of the moduli from the states $\mathcal{V}_i$. It is natural to achieve the second point by computing a correlation function on the Riemann surfaces $\Sigma_{g,n}$. Moreover, Chaps. 2 and 3 indicate that the ghosts are part of the definition of the measure. Hence, one can expect the $\omega_{M_{g,n}}^{g,n}$ to have the form:

$$\omega_{M_{g,n}}^{g,n}(\mathcal{V}_1, \ldots, \mathcal{V}_n) = \left\langle \text{ghosts} \times \prod_{i=1}^n \mathcal{V}_i \right\rangle_{\Sigma_{g,n}} \times \bigwedge_{\lambda=1}^{M_{g,n}} dt_\lambda. \tag{11.55}$$

We will motivate an expression in Chap. 14 before checking that it has the correct properties. The ghost insertions are necessary to saturate the number of zero-modes to obtain a non-vanishing result. By convention, the ghosts are inserted on the left: while this does not make difference for on-shell closed states, this will for off-shell states and for the other types of strings (open and supersymmetric) since the operators can be Grassmann odd.

11.3.2 Local Coordinates

The next step is to consider off-shell states $\mathcal{V}_i \in \mathcal{H}$. As motivated previously, one needs to introduce local coordinates defined by the maps:

$$z = f_i(w_i), \qquad z_i = f_i(0). \tag{11.56}$$

There is one local coordinate for each operator, which is inserted at the origin. Local coordinates on the surfaces can be seen in two different fashions (Fig. 11.1): either as describing patches on the surface, in which case the maps f_i correspond to transition functions, or one can interpret them by cutting disks centred at the punctures and whose interiors are mapped to complex planes, and the maps f_i tell how to insert the plane inside the disk.

When the amplitude $A_{g,n}$ is defined in terms of local coordinates, it will depend on the maps f_i, and one needs to ensure that this cancels when the A_is are on-shell. But, the choice of the maps f_i is arbitrary: selecting a specific set hides that all choices are physically equivalent and should lead to the same results on-shell. For this reason, the geometry can be enriched with the local coordinates, in the same way that the puncture locations were added as a fibre to the moduli space $\mathcal{M}_g$ to get the marked moduli space $\mathcal{M}_{g,n}$. Hence, the fundamental geometrical object is the fibre bundle $\mathcal{P}_{g,n}$ with $\mathcal{M}_{g,n}$ being the base and the local coordinates of the fibre. Since there is an infinite number of functions, the fibre is infinite-dimensional, and so is the space $\mathcal{P}_{g,n}$.

Every point of $\mathcal{P}_{g,n}$ corresponds to a genus-g Riemann surface with n punctures together with a choice of local coordinates around the punctures. The form $\omega_{\mathsf{M}_{g,n}}^{g,n}$ is defined in this bigger space, and the integration giving the off-shell amplitude (11.54) is performed over a $\mathsf{M}_{g,n}$-dimensional section $\mathcal{S}_{g,n} \subset \mathcal{P}_{g,n}$ (Fig. 11.2):

$$A_{g,n}(\mathcal{V}_1, \ldots, \mathcal{V}_n)_{\mathcal{S}_{g,n}} = \int_{\mathcal{S}_{g,n}} \omega_{\mathsf{M}_{g,n}}^{g,n}(\mathcal{V}_1, \ldots, \mathcal{V}_n)\big|_{\mathcal{S}_{g,n}}. \tag{11.57}$$

The subscript in the LHS indicates that the amplitudes depend on $\mathcal{S}_{g,n}$ through the choice of local coordinates. The on-shell independence of $A_{g,n}$ on the local coordinates translates into the independence on the choice of the section:

$$\forall \mathcal{S}_{g,n}: \quad A_{g,n}(\mathcal{V}_1, \ldots, \mathcal{V}_n)_{\mathcal{S}_{g,n}} = A_{g,n}(\mathcal{V}_1, \ldots, \mathcal{V}_n) \quad \text{(on-shell)}. \tag{11.58}$$

The section is taken to be continuous, which means that two neighbouring surfaces of the moduli space must have close local coordinates.

In order to define the amplitude, one needs to find the expression of the $\mathsf{M}_{g,n}$-form $\omega_{\mathsf{M}_{g,n}}^{g,n}$ on $\mathcal{P}_{g,n}$. It is in fact simpler to define general p-forms $\omega_p^{g,n}$ on $\mathcal{P}_{g,n}$, in particular, for proving general properties about the forms and the amplitudes. Given a manifold, a p-form is an element of the cotangent space, and it can be defined through its contraction with vectors (tangent space). Since vectors correspond to small variation of the manifold coordinates, it is necessary to find a parametrization of $\mathcal{P}_{g,n}$. The geometry of $\mathcal{P}_{g,n}$—and of its relevant subspaces and tangent space—is studied in the next chapter. Then, we will come back on the construction of the amplitudes in Chap. 13.

Fig. 11.1 Usage of local
coordinates for a Riemann
surface. (**a**) Original surface
with three punctures. (**b**)
Disks delimiting the local
coordinate patches. (**c**)
Complex plane mapped to
disks centred around the
previous puncture location

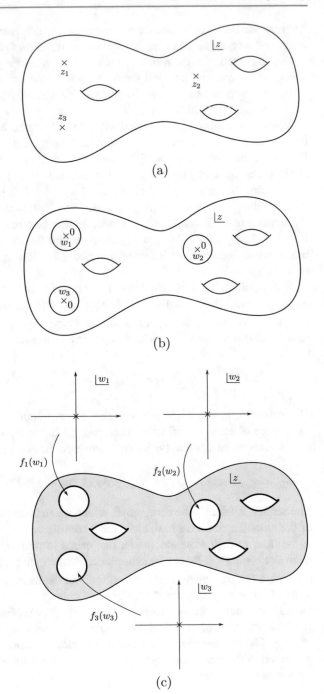

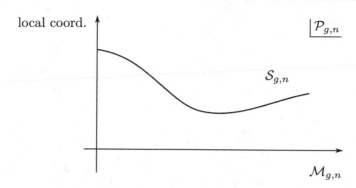

Fig. 11.2 Section $\mathcal{S}_{g,n}$ of the fibre bundle $\mathcal{P}_{g,n}$, the latter having $\mathcal{M}_{g,n}$ as a base and the local coordinates as a fibre

11.4   Suggested Readings

- General references on off-shell string theory include [4, sec. 7, 3, sec. 2, 1, 2].
- Interpretation of local coordinates [2, sec. 5.2].

References

1. C. de Lacroix, H. Erbin, S.P. Kashyap, A. Sen, M. Verma, Closed super-string field theory and its applications. Int. J. Mod. Phys. A **32**(28n29), 1730021 (2017). https://doi.org/10.1142/S0217751X17300216. arXiv: 1703.06410
2. J. Polchinski, *String Theory: Volume 1, An Introduction to the Bosonic String* (Cambridge University Press, Cambridge, 2005)
3. A. Sen, Off-shell amplitudes in superstring theory. Fortschritte der Physik **63**(3–4), 149–188 (2015). https://doi.org/10.1002/prop.201500002. arXiv: 1408.0571
4. B. Zwiebach, Closed string field theory: quantum action and the BV master equation. Nucl. Phys. B **390**(1), 33–152 (1993). https://doi.org/10.1016/0550-3213(93)90388-6. arXiv:hep-th/9206084

Geometry of Moduli Spaces and Riemann Surfaces

12

Abstract

In this chapter, we describe how to parametrize the moduli space $\mathcal{M}_{g,n}$ and the local coordinates that together form the fibre bundle $\mathcal{P}_{g,n}$ introduced in the previous chapter. Then, we can characterize the tangent space that we will need in the next chapter to write the p-forms on $\mathcal{P}_{g,n}$ necessary to write the amplitudes. Finally, we introduce the notion of plumbing fixture, an operation that glues together punctures located on the same or different surfaces.

12.1 Parametrization of $\mathcal{P}_{g,n}$

The first step is to find a parametrization of the Riemann surfaces. As we have seen (Chap. 11), the dependence of the surface on the punctures can be described by local coordinates, that is, transition functions. The patch is defined by cutting a disk around each puncture, and the transition functions are defined on the circle given by the intersection of the disk with the rest of the surface. The number of disks is simply:

$$\#\text{disks} = n. \tag{12.1}$$

It makes sense to look for a similar description of the other moduli (associated to the genus) by introducing additional coordinate patches. One can imagine that all the dependence of the moduli and punctures will reside in the transition functions between patches if the different patches are isomorphic to a surface without any moduli: the 3-punctured sphere $\Sigma_{0,3}$. Hence, one can look for a decomposition of the surface by cutting disks such that one is left with 3-punctured spheres only, and transition functions are defined on the circles at the intersections of the spheres.

Next, we need to find the number of spheres with 3 holes (or punctures). We start first with $\Sigma_{0,n}$: in this case, it is straightforward to find that there will be $n - 2$

© Springer Nature Switzerland AG 2021
H. Erbin, *String Field Theory*, Lecture Notes in Physics 980,
https://doi.org/10.1007/978-3-030-65321-7_12

spheres. Indeed, for each additional puncture beyond $n = 3$, an additional sphere is created by cutting a circle. For $g \geq 1$, natural places to split the surface are handles: two circles can be cut for each of them. By inspection, one finds that it leads to 2 spheres for each handle (one on the right and the other on the left).[1] This shows that the number of spheres is

$$\text{\#spheres} = 2g - 2 + n. \tag{12.2}$$

The number of circles corresponds to the number of boundaries divided by two since the boundaries are glued pairwise: each disk has one boundary and each sphere has 3, which leads to

$$\text{\#circles} = \frac{n + 3(2g - 2 + n)}{2} = 3g - 3 + 2n. \tag{12.3}$$

The idea of the construction is to split the surface into elementary objects (spheres and disks) such that the full surface is seen as the union of all of them (gluing along circles), and no information is left in the individual geometries. This parametrization is particularly useful because there are simple coordinate systems on spheres and disks and these surfaces are easy to visualize and to work with. For example, they can be easily mapped to the complex plane.

To conclude, a genus-g Riemann surface $\Sigma_{g,n}$ with n punctures can be seen as the collection of:

- $2g - 2 + n$ three-punctured spheres $\{S_a\}$ with coordinates z_a,
- n disks $\{D_i\}$ with coordinates w_i around each puncture,
- $3g - 3 + 2n$ circles $\{C_\alpha\}$ at the intersections of the spheres and disks.

Examples for $\Sigma_{0,4}$ and $\Sigma_{2,2}$ are given in Figs. 12.1 and 12.2, respectively.

There are two types of circles: respectively, the ones at the overlap between two spheres, and between a disk and a 3-sphere:

$$C_{\Lambda(ab)} := S_a \cap S_b, \qquad C_{i(a)} := S_a \cap D_i, \qquad \{C_\alpha\} = \{C_\Lambda, C_i\}, \tag{12.4}$$

where Λ counts in fact the number of moduli:

$$\Lambda = 1, \ldots, \mathsf{M}_{g,n}^c, \qquad \mathsf{M}_{g,n}^c = 3g - 3 + n. \tag{12.5}$$

On the overlap circles, the coordinate systems are related by transition functions:

$$\begin{aligned} \text{on } C_{\Lambda(ab)} : \qquad z_a &= F_{ab}(z_b), \\ \text{on } C_{i(a)} : \qquad z_a &= f_{ai}(w_i). \end{aligned} \tag{12.6}$$

---

[1] The simplest way to find this result is to consider $\Sigma_{g,2}$ and to write one puncture at each side of the surface (as in Fig. 12.2). To generalize further, one can consider a generic n and put all punctures but one on one side of the surfaces.

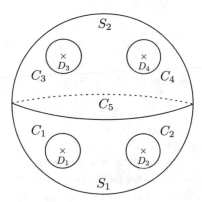

Fig. 12.1 Parametrization of $\Sigma_{0,4}$

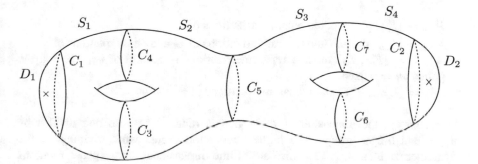

Fig. 12.2 Parametrization of $\Sigma_{2,2}$

Then, the set of functions $\{F_{ab}, f_{ai}\}$ completely specifies the Riemann surface $\Sigma_{g,n}$ together with the choice of the local coordinate systems around the punctures. The transition functions can thus be used to parametrize the moduli space $\mathcal{M}_{g,n}$ and the fibre bundle $\mathcal{P}_{g,n}$, but it is highly redundant because many different functions lead to the same Riemann surface. A unique characterization of the different spaces is obtained by making identifications up to symmetries.

In the previous chapter, we have seen that the metric in the local coordinate system is flat, $ds^2 = |dw|^2$. This means that two systems differing by a global phase rotation

$$w_i \longrightarrow \tilde{w}_i = e^{i\alpha_i} w_i \tag{12.7}$$

lead to surfaces with local coordinates that cannot be distinguished. Correspondingly, the two maps f_i and $\tilde{f}_i$ that relate the local coordinates w_i and $\tilde{w}_i$ to the coordinate z

$$z = f_i(w_i), \qquad z = \tilde{f}_i(\tilde{w}_i) \tag{12.8}$$

are related as

$$f_i(w_i) = \tilde{f}_i(e^{i\alpha_i} w_i). \tag{12.9}$$

Hence, this motivates to consider the smaller space

$$\hat{\mathcal{P}}_{g,n} = \mathcal{P}_{g,n}/\mathrm{U}(1)^n, \tag{12.10}$$

where the action of each $\mathrm{U}(1)$ is defined by the equivalence (12.9). The necessity to consider this subspace will be strengthened further later and will correspond to the level-matching condition. Below, global phase rotations are also interpreted in terms of the plumbing fixture, see (12.57).

The different spaces that we need are parametrized by the transition functions up to the following identifications:

- $\mathcal{P}_{g,n} = \{F_{ab}, f_{ai}\}$ modulo reparametrizations of z_a,
- $\hat{\mathcal{P}}_{g,n} = \{F_{ab}, f_{ai}\}$ modulo reparametrizations of z_a and phase rotations of w_i,
- $\mathcal{M}_{g,n} = \{F_{ab}, f_{ai}\}$ modulo reparametrizations of z_a and of w_i, keeping the points $w_i = 0$ fixed,
- $\mathcal{M}_g = \{F_{ab}, f_{ai}\}$ modulo reparametrizations of z_a and w_i.

At each step, the dimension of the space is reduced because one divides by bigger and bigger groups. The highest reduction occurs when dividing by the reparametrizations of w_i that form an infinite-dimensional group (phase rotations form a finite-dimensional subgroup of them).

For concreteness, it is useful to introduce explicit coordinates x_s on $\mathcal{P}_{g,n}$ ($s \in \mathbb{N}$ since the space is infinite dimensional). The transition functions on the Riemann surface depend on the x_s, which explains why they can be used to parametrize the moduli spaces. Describing the spheres S_a by complex planes with punctures located at $z_{a,1}$, $z_{a,2}$ and $z_{a,3}$, the transition functions on the $\mathsf{M}_{g,n}$ circles $C_{\Lambda(ab)} = S_a \cap S_b$ for all $a < b$ can be taken to be

$$\text{on } C_{\Lambda(ab)}: \qquad z_a - z_{a,m} = \frac{q_\Lambda}{z_b - z_{b,n}}, \tag{12.11}$$

where $z_{a,m}$ and $z_{b,n}$ denote the punctures of S_a and S_b lying in C_Λ. Then, the complex parameters q_Λ with $\Lambda = 1, \ldots, \mathsf{M}_{g,n}^c$ are coordinates on the moduli space $\mathcal{M}_{g,n}$. On the remaining n circles $C_{i(a)} = S_a \cap D_i$, the transition functions can be expanded in series

$$\text{on } C_{i(a)}: \qquad z_a - z_{a,m} = w_i + \sum_{N=1}^{\infty} p_{i,N} w_i^N, \tag{12.12}$$

where $z_{a,m}$ is the puncture of S_a lying in C_i. There is no negative index in the series because the RHS must vanish for $w_i = 0$ that maps to $z_a = z_{a,m}$ (puncture location). The complex coefficients of the series $p_{i,N}$ ($i = 1, \ldots, n$ and $N \in \mathbb{N}^*$) provide coordinates for the fibre. Thus, coordinates for $\mathcal{P}_{g,n}$ are

$$\{x_s\} = \{q_\Lambda, p_{i,N}\}. \tag{12.13}$$

As usual, derivatives with respect to x_s are abbreviated by ∂_s.

When the dependence in the x_s must be stressed, the transition functions (12.6) are denoted by

$$z_a = F_{ab}(z_b; x_s), \qquad z_a = f_{ai}(w_i; x_s). \tag{12.14}$$

In this coordinate system, each parameter appears in only one transition function and it looks like one can separate the fibre from the basis. But, this is not an invariant statement as this would not hold in other coordinate systems. For example, one can rescale the coordinates to lump all dependence on q_Λ in a single circle.

Since both cases are formally identical, it is convenient to fix the orientation of each C_α and to denote by σ_α (resp., τ_α) the coordinate on the left (resp., right) of the contour, such that the transition functions read

$$\text{on } C_\alpha: \qquad \sigma_\alpha = F_\alpha(\tau_\alpha; x_s). \tag{12.15}$$

Now that we have coordinates on $\mathcal{P}_{g,n}$, it is possible to construct tangent vectors.

12.2   Tangent Space

A tangent vector $V_s \in T\mathcal{P}_{g,n}$ corresponds to an infinitesimal variation of the coordinates on the manifold

$$\delta x_s = \epsilon \, V_s, \tag{12.16}$$

where ϵ is a small parameter, such that functions of x_s vary as

$$\epsilon \, V_s \, \partial_s f = f(x_s + \epsilon \, V_s) - f(x_s). \tag{12.17}$$

The transition functions F_α provide an equivalent (but redundant) set of coordinates for $\mathcal{P}_{g,n}$. Hence, vectors in $T\mathcal{P}_{g,n}$ can also be obtained by considering small variations of the transition functions F_α:

$$F_\alpha \longrightarrow F_\alpha + \epsilon \, \delta F_\alpha. \tag{12.18}$$

Considering an overlap circle C_α, a deformation of the transition function

$$\sigma_\alpha = F_\alpha(\tau_\alpha) \tag{12.19}$$

for fixed τ_α can be interpreted as a change of the coordinate σ_α:

$$\sigma'_\alpha = F_\alpha(\tau_\alpha) + \epsilon\,\delta F_\alpha(\tau_\alpha) = \sigma_\alpha + \epsilon\,\delta F_\alpha(\tau_\alpha) = \sigma_\alpha + \epsilon\,\delta F_\alpha\big(F_\alpha^{-1}(\sigma_\alpha)\big). \qquad (12.20)$$

This transformation is generated by a vector field $v^{(\alpha)}$ on the Riemann surface $\Sigma_{g,n}$:

$$\sigma'_\alpha = \sigma_\alpha + \epsilon v^{(\alpha)}(\sigma_\alpha), \qquad v^{(\alpha)} = \delta F_\alpha \circ F_\alpha^{-1}. \qquad (12.21)$$

The situation is symmetrical, and one can obviously fix σ_α and vary τ_α. The vector field is regular around the circle C_α (to have a well-defined change of coordinates), but it can have singularities away from the circle C_α. Hence, the vector field $v^{(\alpha)}$ together with the circle C_α defines a vector of $\mathcal{P}_{g,n}$:

$$V^{(\alpha)} \sim \big(v^{(\alpha)}, C_{(\alpha)}\big). \qquad (12.22)$$

This provides a basis of $T\mathcal{P}_{g,n}$. This is sufficient when using the coordinate system (12.13), but, in more general situations, one needs to consider linear combinations. For example, if a modulus appears in several transition functions, then the associated vector field will be defined on the corresponding circles. A general vector V is described by a vector field v with support on a subset C of the circles C_α:

$$V \sim (v, C), \qquad C \subseteq \bigcup_\alpha C_\alpha, \qquad (12.23)$$

and the restriction of v on the various circles is written as

$$v|_{C_\alpha} = v^{(\alpha)}. \qquad (12.24)$$

Note that the vector field $v^{(\alpha)}$ and its complex conjugate $\bar{v}^{(\alpha)}$ are independent and are associated to different tangent vectors. This construction is called the Schiffer variation.

The simplest tangent vectors ∂_s are given by varying one coordinate of $\mathcal{P}_{g,n}$ while keeping the other fixed:

$$x_s \longrightarrow x_s + \epsilon\,\delta x_s. \qquad (12.25)$$

On each circle C_α, this gives a deformation of the transition functions

$$C_\alpha: \quad F_\alpha \longrightarrow F_\alpha + \epsilon\,\delta F_\alpha, \qquad \delta F_\alpha = \frac{\partial F_\alpha}{\partial x_s}\,\delta x_s \qquad (12.26)$$

(no sum over s), such that the change of coordinates reads

$$\sigma'_\alpha = \sigma_\alpha + \epsilon\, v_s^{(\alpha)}(\sigma_\alpha)\,\delta x_s, \qquad v_s^{(\alpha)}(\sigma_\alpha) = \frac{\partial F_\alpha}{\partial x_s}\big(F_\alpha^{-1}(\sigma_\alpha)\big). \qquad (12.27)$$

If x_s are given by (12.13), the vectors have support in only one circle.

There is, however, a redundancy in these vectors. Not all of them lead to a motion in $\mathcal{P}_{g,n}$ because some modifications can be absorbed with a reparametrization of the z_a. For example, if a given $v^{(\alpha)}$ can be extended holomorphically outside the circle C_α in the neighbour sphere, then its effect can be undone by reparametrizing the corresponding coordinate. A similar discussion holds for the other spaces, and relations can be found by restricting the vector on subspaces. A non-trivial vector $(v^{(i)}, C_i)$ of $\mathcal{P}_{g,n}$ becomes trivial on $\mathcal{M}_{g,n}$ if it can be cancelled with a reparametrization of w_i that leaves the origin fixed.

12.3 Plumbing Fixture

The plumbing fixture is a way to glue together two Riemann surfaces (separating case) or two parts of the same surface (non-separating case), in order to build a surface with a higher number of holes and punctures. This geometric operation will correspond precisely to the concept of gluing two Feynman graphs with a propagator in Siegel gauge.

The plumbing fixture depends on a (complex) one parameter, which leads to a family of surfaces. This provides the correct number of moduli for the surface obtained after gluing. This brings to the question of describing the moduli spaces $\mathcal{M}_{g,n}$ in terms of the moduli spaces with lower genus and number of punctures.

12.3.1 Separating Case

Consider two Riemann surfaces Σ_{g_1,n_1} and Σ_{g_2,n_2} with local coordinates $w_1^{(1)}, \ldots, w_{n_1}^{(1)}$ and $w_1^{(2)}, \ldots, w_{n_2}^{(2)}$.

The first step is to cut two disks $D_q^{(1)}$ and $D_q^{(2)}$ of radius $|q|^{1/2}$ around a puncture on each surface, taken to be the n_1-th and n_2-th for definiteness:

$$D_q^{(1)} = \{|w_{n_1}^{(1)}| \leq |q|^{1/2}\}, \qquad D_q^{(2)} = \{|w_{n_2}^{(2)}| \leq |q|^{1/2}\}, \qquad (12.28)$$

where $q \in \mathbb{C}$ is fixed (Fig. 12.3).[2] Then, both surfaces can be glued (indicated by the binary operation #) together into a new surface

$$\Sigma_{g,n} = \Sigma_{g_1,n_1} \# \Sigma_{g_2,n_2}, \qquad \begin{cases} g = g_1 + g_2, \\ n = n_1 + n_2 - 2 \end{cases} \qquad (12.29)$$

---

[2] The disks $D_q^{(i)}$ should be equal or smaller than the disks $D_{n_1}^{(1)}$ and $D_{n_2}^{(2)}$.

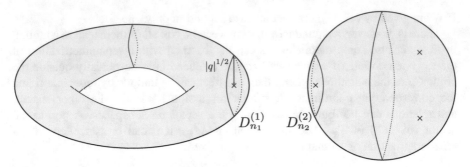

Fig. 12.3 Disks around one puncture of the surfaces $\Sigma_{1,1}$ and $\Sigma_{0,3}$. The disks appear as a cap because it is on top of the surface, which is curved

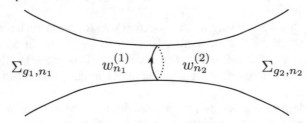

Fig. 12.4 Integration contour on the circle between the two local coordinates that are glued together

by removing the disks $D_q^{(1)}$ and $D_q^{(2)}$ and by identifying the circles $\partial D_q^{(1)}$ and $\partial D_q^{(2)}$. At the level of the coordinates, this is achieved by the *plumbing fixture* operation:

$$w_{n_1}^{(1)} w_{n_2}^{(2)} = q, \qquad |q| \le 1. \tag{12.30}$$

The restriction on q arises because we have $|w_{n_1}^{(1)}|, |w_{n_2}^{(2)}| \le 1$ (for a discussion, see [1]). This case is called *separating* because cutting the new tube splits the surface into two components. Locally, the new surface looks like Fig. 12.4. It is also convenient to parametrize q as

$$q = e^{-s+i\theta}, \qquad s \in \mathbb{R}_+, \qquad \theta \in [0, 2\pi). \tag{12.31}$$

The parameters s and θ are interpreted below as moduli of the Riemann surface. The geometry of the new surface can be viewed in three different ways:

1. both surfaces Σ_{g_1,n_1} and Σ_{g_2,n_2} (with the disks removed) are connected directly at their boundaries (Fig. 12.5a);
2. both surfaces Σ_{g_1,n_1} and Σ_{g_2,n_2} (with the disks removed) are connected by a cylinder of finite size (Fig. 12.5b);
3. the surface Σ_{g_2,n_2} is inserted inside the disk $D_q^{(1)}$, or conversely Σ_{g_1,n_1} inside $D_q^{(2)}$ (Fig. 12.5c, d).

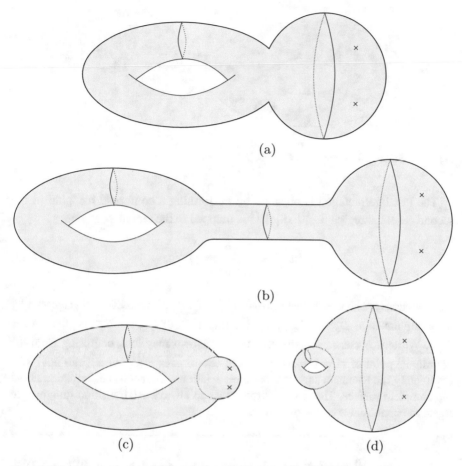

Fig. 12.5 Different representations of the surface $\Sigma_{1,2}$ obtained after gluing $\Sigma_{1,1}$ and $\Sigma_{0,3}$ through the plumbing fixture. (**a**) Direct gluing of circles. (**b**) Connection by a long tube. (**c**) Insertion of the second surface into the first one. (**d**) Insertion of the first surface into the second one

The first interpretation is the most direct one: the disks are simply removed and the boundaries are glued together by a small tube of radius $|w_1^{(1)}| < |q|^{1/2}$. The connection between both surfaces can be smoothed (and figures are often drawn in this way—for example Fig. 12.6), but this is not necessary (the smoothing is achieved by cutting disks of size $|q|^{1/2} - \epsilon$ and gluing the boundaries to the disks of radius $|q|^{1/2}$).

In the second interpretation, one rescales the local coordinates in order to bring the radius of the disk to 1 instead of $|q|^{1/2}$. In terms of these new coordinates, the surfaces are connected by a tube of length $s = -\ln|q|$ after a Weyl transformation (see [2, sec. 9.3] for a longer discussion).

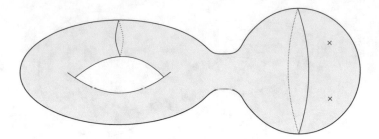

Fig. 12.6 Smoothed connection between both surfaces

The last interpretation is obtained by performing a conformal mapping of the second case: the region $|w_{n_1}^{(1)}| < |q|^{1/2}$ is mapped to the region

$$|w_{n_2}^{(2)}| = \frac{|q|}{|w_{n_1}^{(1)}|} > |q|^{1/2}, \tag{12.32}$$

and conversely. The idea is that the disk $D_q^{(1)}$ of Σ_{g_1,n_1} is removed and replaced by the complement of $D_q^{(2)}$ in Σ_{g_2,n_2}, i.e. the full surface $\Sigma_{g_2,n_2} - D_q^{(2)}$ is glued inside $D_q^{(1)}$. While it is clear geometrically, this statement may look confusing from the coordinate point of view because the local coordinates $w_{n_1}^{(1)}$ and $w_{n_2}^{(2)}$ do not cover completely the Riemann surfaces, but their relation still encodes information about the complete surface. The reason is that one can always use transition functions to relate the coordinates on the two surfaces.

Example 12.1

Denote by $S_a^{(1)}$ and $S_b^{(2)}$ the spheres sharing a boundary with $D_{n_1}^{(1)}$ and $D_{n_2}^{(2)}$, and write the corresponding coordinates by $z_a^{(1)}$ and $z_b^{(2)}$ such that the transition functions are

$$z_a^{(1)} = f_{an_1}^{(1)}(w_{n_1}^{(1)}), \qquad z_b^{(2)} = f_{bn_2}^{(2)}(w_{n_2}^{(2)}). \tag{12.33}$$

Then the coordinates z_a and z_b are related by

$$z_a^{(1)} = f_{an_1}^{(1)}(w_{n_1}^{(1)}) = f_{an_1}^{(1)}\left(\frac{q}{w_{n_2}^{(2)}}\right) = f_{an_1}^{(1)}\left(\frac{q}{f_{bn_2}^{(2)-1}(z_b^{(2)})}\right) \tag{12.34}$$

such that the new transition function reads

$$z_a = F_{ab}(z_b), \qquad F_{ab} = f_{an_1}^{(1)} \circ (q \cdot I) \circ f_{bn_2}^{(2)-1}, \tag{12.35}$$

where I is the inversion (the superscript on the coordinates z_a and z_b has been removed to indicate that they are now seen as coordinates on the same surface $\Sigma_{g,n}$). ◀

The Riemann surface $\Sigma_{g,n}$ is a point of $\mathcal{M}_{g,n}$. By varying the moduli parameters of Σ_{g_1,n_1} and Σ_{g_2,n_2}, one obtains other surfaces in $\mathcal{M}_{g,n}$. But the number of parameters furnished by Σ_{g_1,n_1} and Σ_{g_2,n_2} does not match the dimension (11.53) of $\mathcal{M}_{g,n}$:

$$\mathsf{M}_{g_1,n_1} + \mathsf{M}_{g_2,n_2} = 6g_1 - 6 + 2n_1 + 6g_2 - 6 + 2n_2 = \mathsf{M}_{g,n} - 2. \tag{12.36}$$

This means that the subspace of $\mathcal{M}_{g,n}$ obtained by gluing all the possible surfaces in $\mathcal{M}_{g_1,n_1}$ and $\mathcal{M}_{g_2,n_2}$ is of codimension 2. The missing complex parameter is q: in writing the plumbing fixture, it was taken to be fixed, but it can be varied to generate a 2-parameter family of Riemann surfaces in $\mathcal{M}_{g,n}$, with the moduli of the original surfaces held fixed.

The surface $\Sigma_{g,n}$ is equipped with local coordinates inherited from the original surfaces Σ_{g_1,n_1} and Σ_{g_2,n_2}. Hence, the plumbing fixture of points in $\mathcal{P}_{g_1,n_1}$ and $\mathcal{P}_{g_2,n_2}$ automatically leads to a point of $\mathcal{P}_{g,n}$. The fact that the local coordinates are inherited from lower-order surfaces is called *gluing compatibility*. It is also not necessary to add parameters to describe the fibre direction.

12.3.2 Non-separating Case

In the previous section, the plumbing fixture was used to glue punctures on two different surfaces. In fact, one can also glue two punctures on the same surface to get a new surface with an additional handle:

$$\Sigma_{g,n} = \#\Sigma_{g_1,n_1}, \qquad \begin{cases} g = g_1 + 1, \\ n = n_1 - 2, \end{cases} \tag{12.37}$$

defining # as a unary operator. This gluing is called *non-separating* because there is a single surface before the identification of the disks.

In terms of the local coordinates, the gluing relation reads

$$w^{(1)}_{n_1-1} w^{(1)}_{n_1} = q, \tag{12.38}$$

where we consider the last two punctures for definiteness.

The dimensions of both moduli spaces are related by

$$\mathsf{M}_{g_1,n_1} = \mathsf{M}_{g,n} - 2. \tag{12.39}$$

Again, the two missing parameters are provided by varying q, and we obtain a $M_{g,n}$-dimensional subspace of $\mathcal{M}_{g,n}$.

Example 12.2

Here are some examples of surfaces obtained by gluing:

- $\Sigma_{0,4} = \Sigma_{0,3} \# \Sigma_{0,3}$
- $\Sigma_{0,5} = \Sigma_{0,3} \# \Sigma_{0,3} \# \Sigma_{0,3}, \Sigma_{0,3} \# \Sigma_{0,4}$
- $\Sigma_{1,1} = \# \Sigma_{0,3}$
- $\Sigma_{1,2} = \# \Sigma_{0,4}, \Sigma_{1,1} \# \Sigma_{0,3}$

Note that the moduli on the LHS and RHS are fixed (we will see later that not all surfaces can be obtained by gluing). ◄

12.3.3 Decomposition of Moduli Spaces and Degeneration Limit

We have seen that the separating and non-separating plumbing fixtures yield a family of surfaces in $\mathcal{M}_{g,n}$ described in terms of lower-dimensional moduli spaces. The question is whether all points in $\mathcal{M}_{g,n}$ can be obtained in this way by looking at all the possible gluing (varying g_1, n_1, g_2 and n_2). It turns out that this is not possible, which is at the core of the difficulties to construct a string field theory.

Which surfaces are obtained from this construction? In order to interpret the regions of $\mathcal{M}_{g,n}$ covered by the plumbing fixture, the parametrization (12.31) is the most useful. Previously, we explained that s gives the size of the tube connecting the two surfaces. Since the latter is like a sphere with two punctures, it corresponds to a cylinder (interpreted as a propagating intermediate closed string). The angle θ in (12.31) is the twist of the cylinder connecting both components. This amounts to start with $\theta = 0$, then to cut the cylinder, to twist it by an angle θ and to glue again.

The limit $s \to \infty$ ($|q| \to 0$) is called the *degeneration limit*: the degenerate surface $\Sigma_{g,n}$ reduces to Σ_{g_1,n_1} and Σ_{g_2,n_2} connected by a very long tube attached to two punctures (separating case), or to $\Sigma_{g-1,n+2}$ with a very long handle (non-separating case). So it means that the family of surfaces described by the plumbing fixture are "close" to degeneration. Another characterization (for the separating case) is that the punctures on Σ_{g_1,n_1} are closer (according to some distance, possibly after a conformal transformation) to each other than to the punctures on Σ_{g_2,n_2}.

Conversely, there are surfaces that cannot be described in this way: the plumbing fixture does not cover all the possible values of the moduli. For a given $\mathcal{M}_{g,n}$, we denote the surfaces that cannot be obtained by the plumbing fixture by $\mathcal{V}_{g,n}$. This space does not contain any surface arbitrarily close to degeneration (i.e. with long handles or tubes). In terms of punctures, it also means that there is no conformal frame where the punctures split into two sets.

In the previous subsection, we considered two specific punctures, but any other punctures could be chosen. Hence, there are many ways to split $\Sigma_{g,n}$ into two surfaces Σ_{g_1,n_1} and Σ_{g_2,n_2} (with fixed g_1, g_2, n_1 and n_2): every partition of the punctures and holes in two sets leads to different degeneration limits (because they

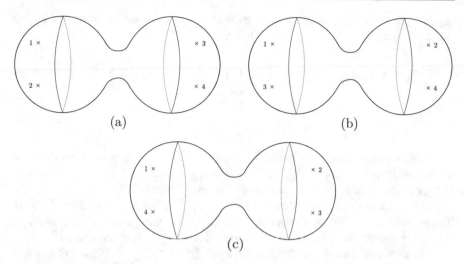

Fig. 12.7 Permutations of punctures while gluing two spheres: they correspond to different (disconnected) parts of $\mathcal{M}_{0,4}$. (**a**) Degeneration $12 \rightarrow 34$. (**b**) Degeneration $13 \rightarrow 24$. (**c**) Degeneration $14 \rightarrow 23$

are associated to different moduli—Fig. 12.7). Since each puncture is described by a modulus, choosing different punctures for gluing gives different set of moduli for $\Sigma_{g,n}$, such that each possibility covers a different subspace of $\mathcal{M}_{g,n}$. The part of the moduli space $\mathcal{M}_{g,n}$ covered by the plumbing fixture of all surfaces Σ_{g_1,n_1} and Σ_{g_2,n_2} (with fixed g_1, g_2, n_1, n_2) is denoted by $\mathcal{M}_{g_1,n_1} \# \mathcal{M}_{g_2,n_2}$:

$$\mathcal{M}_{g_1,n_1} \# \mathcal{M}_{g_2,n_2} \subset \mathcal{M}_{g,n}, \tag{12.40}$$

where the operation # includes the plumbing fixture for all values of q and all pairs of punctures. Similarly, the part covered by the non-separating plumbing fixture is written as $\# \mathcal{M}_{g_1,n_1}$:

$$\# \mathcal{M}_{g_1,n_1} \subset \mathcal{M}_{g,n}. \tag{12.41}$$

Importantly, the regions covered by the plumbing fixture depend on the choice of the local coordinates because (12.30) is written in terms of local coordinates. The subspaces $\mathcal{M}_{g_1,n_1} \# \mathcal{M}_{g_2,n_2}$ and $\# \mathcal{M}_{g_1,n_1}$ are not necessarily connected (in the topological sense).

The moduli space $\mathcal{M}_{g,n}$ cannot be completely covered by the plumbing fixture of lower-dimensional surfaces. We define the propagator and fundamental vertex regions $\mathcal{F}_{g,n}$ and $\mathcal{V}_{g,n}$ as the subspaces that can and cannot be described by the plumbing fixture:

$$\mathcal{F}_{g,n} := \#\mathcal{M}_{g-1,n+2} \; \bigcup \; \left(\bigcup_{\substack{n_1+n_2=n+2 \\ g_1+g_2=g}} \mathcal{M}_{g_1,n_1} \#\mathcal{M}_{g_2,n_2} \right), \qquad (12.42a)$$

$$\mathcal{V}_{g,n} := \mathcal{M}_{g,n} - \mathcal{F}_{g,n}. \qquad (12.42b)$$

In the RHS, it is not necessary to consider multiple non-separating plumbing fixtures for the first term because $\#\mathcal{M}_{g-2,n+4} \subset \mathcal{M}_{g-1,n+2}$, etc. For the same reason, it is sufficient to consider a single separating plumbing fixture. Note that $\mathcal{V}_{g,n}$ and $\mathcal{F}_{g,n}$ are in general not connected subspaces. A simple illustration is given in Fig. 12.8. The actual decomposition of $\mathcal{M}_{0,4}$ is given in Fig. 12.9. Importantly, $\mathcal{F}_{g,n}$ and $\mathcal{V}_{g,n}$ depend on the choice of the local coordinates for all $\mathcal{V}_{g',n'}$ appearing in the RHS.

It is also useful to define the subspaces $\mathcal{F}_{g,n}^{\mathrm{1PR}}$ and $\mathcal{V}_{g,n}^{\mathrm{1PI}}$ of $\mathcal{M}_{g,n}$ that can and cannot be described with the separating plumbing fixture only:

$$\mathcal{F}_{g,n}^{\mathrm{1PR}} := \bigcup_{\substack{n_1+n_2=n+2 \\ g_1+g_2=g}} \mathcal{M}_{g_1,n_1} \#\mathcal{M}_{g_2,n_2}, \qquad (12.43a)$$

$$\mathcal{V}_{g,n}^{\mathrm{1PI}} := \mathcal{M}_{g,n} - \mathcal{F}_{g,n}^{\mathrm{1PR}}. \qquad (12.43b)$$

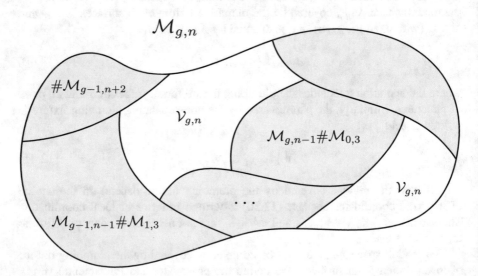

Fig. 12.8 Schematic illustration of the covering of $\mathcal{M}_{g,n}$ from the plumbing fixture of lower-dimensional spaces. The fundamental region $\mathcal{V}_{g,n}$ (usually disconnected) is not covered by the plumbing fixture

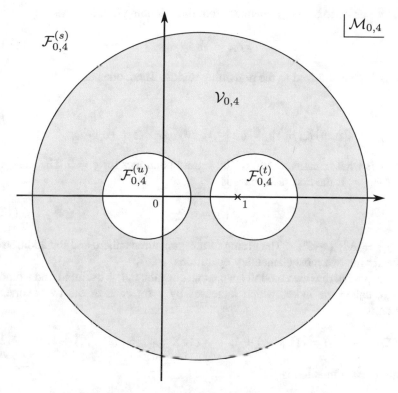

Fig. 12.9 In white are the subspaces of the moduli space $\mathcal{M}_{0,4}$ covered by the plumbing fixture. The three different regions correspond to the three different ways to pair the punctures (see Fig. 12.7). In grey is the fundamental vertex region $\mathcal{V}_{0,4}$

1PR (1PI) stands for 1-particle (ir)reducible, a terminology that will become clear later. Note the relation:

$$\mathcal{V}_{g,n}^{\text{1PI}} = \mathcal{V}_{g,n} \ \cup \ \left(\bigcup_{g'} \#\mathcal{M}_{g-g',n+g'} \right). \tag{12.44}$$

The two plumbing fixtures behave as follow:

- separating: increases both n and g (if both surfaces have a non-vanishing g);
- non-separating plumbing: increases g but decreases n.

The construction is obviously recursive: starting from the lowest-dimensional moduli space, which is $\mathcal{M}_{0,3}$ (no moduli), one has

$$\mathcal{V}_{0,3} = \mathcal{M}_{0,3}, \qquad \mathcal{F}_{0,3} = \emptyset. \tag{12.45}$$

Next, the subspace of $\mathcal{M}_{0,4}$ obtained from the plumbing fixture is

$$\mathcal{F}_{0,4} = \mathcal{V}_{0,3} \# \mathcal{V}_{0,3}, \tag{12.46}$$

and $\mathcal{V}_{0,4}$ is characterized as the remaining region. Then, one has

$$
\begin{aligned}
\mathcal{F}_{0,5} &= \mathcal{M}_{0,4} \# \mathcal{M}_{0,3} \\
&= \mathcal{F}_{0,4} \# \mathcal{V}_{0,3} + \mathcal{V}_{0,4} \# \mathcal{V}_{0,3} = \mathcal{V}_{0,3} \# \mathcal{V}_{0,3} \# \mathcal{V}_{0,3} + \mathcal{V}_{0,4} \# \mathcal{V}_{0,3},
\end{aligned}
\tag{12.47}
$$

and $\mathcal{V}_{0,5}$ is what remains of $\mathcal{M}_{0,5}$. The pattern continues for $g = 0$. The same story holds for $g \geq 1$: the first such space is

$$\mathcal{F}_{1,1} = \# \mathcal{V}_{0,3}, \tag{12.48}$$

and $\mathcal{V}_{1,1} = \mathcal{M}_{1,1} - \mathcal{F}_{1,1}$. The gluing of a 3-punctured sphere and the addition of a handle are the two most elementary operations.

To keep track of which moduli spaces can contribute, it is useful to find a function of $\Sigma_{g,n}$, called the index, which increases by 1 for each of the two elementary operations:

$$r(\Sigma_{g_1,n_1} \# \Sigma_{0,3}) = r(\Sigma_{g_1,n_1}) + 1, \qquad r(\# \Sigma_{g_1,n_1}) = r(\Sigma_{g_1,n_1}) + 1. \tag{12.49}$$

An appropriate function is

$$r(\Sigma_{g,n}) = 3g + n - 2 \in \mathbb{N}^*, \tag{12.50}$$

which is normalized such that

$$r(\Sigma_{0,3}) = 1. \tag{12.51}$$

For a generic separating plumbing fixture, we find

$$r(\Sigma_{g_1,n_1} \# \Sigma_{g_2,n_2}) = r(\Sigma_{g_1,n_1}) + r(\Sigma_{g_2,n_2}). \tag{12.52}$$

Since the index increases, surfaces with a given r can be obtained by considering all the gluings of surfaces with $r' < r$.

12.3.4 Stubs

To conclude this chapter, we introduce the concept of *stubs*. Previously in (12.31), the range of the parameter s was the complete line of positive numbers, $s \in \mathbb{R}_+$. This means that tubes of all lengths were considered to glue surfaces. But, we could also introduce a minimal length $s_0 > 0$, called the *stub parameter*, for the tube. In this case, the plumbing fixture parameter is generalized to

$$q = e^{-s+i\theta}, \qquad s \in [s_0, \infty), \qquad \theta \in [0, 2\pi), \qquad s_0 \geq 0. \qquad (12.53)$$

What is the effect on the subspaces $\mathcal{F}_{g,n}(s_0)$ and $\mathcal{V}_{g,n}(s_0)$? Obviously, less surfaces can be described by the plumbing fixture if $s_0 > 0$ than if $s_0 = 0$, since the plumbing fixture cannot describe anymore surfaces that contain a tube of length less than s_0. Equivalently, the values of the moduli described by the plumbing fixture is more restricted when $s_0 > 0$. More generally, one has

$$s_0 < s_0' : \qquad \mathcal{F}_{g,n}(s_0') \subset \mathcal{F}_{g,n}(s_0) \qquad \mathcal{V}_{g,n}(s_0) \subset \mathcal{V}_{g,n}(s_0'). \qquad (12.54)$$

This is illustrated in Fig. 12.10. Even if s_0 is very large, $\mathcal{V}_{g,n}$ still does not include surfaces arbitrarily close to degeneracy. In general, we omit the dependence in s_0 except when it is necessary.

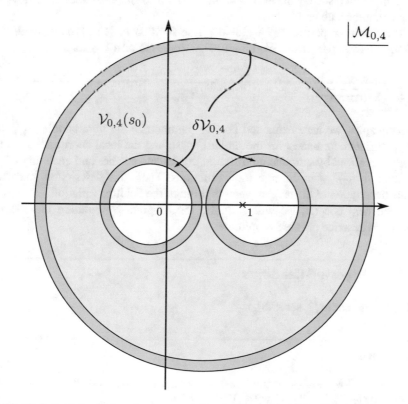

Fig. 12.10 In light grey is the subspace covered by the $\mathcal{V}_{0,4}(s_0)$ as in Fig. 12.9. In dark grey is the difference $\delta\mathcal{V}_{0,4} = \mathcal{V}_{0,4}(s_0 + \delta s_0) - \mathcal{V}_{0,4}(s_0)$ with $\delta s_0 > 0$

To interpret the stub parameter, consider two local coordinates w_1 and w_2 and rescale them by $\lambda \in \mathbb{C}$ with $\mathrm{Re}\,\lambda > 0$:

$$w_1 = \lambda\,\tilde{w}_1, \qquad w_2 = \lambda\,\tilde{w}_2. \tag{12.55}$$

Then, the plumbing fixture (12.30) becomes

$$\tilde{w}_1\tilde{w}_2 = \mathrm{e}^{-\tilde{s}+\mathrm{i}\tilde{\theta}}, \tag{12.56}$$

with

$$\tilde{s} = s + 2\ln|\lambda|, \qquad \tilde{\theta} = \theta + \mathrm{i}\ln\frac{\lambda}{\bar{\lambda}}. \tag{12.57}$$

If $s \in \mathbb{R}_+$, the corresponding range of $\tilde{s}$ is

$$\tilde{s} \in [s_0, \infty), \qquad s_0 := 2\ln|\lambda|. \tag{12.58}$$

This shows that rescaling the local coordinates by a constant parameter is equivalent to changing the stub parameter.

Note also how performing a global phase rotation in (12.57) is equivalent to shifting the twist parameter. Working in $\hat{\mathcal{P}}_{g,n}$ forces to take $\lambda \in \mathbb{R}_+$.

12.4   Summary

In this chapter, we have explained how to parametrize the fibre bundle $\mathcal{P}_{g,n}$, that is, appropriate coordinates for the moduli space and the local coordinate systems. This was realized by introducing different coordinate patches and encoding all the information of $\mathcal{P}_{g,n}$ in the transition functions. Then, this description leads to a simple description of the tangent vectors through the Schiffer variation.

In the next chapter, we will continue the program by building the p-forms required to describe off-shell amplitudes.

12.5   Suggested Readings

- Plumbing fixture [2, sec. 9.3].

References

1. T. Erler, Four lectures on closed string field theory. Phys. Rep. **851**, 1–36 (2020). https://doi.org/10.1016/j.physrep.2020.01.003. arXiv: 1905.06785
2. J. Polchinski, *String Theory: Volume 1, An Introduction to the Bosonic String* (Cambridge University Press, Cambridge, 2005)

Off-Shell Amplitudes

13

Abstract

While the previous chapter was purely geometrical, this one makes contact with string theory through the worldsheet CFT. We continue the description of $\mathcal{P}_{g,n}$ by constructing p-forms. The reason why we need to consider the CFT is that ghosts are necessary to build the p-forms: this can be understood from Chap. 2, where we found that the ghosts must be interpreted as part of the measure on the moduli space. Then, we build the off-shell amplitudes and discuss some properties.

13.1 Cotangent Spaces and Amplitudes

In this section, we construct the p-forms on $\mathcal{P}_{g,n}$ that are needed for the amplitudes. We first motivate the expressions from general ideas and check later that they have the correct properties.

13.1.1 Construction of Forms

A p-form $\omega_p^{(g,n)} \in \bigwedge^p T^* \mathcal{P}_{g,n}$ is a multilinear antisymmetric map from $\bigwedge^p T\mathcal{P}_{g,n}$ to a function of the moduli parameters. The superscript on the form is omitted when there is no ambiguity about the space considered. The components $\omega_{i_1 \cdots i_p}$ of the p-form are defined by inserting p basis vectors $\partial_{s_1}, \ldots, \partial_{s_p}$

$$\omega_{i_1 \cdots i_p} := \omega_p(\partial_{s_1}, \ldots, \partial_{s_p}), \tag{13.1}$$

© Springer Nature Switzerland AG 2021
H. Erbin, *String Field Theory*, Lecture Notes in Physics 980,
https://doi.org/10.1007/978-3-030-65321-7_13

where $\partial_s = \frac{\partial}{\partial x_s}$ and x_s are the coordinates (12.13). It is antisymmetric in any pair of two indices

$$\omega_{i_1 i_2 \cdots i_p} = -\omega_{i_2 i_1 \cdots i_p},\tag{13.2}$$

and multilinearity implies that

$$\omega_p\bigl(V^{(1)}, \ldots, V^{(p)}\bigr) = \omega_p\bigl(V^{(1)}_{s_1}\partial_{s_1}, \ldots, V^{(p)}_{s_p}\partial_{s_p}\bigr) = \omega_{i_1 \cdots i_p} V^{(1)}_{s_1} \cdots V^{(p)}_{s_p},\tag{13.3}$$

given vectors $V^{(\alpha)} = V^{(\alpha)}_s \partial_s$.

The p-forms that are needed to define off-shell amplitudes depend on the external states $\mathcal{V}_i$ $(i = 1, \ldots, n)$ inserted at the punctures z_i. They are maps from $\bigwedge^p T\mathcal{P}_{g,n} \times \mathcal{H}^n$ to a function on $\mathcal{P}_{g,n}$. The dependence on the states is denoted equivalently as

$$\omega_p(\mathcal{V}_1, \ldots, \mathcal{V}_n) := \omega_p(\otimes_i \mathcal{V}_i).\tag{13.4}$$

The simplest way to get a function on $\mathcal{P}_{g,n}$ from the states $\mathcal{V}_i$ is to compute a CFT correlation function of the operators inserted at the points $z_i = f_i(0)$ on the surface $\Sigma_{g,n}$ described by the point in $\mathcal{M}_{g,n}$.

The 0-form is just a function and is defined by

$$\omega_0 = (2\pi i)^{-M^c_{g,n}} \left\langle \prod_{i=1}^n f_i \circ \mathcal{V}_i(0) \right\rangle_{\Sigma_{g,n}}.\tag{13.5}$$

For simplicity, the dependence in the local coordinates f_i is kept implicit in the rest of the chapter.

A natural approach for constructing p-forms is to build them from elementary 1-forms and to use ghosts to enforce the antisymmetry. Remembering the Beltrami differentials found in Chap. 2, the contour integral of ghosts $b(z)$ weighted by some vector field is a good starting point. In the current language, it is defined by its contraction with a vector $V = (v, C) \in T\mathcal{P}_{g,n}$ defined in (12.23)

$$B(V) := \oint_C \frac{dz}{2\pi i} b(z)v(z) + \oint_C \frac{d\bar{z}}{2\pi i} \bar{b}(\bar{z})\bar{v}(\bar{z}),\tag{13.6}$$

where $b(z)$ and $\bar{b}(\bar{z})$ are the b ghost components, and v is the vector field on $\Sigma_{g,n}$ defining V. The contours run anti-clockwise. If the contour C includes several circles $(C = \cup_\alpha C_\alpha)$, $B(V)$ is defined as the sum of the contour integrals on each circle

$$B(V) := \sum_\alpha \oint_{C_\alpha} \frac{dz}{2\pi i} b(z)v(z) + \text{c.c.}.\tag{13.7}$$

It is also useful to define another object built from the energy–momentum tensor

$$T(V) := \oint_C \frac{dz}{2\pi i} T(z)v(z) + \oint_C \frac{d\bar{z}}{2\pi i} \bar{T}(\bar{z})\bar{v}(\bar{z}), \tag{13.8}$$

where T and $\bar{T}$ are the components of the energy–momentum tensor. It is defined such that

$$T(V) = \{Q_B, B(V)\}. \tag{13.9}$$

Considering the coordinate system (12.13), the Beltrami form can be decomposed as

$$B = B_s dx_s, \qquad B_s := B(\partial_s), \tag{13.10a}$$

$$B_s = \sum_\alpha \oint_{C_\alpha} \frac{d\sigma_\alpha}{2\pi i} b(\sigma_\alpha) \frac{\partial F_\alpha}{\partial x_s}(F_\alpha^{-1}(\sigma_\alpha)) + \sum_\alpha \oint_{C_\alpha} \frac{d\bar{\sigma}_\alpha}{2\pi i} \bar{b}(\bar{\sigma}_\alpha) \frac{\partial \bar{F}_\alpha}{\partial x_s}(\bar{F}_\alpha^{-1}(\bar{\sigma}_\alpha)), \tag{13.10b}$$

where the contour orientations are defined by having the σ_α coordinate system on the left.

We define the p-form contracted with a set of vectors $V^{(1)}, \ldots, V^{(p)}$ by

$$\omega_p(V^{(1)}, \ldots, V^{(p)})(\mathscr{V}_1, \ldots, \mathscr{V}_n) := (2\pi i)^{-M_{g,n}^c} \left\langle B(V^{(1)}) \cdots B(V^{(p)}) \prod_{i=1}^n \mathscr{V}_i \right\rangle_{\Sigma_{g,n}}, \tag{13.11}$$

and the corresponding p-form reads

$$\omega_p = \omega_{p,s_1 \cdots s_p} dx_{s_1} \wedge \cdots \wedge dx_{s_p} \tag{13.12a}$$

$$= (2\pi i)^{-M_{g,n}^c} \left\langle B_{s_1} dx_{s_1} \wedge \cdots \wedge B_{s_p} dx_{s_p} \prod_{i=1}^n \mathscr{V}_i \right\rangle_{\Sigma_{g,n}}. \tag{13.12b}$$

In this expression, the form contains an infinite number of components $\omega_{p,s_1 \cdots s_p}$ since there is an infinite number of coordinates. Note that the normalization is independent of p.

In practice, one is not interested in $\mathcal{P}_{g,n}$, but rather in a subspace of it. Given a q-dimensional subspace $\mathcal{S}$ of $\mathcal{P}_{g,n}$ parametrized by q real coordinates $t_1, \ldots, t_q$

$$x_s = x_s(t_1, \ldots, t_q), \tag{13.13}$$

the restriction of a p-form to this subspace is obtained by the chain rule

$$\forall p \le q: \quad \omega_p|_S = (2\pi i)^{-M_{g,n}^c} \left\langle B_{r_1} \frac{\partial x_{s_1}}{\partial t_{r_1}} \, dt_{r_1} \wedge \cdots \wedge B_{r_p} \frac{\partial x_{s_p}}{\partial t_{r_p}} \, dt_{r_p} \prod_{i=1}^{n} \mathscr{V}_i \right\rangle_{\Sigma_{g,n}},$$

$$\forall p > q: \quad \omega_p|_S = 0.$$

$$(13.14)$$

We will often write the expression directly in terms of the coordinates of S and abbreviate the notation as

$$B_r := \frac{\partial x_s}{\partial t_r} B_s.$$

$$(13.15)$$

13.1.2 Amplitudes and Surface States

It is now possible to write the amplitude more explicitly. An on-shell amplitude is defined as an integral over $\mathcal{M}_{g,n}$. Off-shell, one needs to consider local coordinates around each puncture, that is, a point of the fibre for each point of the base $\mathcal{M}_{g,n}$. This defines a $\mathsf{M}_{g,n}$-dimensional section $\mathcal{S}_{g,n}$ of $\mathcal{P}_{g,n}$ (Fig. 11.2). The g-loop n-point off-shell amplitude of the states $\mathscr{V}_1, \ldots, \mathscr{V}_n$ reads

$$A_{g,n}(\mathscr{V}_1, \ldots, \mathscr{V}_n)_{\mathcal{S}_{g,n}} := \int_{\mathcal{S}_{g,n}} \omega_{\mathsf{M}_{g,n}}^{g,n}(\mathscr{V}_1, \ldots, \mathscr{V}_n)\Big|_{\mathcal{S}_{g,n}}, \qquad (13.16a)$$

$$\omega_{\mathsf{M}_{g,n}}^{g,n}(\mathscr{V}_1, \ldots, \mathscr{V}_n)\Big|_{\mathcal{S}_{g,n}} = (2\pi i)^{-M_{g,n}^c} \left\langle \bigwedge_{\lambda=1}^{\mathsf{M}_{g,n}} B_s \frac{\partial x_s}{\partial t_\lambda} \, dt_\lambda \prod_{i=1}^{n} f_i \circ \mathscr{V}_i(0) \right\rangle_{\Sigma_{g,n}},$$

$$(13.16b)$$

where the choice of the f_i is dictated by the section $\mathcal{S}_{g,n}$. From now on, we stop to write the restriction of the form to the section. We also restrict to the cases where $\chi_{g,n} = 2 - 2g - n < 0$.

The complete (perturbative) n-point amplitude is the sum of contributions from all loops

$$A_n(\mathscr{V}_1, \ldots, \mathscr{V}_n) := \sum_{g \ge 0} A_{g,n}(\mathscr{V}_1, \ldots, \mathscr{V}_n). \qquad (13.17)$$

More generally, we define the integral over a section $\mathcal{R}_{g,n}$ whose projection on the base is a subspace of $\mathcal{M}_{g,n}$ (and not the full space as for the amplitude) as

$$\mathcal{R}_{g,n}(\mathscr{V}_1, \ldots, \mathscr{V}_n) := \int_{\mathcal{R}_{g,n}} \omega_{\mathsf{M}_{g,n}}^{g,n}(\mathscr{V}_1, \ldots, \mathscr{V}_n). \qquad (13.18)$$

For simplicity, we will sometimes use the same notation for the section of $\mathcal{P}_{g,n}$ and its projection on the base $\mathcal{M}_{g,n}$. For this reason, the reader should assume that some choice of local coordinates around the punctures is made except otherwise stated.

Given sections $\mathcal{R}_{g,n}$, the sum over all genus contribution is written formally as

$$\mathcal{R}_n := \sum_{g \geq 0} \mathcal{R}_{g,n}, \tag{13.19}$$

such that

$$\mathcal{R}_n(\mathcal{V}_1, \ldots, \mathcal{V}_n) := \sum_{g \geq 0} \mathcal{R}_{g,n}(\mathcal{V}_1, \ldots, \mathcal{V}_n) = \sum_{g \geq 0} \int_{\mathcal{R}_{g,n}} \omega_{\mathsf{M}_{g,n}}^{g,n}(\mathcal{V}_1, \ldots, \mathcal{V}_n). \tag{13.20}$$

A *surface state* is defined as a n-fold bra that reproduces the expression of a given function when contracted with n states A_i. The surface $\langle \Sigma^{g,n}|$, form $\langle \omega^{g,n}|$, section $\langle \mathcal{R}_{g,n}|$ and amplitude $\langle A^{g,n}|$ n-fold states are defined by the following expressions:

$$\langle \Sigma_{g,n}| B_{s_1} \cdots B_{s_p} | \otimes_i \mathcal{V}_i \rangle := \omega_{s_1 \cdots s_p}(\mathcal{V}_1, \ldots, \mathcal{V}_n), \tag{13.21a}$$

$$\langle \omega_p^{g,n}| \otimes_i \mathcal{V}_i \rangle := \omega_p(\mathcal{V}_1, \ldots, \mathcal{V}_n), \tag{13.21b}$$

$$\langle A_{g,n}| \otimes_i \mathcal{V}_i \rangle := A_{g,n}(\mathcal{V}_1, \ldots, \mathcal{V}_n). \tag{13.21c}$$

The last relation is generalized to any section $\mathcal{R}_{g,n}$

$$\langle \mathcal{R}_{g,n}| \otimes_i \mathcal{V}_i \rangle := \mathcal{R}_{g,n}(\mathcal{V}_1, \ldots, \mathcal{V}_n). \tag{13.21d}$$

The reason for introducing these objects is that the form (13.12) is a linear map from $\mathcal{H}^{\otimes n}$ to a form on $\mathcal{M}_{g,n}$—see (13.4). Thus, there is always a state $\langle \Sigma_{g,n}|$ such that its BPZ product with the states reproduces the form. In particular, the state $\langle \Sigma_{g,n}|$ contains all the information about the local coordinates and the moduli (the dependence is kept implicit). The definitions of the other states follow similarly. These states are defined as bras, but they can be mapped to kets.

One finds the obvious relations

$$\langle \omega_p^{g,n}| = \langle \Sigma_{g,n}| B_{s_1} dx^{s_1} \cdots B_{s_p} dx^{s_p}, \qquad \langle A_{g,n}| = \int_{\mathcal{M}_{g,n}} \langle \omega_p^{g,n}|. \tag{13.22}$$

The surface states do not contain information about the matter CFT: they collect the universal data (like local coordinates) needed to describe amplitudes. Hence, it is an important step in the description of off-shell string theory to characterize this data. However, note that the relation between a surface state and the corresponding form *does depend* on the CFT.

Example 13.1: On-Shell Amplitude $A_{0,4}$

The transition functions are given by (see Fig. 12.1):

$$
\begin{aligned}
&C_1 : w_1 = z_1 - y_1, && C_3 : w_3 = z_2 - y_3, && C_5 : z_1 = z_2, \\
&C_2 : w_2 = z_1 - y_2, && C_4 : w_4 = z_2 - y_4.
\end{aligned}
\tag{13.23}
$$

Three of the parameters (y_1, y_2 and y_3) are fixed, while the single complex modulus of $\mathcal{M}_{0,4}$ is taken to be y_4. Since we are interested in the on-shell amplitude, it is not necessary to introduce local coordinates and the associated parameters.

A variation of the modulus

$$
y_4 \longrightarrow y_4 + \delta y_4, \qquad \bar{y}_4 \longrightarrow \bar{y}_4 + \delta \bar{y}_4
\tag{13.24}
$$

is equivalent to a change in the transition function of C_4. This translates in turn into a transformation of z_2

$$
z_2' = z_2 + \delta y_4, \qquad \bar{z}_2' = \bar{z}_2 + \delta \bar{y}_4.
\tag{13.25}
$$

Then, the tangent vector $V = \partial_{y_4}$ is associated to the vector field

$$
v = 1, \qquad \bar{v} = 0,
\tag{13.26}
$$

with support on C_4. For $V = \partial_{\bar{y}_4}$, one finds

$$
v = 0, \qquad \bar{v} = 1.
\tag{13.27}
$$

The Beltrami 1-forms for the unit vectors are

$$
B(\partial_{y_4}) = \oint_{C_4} dz_2 \, b(z_2)(+1), \qquad B(\partial_{\bar{y}_4}) = \oint_{C_4} d\bar{z}_2 \, \bar{b}(\bar{z}_2)(+1),
\tag{13.28}
$$

with both contours running anti-clockwise.

The components of the 2-form read

$$
\omega_2(\partial_{y_4}, \partial_{\bar{y}_4}) = \frac{1}{2\pi i} \left\langle B(\partial_{y_4}) B(\partial_{\bar{y}_4}) \prod_{i=1}^{4} \mathcal{V}_i \right\rangle_{\Sigma_{0,4}}
$$

$$
= \frac{1}{2\pi i} \left\langle \oint_{C_4} dz_2 \, b(z_2) \oint_{C_4} d\bar{z}_2 \, \bar{b}(\bar{z}_2) \prod_{i=1}^{4} \mathcal{V}_i \right\rangle_{\Sigma_{0,4}}.
$$

For on-shell states $\mathscr{V}_i = c\bar{c}V_i(y_i, \bar{y}_i)$, this becomes

$$\omega_2(\partial_{y_4}, \partial_{\bar{y}_4}) = \frac{1}{2\pi i} \left\langle \prod_{i=1}^{3} c\bar{c}V_i(y_i, \bar{y}_i) \oint_{C_4} dz_2\, b(z_2) \right.$$

$$\left. \times \oint_{C_4} d\bar{z}_2\, \bar{b}(\bar{z}_2)\bar{c}(\bar{y}_4)c(y_4)V_4(y_4, \bar{y}_4) \right\rangle \Bigg|_{\Sigma_{0,4}}.$$

The first three operators could be moved to the left because they are not encircled by the integration contour. Note the difference with the example discussed in Sect. 11.1.2: here, the contour encircles z_3, while it was encircling y_3 for the s-channel.

Using the OPE

$$\oint_{C_4} dz_2\, b(z_2)c(y_4) \sim \oint_{C_4} dz_2\, \frac{1}{z_2 - y_4} \tag{13.29}$$

to simplify the product of b and c gives the amplitude

$$A_{0,4} = \frac{1}{2\pi i} \int dy_4 \wedge d\bar{y}_4 \left\langle \prod_{i=1}^{3} c\bar{c}V_i(y_i)\, V_4(y_4) \right\rangle \Bigg|_{\Sigma_{0,4}}. \tag{13.30}$$

This is the standard formula for the 4-point function derived from the Polyakov path integral. ◄

13.2 Properties of Forms

In this section, we check that the form (13.12) has the correct properties

- antisymmetry under exchange of two vectors;
- given a trivial vector of (a subspace of) $\mathcal{P}_{g,n}$ (Sect. 12.1), its contraction with the form vanishes: $\omega_p(V^{(1)}, \ldots, V^{(p)}) = 0$ if any of the $V^{(i)}$ generates:
  - reparametrizations of z_a for $V^{(i)} \in T\mathcal{P}_{g,n}$,
  - rotation $w_i \to (1 + i\alpha_i)w_i$ for $V^{(i)} \in T\hat{\mathcal{P}}_{g,n}$,
  - reparametrizations of w_i keeping $w_i = 0$ if the states are on-shell for $V^{(i)} \in T\mathcal{M}_{g,n}$;
- BRST identity, which is necessary to prove several properties of the amplitudes.

The first property is obvious. Indeed, the form is correctly antisymmetric under the exchange of two vectors $V^{(i)}$ and $V^{(j)}$ due to the ghost insertions.

13.2.1 Vanishing of Forms with Trivial Vectors

Reparametrization of z_a Consider the sphere S_a with coordinate z_a, and denote by C_1, C_2 and C_3 the three boundaries. Then, a reparametrization

$$z_a \longrightarrow z_a + \phi(z_a) \tag{13.31}$$

is generated by a vector field $\phi(z)$ that is regular on S_a. This transformation modifies the transition functions on the three circles and is thus associated to a tangent vector V described by a vector field v with support on the three circles

$$C_i : \quad v^{(i)} = \phi|_{C_i}. \tag{13.32}$$

The Beltrami form then reads

$$B(V) = \sum_{i=1}^{3} \oint_{C_i} dz_a \, b(z_a)\phi(z_a) + \text{c.c.}, \tag{13.33}$$

where the orientations of the contours are such that S_a is on the left. Since the vector field ϕ is regular in S_a, two of the contours can be deformed until they merge together. The resulting orientation is opposite to the one of the last contours (Fig. 13.1). As a consequence, both cancel and the integral vanishes.

Rotation of w_i Consider an infinitesimal phase rotation of the local coordinate w_i in the disk D_i

$$w_i \longrightarrow (1 + i\alpha_i)w_i, \qquad \bar{w}_i \longrightarrow (1 - i\alpha_i)\bar{w}_i, \tag{13.34}$$

with $\alpha_i \in \mathbb{R}$. The tangent vector is defined by the circle C_i and the vector field by

$$v = iw_i, \qquad \bar{v} = -i\bar{w}_i. \tag{13.35}$$

Fig. 13.1 Deformation of the contour of integration defining the Beltrami form for a reparametrization of z_a. The figure is drawn for two circles at a hole, but the proof is identical for other types of circles

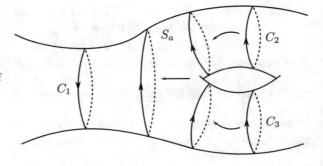

The Beltrami form for this vector is

$$B(V) = i \oint_{C_i} dw_i \, w_i \, b(w_i) - i \oint_{C_i} d\bar{w}_i \, \bar{w}_i \, \bar{b}(\bar{w}_i), \tag{13.36}$$

where D_i is kept to the left.

In the p-form (13.11), the ith operator $\mathscr{V}_i$ is inserted in D_i and encircled by C_i. Because there is no other operator inside D_i, the contribution of this disk to the form is

$$B(V)\mathscr{V}_i(0) = i \oint_{C_i} dw_i \, w_i \, b(w_i)\mathscr{V}_i(0) - i \oint_{C_i} d\bar{w}_i \, \bar{w}_i \, \bar{b}(\bar{w}_i)\mathscr{V}_i(0). \tag{13.37}$$

The state–operator correspondence allows to rewrite this result as

$$i(b_0 - \bar{b}_0) \, |\mathscr{V}_i\rangle, \tag{13.38}$$

since the contour integral picks the zero-modes of b and of $\bar{b}$. Requiring that the form vanishes implies the ghost counter-part of the level-matching condition

$$b_0^- \, |\mathscr{V}_i\rangle = 0. \tag{13.39}$$

Hence, consistency of off-shell amplitudes implies that

$$\mathscr{V}_i \in \mathcal{H}^-, \tag{13.40}$$

where $\mathcal{H}^-$ is defined in (11.38).

Reparametrization of w_i A reparametrization of the local coordinate w_i keeping the origin of D_i fixed reads

$$w_i \longrightarrow f(w_i), \qquad f(0) = 0. \tag{13.41}$$

The function can be expanded in series

$$f(w_i) = \sum_{m \geq 0} p_m w_i^{m+1}. \tag{13.42}$$

Because the transformation is holomorphic, it can be extended on C_i. Each parameter p_m provides a coordinate of $\mathcal{P}_{g,n}$ and whose deformation corresponds to a vector field

$$v_m = w_i^{m+1}, \qquad \bar{v}_m = 0. \tag{13.43}$$

The corresponding Beltrami differential is

$$B(\partial_{p_m}) = \oint_{C_i} dw_i\, b(w_i) w_i^{m+1}.$$ (13.44)

Since only the operator $\mathcal{V}_i$ is inserted in the disk, the state–operator correspondence gives $b_m |\mathcal{V}_i\rangle$. Requiring that the form vanishes on $\mathcal{M}_{g,n}$ for all m and also for the anti-holomorphic vectors gives the conditions

$$\forall m \geq 0: \qquad b_m |\mathcal{V}_i\rangle = 0, \qquad \bar{b}_m |\mathcal{V}_i\rangle = 0.$$ (13.45)

This holds automatically for on-shell states $\mathcal{V}_i = c\bar{c}V_i$.

13.2.2 BRST Identity

The BRST identity for the p-form (13.12) reads

$$\omega_p\Big(\sum_i Q_B^{(i)} \otimes_i \mathcal{V}_i\Big) = (-1)^p d\omega_{p-1}(\otimes\mathcal{V}_i),$$ (13.46)

using the notation (13.4). The BRST operator acting on the ith Hilbert space is written as

$$Q_B^{(i)} = 1_{i-1} \otimes Q_B \otimes 1_{n-i}$$ (13.47)

and acts as

$$Q_B \mathcal{V}_i(z, \bar{z}) = \frac{1}{2\pi i} \oint dw\, j_B(w) \mathcal{V}_i(z, \bar{z}) + \text{c.c.}.$$ (13.48)

More explicitly, the LHS corresponds to

$$\omega_p\Big(\sum_i Q_B^{(i)} \otimes_i \mathcal{V}_i\Big) = \omega_p(Q_B\mathcal{V}_1, \mathcal{V}_2, \ldots, \mathcal{V}_n) + (-1)^{|\mathcal{V}_1|}\omega_p(\mathcal{V}_1, Q_B\mathcal{V}_2, \ldots, \mathcal{V}_n)$$

$$+ \cdots + (-1)^{|\mathcal{V}_1|+\cdots+|\mathcal{V}_{n-1}|}\omega_p(\mathcal{V}_1, \mathcal{V}_2, \ldots, Q_B\mathcal{V}_n).$$ (13.49)

We give just an hint of this identity, and the complete proof can be found in [9, pp. 85–89, 6, sec. 2.5].

The contour of the BRST current around each puncture can be deformed, picking singularities due to the presence of the Beltrami forms. Using (13.9), we find that anti-commuting the BRST charge with the Beltrami form B_s leads to an insertion of

$$T_s = \{Q_B, B_s\}.$$ (13.50)

The energy–momentum tensor generates changes of coordinates. Hence, $T_s = T_{\partial_s}$ is precisely the generator associated to an infinitesimal change of the coordinate x_s on $\mathcal{P}_{g,n}$. The latter is given by the vector ∂_s. For this reason, one can write

$$\mathrm{d}x_s \{Q_B, B_s\} = \mathrm{d}x_s\, T_s = \mathrm{d}x_s\, \partial_s = \mathrm{d}, \tag{13.51}$$

where d is the exterior derivative on $\mathcal{P}_{g,n}$. The minus signs arise if the states $\mathcal{V}_i$ are Grassmann odd.

13.3   Properties of Amplitudes

In order for the p-form (13.12) to be non-vanishing, its total ghost number should match the ghost number anomaly

$$N_{\mathrm{gh}}\big(\omega_p(\mathcal{V}_1, \ldots, \mathcal{V}_n)\big) = \sum_{i=1}^{n} N_{\mathrm{gh}}(\mathcal{V}_i) - p = 6 - 6g, \tag{13.52}$$

using $N_{\mathrm{gh}}(B) = -1$. For an amplitude, one has $p = \mathsf{M}_{g,n} = 6g - 6 + 2n$, and thus,

$$N_{\mathrm{gh}}(\omega_{\mathsf{M}_{g,n}}) = 6 - 6g \quad \Longrightarrow \quad \sum_{i=1}^{n} N_{\mathrm{gh}}(\mathcal{V}_i) = 2n. \tag{13.53}$$

This condition holds automatically for on-shell states since $N_{\mathrm{gh}}(c\bar{c}V_i) = 2$.

13.3.1   Restriction to $\hat{\mathcal{P}}_{g,n}$

The goal of this section is to explain why amplitudes must be described in terms of a section of $\hat{\mathcal{P}}_{g,n}$ (12.10) instead of $\mathcal{P}_{g,n}$. This means that one should identify local coordinates differing by a global phase rotation.

The off-shell amplitudes (13.16) are multi-valued on $\mathcal{P}_{g,n}$. Indeed, the amplitude depends on the local coordinates[1] and changes by a factor under a global phase rotation of any local coordinate $w_i \to e^{i\alpha} w_i$. However, such a global rotation leaves the surface unchanged, since the flat metric $|\mathrm{d}w_i|^2$ is invariant. This means that the same surface leads to different values for the amplitude. To prevent this multi-valuedness of the amplitudes, it is necessary to identify local coordinates differing by a constant phase.

A second way to obtain this condition is to require that the section $\mathcal{S}_{g,n}$ is globally defined: every point of the section should correspond to a single point of the moduli space $\mathcal{M}_{g,n}$. However, there is a topological obstruction that prevents finding a

---

[1]The current argument does not apply for on-shell amplitudes.

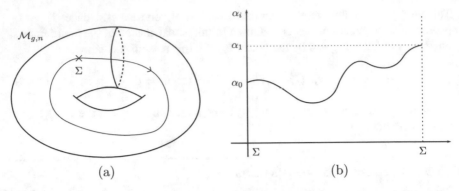

Fig. 13.2 Schematic plot of the change in the phase of the local coordinate w_i as one follows a closed curve in $\mathcal{M}_{g,n}$. If the original phase at Σ is α_0 and if the phase varies continuously along the path, then $\alpha_1 \neq \alpha_0$ when returning back to Σ by continuity. (a) Closed curve in $\mathcal{M}_{g,n}$. (b) Change in the phase of w_i

global section in $\mathcal{P}_{g,n}$ in general. One hint [1, sec. 2, 3, sec. 3] is to exhibit a nowhere vanishing 1-form if $\mathcal{S}_{g,n}$ is globally defined: this leads to a contradiction since such a 1-form does not generally exist (see for example [2, sec. 6.3.2, ch. 7]). Then, consider a closed curve in the moduli space (such curves exist since $\mathcal{M}_{g,n}$ is compact). Starting at a given point Σ of the curve, one finds that the local coordinates typically change by a global phase when coming back to the point Σ (Fig. 13.2), since this describes the same surface and there is no reason to expect the phase to be invariant. Up to this identification, it is possible to find a global section. The latter corresponds to a section of $\hat{\mathcal{P}}_{g,n}$.

Remark 13.1 (Degeneracy of the Antibracket) It is possible to define a BV structure on Riemann surfaces [7,8]. The antibracket is degenerate in $\mathcal{P}_{g,n}$ but not in $\hat{\mathcal{P}}_{g,n}$ [8].

Global phase rotations of the local coordinates are generated by L_0^-. Hence, identifying the local coordinates $w_i \rightarrow e^{i\alpha_i} w_i$ amounts to require that the amplitude is invariant under L_0^-. This is equivalent to imposing the level-matching condition

$$L_0^- \, |\mathcal{V}_i\rangle = 0 \tag{13.54}$$

on the off-shell states. This condition was interpreted in Sect. 3.2.2 as a gauge fixing condition for translations along the S^1 of the string. This shows, in agreement with earlier comments, that the level-matching condition should also be imposed off-shell because no gauge symmetry is introduced for the corresponding transformation.

If the generator L_0^- is trivial, this means that the ghost associated to the corresponding tangent vector must be decoupled. According to Sect. 13.2.1, this corresponds to the constraint

$$b_0^- \, |\mathcal{V}_i\rangle = 0. \tag{13.55}$$

This can be interpreted as a gauge fixing condition (Sect. 10.5), which could in principle be relaxed. However, the decoupling of physical states (equivalent to gauge invariance in SFT) happens only after integrating over the moduli space. This requires having a globally defined section.

As a consequence, off-shell states are elements of the semi-relative Hilbert space

$$\mathscr{V}_i \in \mathcal{H}^- \cap \ker L_0^-, \tag{13.56}$$

and the amplitudes are defined by integrating the form $\omega_{M_{g,n}}$ over a section $\mathcal{S}_{g,n} \subset \hat{\mathcal{P}}_{g,n}$.

Computation: Equation (13.54)
The operator associated to the state through $|A_i\rangle = A_i(0)|0\rangle$ transforms as

$$\mathscr{V}_i(0) \longrightarrow (e^{i\alpha_i})^h (e^{-i\alpha_i})^{\bar{h}} \mathscr{V}_i(0), \tag{13.57}$$

which translates into

$$|\mathscr{V}_i\rangle \longrightarrow e^{i\alpha_i(L_0 - \bar{L}_0)} |\mathscr{V}_i\rangle \tag{13.58}$$

for the state, using the fact that the vacuum is invariant under L_0 and $\bar{L}_0$. Then, requiring the invariance of the state leads to (13.54).

13.3.2 Consequences of the BRST Identity

Two important properties of the on-shell amplitudes can be deduced from the BRST identity (13.46): the independence of physical results on the choice of local coordinates and the decoupling of pure gauge states.

Given BRST closed states, the LHS of (13.46) vanishes identically

$$\forall i: \quad Q_B |\mathscr{V}_i\rangle = 0 \quad \Longrightarrow \quad d\omega_{p-1}(\mathscr{V}_1, \ldots, \mathscr{V}_n) = 0. \tag{13.59}$$

Using this result, one can compare the on-shell amplitudes computed for two different sections $\mathcal{S}$ and $\mathcal{S}'$

$$\int_{\mathcal{S}} \omega_{M_{g,n}} - \int_{\mathcal{S}'} \omega_{M_{g,n}} = \int_{\partial \mathcal{T}} \omega_{M_{g,n}-1} = \int_{\mathcal{T}} d\omega_{M_{g,n}-1} = 0, \tag{13.60}$$

using Stokes' theorem and where $\mathcal{T}$ is the surface delimited by the two sections (Fig. 13.3). This implies that on-shell amplitudes do not depend on the section, and thus on the local coordinates. In obtaining the result, one needs to assume that the vertical segments do not contribute. The latter corresponds to boundary

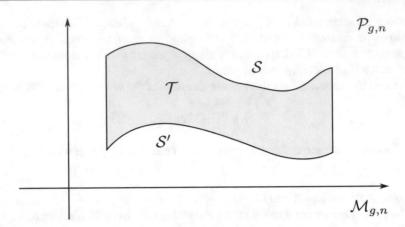

Fig. 13.3 Two sections S and S' of $\mathcal{P}_{g,n}$ delimiting a surface $\mathcal{T}$

contributions of the moduli space. In general, many statements hold up to this condition, which we will not comment more in this book.

Next, we consider a pure gauge state together with BRST closed states

$$|\mathscr{V}_1\rangle = Q_B |\Lambda\rangle, \qquad Q_B |\mathscr{V}_i\rangle = 0. \tag{13.61}$$

The BRST identity (13.46) reads

$$\omega_{\mathrm{M}_{g,n}}(Q_B \Lambda, \mathscr{V}_2, \ldots, \mathscr{V}_n) = \mathrm{d}\omega_{\mathrm{M}_{g,n}-1}(\Lambda, \mathscr{V}_2, \ldots, \mathscr{V}_n), \tag{13.62}$$

which gives the amplitude

$$\int_S \omega_{\mathrm{M}_{g,n}}(Q_B \Lambda, \mathscr{V}_2, \ldots, \mathscr{V}_n) = \int_S \mathrm{d}\omega_{\mathrm{M}_{g,n}-1}(\Lambda, \mathscr{V}_2, \ldots, \mathscr{V}_n)$$

$$= \int_{\partial S} \omega_{\mathrm{M}_{g,n}-1}(\Lambda, \mathscr{V}_2, \ldots, \mathscr{V}_n), \tag{13.63}$$

where the last equality follows from Stokes' theorem. Assuming again that there is no boundary contribution, this vanishes

$$\int_S \omega_{\mathrm{M}_{g,n}}(Q_B \Lambda, \mathscr{V}_2, \ldots, \mathscr{V}_n) = 0. \tag{13.64}$$

This implies that pure gauge states decouple from the physical states.

13.4   Suggested Readings

* Definition of the forms [3, 4, 6, 9].
* Global phase rotation of local coordinates [1, sec. 2, 3, sec. 3, 5, 9, p. 54].

References

1. J. Distler, P. Nelson, Topological couplings and contact terms in 2d field theory. en. Commun. Math. Phys. **138**(2), 273–290 (1991). https://doi.org/10.1007/BF02099493
2. S. Donaldson, *Riemann Surfaces* (English), 1st edn. (Oxford University Press, Oxford, 2011)
3. T. Erler, Four lectures on closed string field theory. Phys. Rep. **851**, 1–36 (2020). https://doi.org/10.1016/j.physrep.2020.01.003. arXiv: 1905.06785
4. T. Erler, S. Konopka, Vertical integration from the large Hilbert space. J. High Energy Phys. **2017**(12), 112 (2017). https://doi.org/10.1007/JHEP12(2017)112. arXiv: 1710.07232
5. P. Nelson, Covariant insertion of general vertex operators. Phys. Rev. Lett. **62**(9), 993–996 (1989). https://doi.org/10.1103/PhysRevLett.62.993
6. A. Sen, Off-shell amplitudes in superstring theory. Fortschritte der Physik **63**(3–4), 149–188 (2015). https://doi.org/10.1002/prop.201500002. arXiv: 1408.0571
7. A. Sen, B. Zwiebach, Quantum background independence of closed string field theory. Nucl. Phys. B **423**(2–3), 580–630 (1994). https://doi.org/10.1016/0550-3213(94)90145-7. arXiv: hep-th/9311009
8. A. Sen, B. Zwiebach, Background independent algebraic structures in closed string field theory. Commun. Math. Phys. **177**(2), 305–326 (1996). https://doi.org/10.1007/BF02101895. arXiv: hep-th/9408053
9. B. Zwiebach, Closed string field theory: quantum action and the BV master equation. Nucl. Phys. B **390**(1), 33–152 (1993). https://doi.org/10.1016/0550-3213(93)90388-6. arXiv: hep-th/9206084

Amplitude Factorization and Feynman Diagrams

14

Abstract

In the previous chapter, we built the off-shell amplitudes by integrating forms on sections of $\mathcal{P}_{g,n}$. Studying their factorizations leads to rewrite them in terms of Feynman diagrams, which allows to identify the fundamental interactions' vertices. We will then be able to write the SFT action in the next chapter.

14.1 Amplitude Factorization

We have seen how to write off-shell amplitudes. The next step is to rewrite them as a sum of Feynman diagrams through factorization of amplitudes.

Factorization consists in writing a g-loop n-point amplitude in terms of lower-order amplitudes in both g and n connected by propagators. Since an amplitude corresponds to a sum over all possible processes, which corresponds to integrating over the moduli space, it is natural to associate Feynman diagrams to different subspaces of the moduli space. One can expect that the plumbing fixture (Sect. 12.3) is the appropriate translation of the factorization at the level of Riemann surfaces. We will assume that it is the case and check that it is correct a posteriori.

To proceed, we consider the contribution to the amplitude $A_{g,n}$ of the family of surfaces obtained by the plumbing fixture of two surfaces (separating case) or a surface with itself (non-separating case).

14.1.1 Separating Case

In this section, we consider the separating plumbing fixture where part of the moduli space $\mathcal{M}_{g,n}$ is covered by $\mathcal{M}_{g_1,n_1} \# \mathcal{M}_{g_2,n_2}$ with $g = g_1 + g_2$ and $n = n_1 + n_2 - 2$ (Sect. 12.3.1). The local coordinates read $w_i^{(1)}$ and $w_j^{(2)}$ for $i = 1, \ldots, n_1$ and $j =$

© Springer Nature Switzerland AG 2021
H. Erbin, *String Field Theory*, Lecture Notes in Physics 980,
https://doi.org/10.1007/978-3-030-65321-7_14

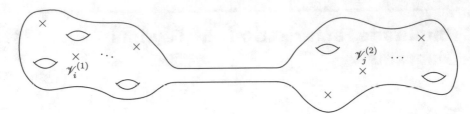

Fig. 14.1 Degeneration limit of $\Sigma_{g,n}$ where the punctures $\mathscr{V}_i^{(1)}$ and $\mathscr{V}_j^{(2)}$ move apart from each other

$1, \ldots, n_2$. By convention, the last coordinate of each set is used for the plumbing fixture:

$$w_{n_1}^{(1)} w_{n_2}^{(2)} = q. \tag{14.1}$$

The g-loop n-point amplitude with external states $\{\mathscr{V}_1^{(1)}, \ldots, \mathscr{V}_{n_1-1}^{(1)},$ $\mathscr{V}_1^{(2)}, \ldots, \mathscr{V}_{n_2-1}^{(2)}\}$ is denoted as

$$A_{g,n} = \int_{\mathcal{S}_{g,n}} \omega_{\mathrm{M}_{g,n}}^{g,n} \left(\mathscr{V}_1^{(1)}, \ldots, \mathscr{V}_{n_1-1}^{(1)}, \mathscr{V}_1^{(2)}, \ldots, \mathscr{V}_{n_2-1}^{(2)} \right). \tag{14.2}$$

We need to study the form $\omega_{\mathrm{M}_{g,n}}^{g,n}$ on $\mathcal{M}_{g_1,n_1} \# \mathcal{M}_{g_2,n_2}$, which means to rewrite it in terms of the data from $\mathcal{M}_{g_1,n_1}$ and $\mathcal{M}_{g_2,n_2}$. This corresponds to the degeneration limit where the two groups of punctures denoted by $\mathscr{V}_i^{(1)}$ and $\mathscr{V}_j^{(2)}$ ($i = 1, \ldots, n_1 - 1$, $j = 1, \ldots, n_2 - 1$) together with g_1 and g_2 holes move apart from each other (Fig. 14.1).

Since q is a coordinate of $\mathcal{P}_{g,n}$, its variation is associated with a tangent vector and a Beltrami 1-form. The latter has to be inserted inside $\omega_{\mathrm{M}_{g,n}}^{g,n}$. A change $q \to q + \delta q$ translates into a change of coordinate

$$w_{n_1}^{\prime(1)} = w_{n_1}^{(1)} + \frac{w_{n_1}^{(1)}}{q} \delta q, \tag{14.3}$$

where $w_{n_2}^{(2)}$ is kept fixed (obviously, this choice is conventional as explained in Sect. 12.2). Thus, the vector field and the Beltrami form are

$$v_q = \frac{w_{n_1}^{(1)}}{q}, \qquad B_q = \frac{1}{q} \oint_{C_q} dw_{n_1}^{(1)} \, b\big(w_{n_1}^{(1)}\big) w_{n_1}^{(1)}. \tag{14.4}$$

Computation: Equation (14.3)

Starting from (14.1), vary $q \to q + \delta q$ while keeping $w_{n_2}^{(2)}$ fixed:

$$w_{n_1}^{\prime(1)} w_{n_2}^{(2)} = q + \delta q$$

$$w_{n_1}^{\prime(1)} = \frac{w_{n_1}^{(1)}}{q}(q + \delta q) = w_{n_1}^{(1)} + \frac{q}{w_{n_1}^{(1)}} \delta q.$$

The second line follows by replacing $w_{n_2}^{(2)}$ using (14.1).

The $\mathsf{M}_{g,n}$-form for the moduli described by the plumbing fixture can be expressed as

$$\omega_{\mathsf{M}_{g,n}}\left(V_1^{(1)}, \ldots, V_{\mathsf{M}_{g_1,n_1}}^{(1)}, \partial_q, \partial_{\bar{q}}, V_1^{(2)}, \ldots, V_{\mathsf{M}_{g_2,n_2}}^{(2)}\right)$$

$$= (2\pi \mathrm{i})^{-\mathsf{M}_{g,n}^c} \left\langle \prod_{\lambda=1}^{\mathsf{M}_{g_1,n_1}} B(V_\lambda^{(1)}) B(\partial_q) B(\partial_{\bar{q}}) \prod_{\kappa=1}^{\mathsf{M}_{g_2,n_2}} B(V_\kappa^{(2)}) \prod_{i=1}^{n_1-1} \mathscr{V}_i^{(1)} \prod_{j=1}^{n_2-1} \mathscr{V}_j^{(2)} \right\rangle_{\Sigma_{g,n}}.$$

$$(14.5)$$

We introduce the surface states Σ_{n_1} and Σ_{n_2} such that the BPZ inner product with the new states $\mathscr{V}_{n_1}^{(1)}$ and $\mathscr{V}_{n_2}^{(2)}$ reproduces the M_{g_1,n_1}- and M_{g_2,n_2}-forms:

$$\langle \Sigma_{n_1} | \mathscr{V}_{n_1}^{(1)} \rangle := \omega_{\mathsf{M}_{g_1,n_1}}(\mathscr{V}_1^{(1)}, \ldots, \mathscr{V}_{n_1}^{(1)})$$

$$= (2\pi \mathrm{i})^{-\mathsf{M}_{g_1,n_1}^c} \left\langle \prod_{\lambda=1}^{\mathsf{M}_{g_1,n_1}} B(V_\lambda^{(1)}) \prod_{i=1}^{n_1-1} \mathscr{V}_i^{(1)} \right\rangle_{\Sigma_{g_1,n_1}}, \qquad (14.6a)$$

$$\langle \Sigma_{n_2} | \mathscr{V}_{n_2}^{(2)} \rangle := \omega_{\mathsf{M}_{g_2,n_2}}(\mathscr{V}_1^{(2)}, \ldots, \mathscr{V}_{n_2}^{(2)})$$

$$= (2\pi \mathrm{i})^{-\mathsf{M}_{g_2,n_2}^c} \left\langle \prod_{\lambda=1}^{\mathsf{M}_{g_1,n_1}} B(V_\lambda^{(2)}) \prod_{j=1}^{n_2-1} \mathscr{V}_j^{(2)} \right\rangle_{\Sigma_{g_2,n_2}}. \qquad (14.6b)$$

As described in Sect. 13.1.2, these states exist since the p-form is linear in each of the external state and the BPZ inner product is non-degenerate. Each of the surface states corresponds to an operator

$$\langle \Sigma_{n_1} | = \langle 0 | \, I \circ \Sigma_{n_1}(0), \qquad \langle \Sigma_{n_2} | = \langle 0 | \, I \circ \Sigma_{n_2}(0), \qquad (14.7)$$

defined from (6.136). Then, the forms can be interpreted as 2-point functions on the complex plane:

$$\langle \Sigma_{n_1} | \mathscr{V}_{n_1}^{(1)} \rangle = \langle I \circ \Sigma_{n_1}(0) \mathscr{V}_{n_1}(0) \rangle_{w_{n_1}^{(1)}}, \qquad \langle \Sigma_{n_2} | \mathscr{V}_{n_2}^{(2)} \rangle = \langle I \circ \Sigma_{n_2}(0) \mathscr{V}_{n_2}(0) \rangle_{w_{n_2}^{(2)}}.$$

(14.8)

All the complexity of the amplitudes has been lumped into the definitions of the surface states that contain information about the surface moduli (including the ghost insertions) and about the $n_1 - 1$ remaining states (including the local coordinate systems). The local coordinates around $\mathscr{V}_{n_1}^{(1)}$ and $\mathscr{V}_{n_2}^{(2)}$ are denoted, respectively, as $w_{n_1}^{(1)}$ and $w_{n_2}^{(2)}$. Correspondingly, the surface operators are inserted in the local coordinates w_1 and w_2 that are related to $w_{n_1}^{(1)}$ and $w_{n_2}^{(2)}$, respectively, through the inversion:

$$w_1 = I\big(w_{n_1}^{(1)}\big), \qquad w_2 = I\big(w_{n_2}^{(2)}\big).$$

(14.9)

In order to rewrite (14.5) in terms of Σ_1 and Σ_2, it is first necessary to express all operators in one coordinate system, for example $w_{n_1}^{(1)}$. Hence, we need to find its relation to w_2. Using the plumbing fixture (14.1), the relation between $w_{n_1}^{(1)}$ and w_2 is:

$$w_{n_1}^{(1)} = \frac{q}{w_{n_2}^{(2)}} = \frac{q}{I(w_2)} = q w_2 := f(w_2).$$

(14.10)

Then, the form (14.5) becomes

$$\omega_{M_{g,n}} = \frac{1}{2\pi i} \langle I \circ \Sigma_{n_1}(0) B_q B_{\bar{q}} \, f \circ \Sigma_{n_2}(0) \rangle_{w_{n_1}^{(1)}} = \frac{1}{2\pi i} \langle \Sigma_{n_1} | B_q B_{\bar{q}} \, q^{L_0} \bar{q}^{\bar{L}_0} | \Sigma_2 \rangle,$$

(14.11)

using that Σ_2 has a well-defined scaling dimension. The factor of $2\pi i$ arises by comparing the contribution from Σ_{n_1} and Σ_{n_2} with the factor in (14.5). The expression can be simplified by using the relation

$$\langle \Sigma_{n_1} | B_q B_{\bar{q}} | \mathscr{V}_{n_1}^{(1)} \rangle = \frac{1}{q\bar{q}} \langle \Sigma_{n_1} | b_0 \bar{b}_0 | \mathscr{V}_{n_1}^{(1)} \rangle$$

(14.12)

using the expression (14.4) for B_q and the state–operator correspondence:

$$B_q \mathscr{V}_{n_1}^{(1)}(z, \bar{z}) = \frac{1}{q} \oint_{C_q} dw_{n_1}^{(1)} \, b\big(w_{n_1}^{(1)}\big) w_{n_1}^{(1)} \mathscr{V}_{n_1}^{(1)}(z, \bar{z}) \longrightarrow \frac{1}{q} b_0 | \mathscr{V}_{n_1}^{(1)} \rangle.$$

(14.13)

Ultimately, the form (14.5) reads

$$\omega_{M_{g,n}} = \frac{1}{2\pi i} \frac{1}{q\bar{q}} \langle \Sigma_{n_1} | b_0 \bar{b}_0 \, q^{L_0} \bar{q}^{\bar{L}_0} | \Sigma_{n_2} \rangle .$$

(14.14)

It is important to remember that the plumbing fixture describes only a patch of the moduli space, and the form defined in this way is valid only locally. As a consequence, the integration over all moduli of $\mathcal{M}_{g_1,n_1} \# \mathcal{M}_{g_2,n_2}$ does *not* describe $\mathcal{M}_{g,n}$, but only a part of it (Sect. 12.3.3). Every degeneration limit with a different puncture distribution in two different groups contributes to a different part of the amplitude.

We denote the contribution to the total amplitude (14.2) from the region of the moduli space connected to this degeneration limit as

$$\mathcal{F}_{g,n}\big(\mathcal{V}_i^{(1)} | \mathcal{V}_j^{(2)}\big)$$

$$:= \frac{1}{2\pi i} \int^{M_{g_1,n_1}} \bigwedge_{\lambda=1} dt_\lambda^{(1)} \bigwedge_{\kappa=1}^{M_{g_2,n_2}} dt_\kappa^{(2)} \wedge \frac{dq}{q} \wedge \frac{d\bar{q}}{\bar{q}} \, \langle \Sigma_{n_1} | b_0 \bar{b}_0 \, q^{L_0} \bar{q}^{\bar{L}_0} | \Sigma_{n_2} \rangle .$$

(14.15)

To proceed, we introduce a basis $\{\phi_\alpha(k)\}$ of eigenstates of L_0 and $\bar{L}_0$, where k^μ is the D-dimensional momentum and α denotes the remaining quantum number. Then, introducing twice the resolution of the identity (11.36) gives

$$\mathcal{F}_{g,n}\big(\mathcal{V}_i^{(1)} | \mathcal{V}_j^{(2)}\big) = \frac{1}{2\pi i} \int \frac{d^D k}{(2\pi)^D} \frac{d^D k'}{(2\pi)^D} (-1)^{|\phi_\alpha|}$$

$$\times \int \frac{dq}{q} \wedge \frac{d\bar{q}}{\bar{q}} \, \langle \phi_\alpha(k)^c | b_0 \bar{b}_0 \, q^{L_0} \bar{q}^{\bar{L}_0} | \phi_\beta(k')^c \rangle$$

(14.16)

$$\times \int^{M_{g_1,n_1}} \bigwedge_{\lambda=1} dt_\lambda^{(1)} \langle \Sigma_{n_1} | \phi_\alpha(k) \rangle \int^{M_{g_2,n_2}} \bigwedge_{\kappa=1} dt_\kappa^{(2)} \langle \phi_\beta(k') | \Sigma_{n_2} \rangle$$

(with implicit sums over α and β). In the last line, one recognizes the expressions of the g_1-loop n_1-point amplitude with external states $\{\mathcal{V}_1^{(1)}, \ldots, \mathcal{V}_{n_1-1}^{(1)}, \phi_\alpha\}$ and of the g_2-loop and n_2-point amplitudes with external states $\{\mathcal{V}_1^{(2)}, \ldots, \mathcal{V}_{n_2-1}^{(2)}, \phi_\beta\}$:

$$A_{g_1,n_1}\big(\mathcal{V}_1^{(1)}, \ldots, \mathcal{V}_{n_1-1}^{(1)}, \phi_\alpha(k)\big) = \int_{\mathcal{S}_{g_1,n_1}} \omega_{M_{g_1,n_1}}\big(\mathcal{V}_1^{(1)}, \ldots, \mathcal{V}_{n_1-1}^{(1)}, \phi_\alpha(k)\big)$$

$$= \int_{\mathcal{S}_{g_1,n_1}} \bigwedge_{\lambda=1}^{M_{g_1,n_1}} dt_\lambda^{(1)} \langle \Sigma_{n_1} | \phi_\alpha(k) \rangle ,$$

(14.17a)

$$A_{g_2,n_2}\left(\mathcal{V}_1^{(2)},\dots,\mathcal{V}_{n_2-1}^{(2)},\phi_\beta(k')\right) = \int_{\mathcal{S}_{g_2,n_2}} \omega_{\mathsf{M}_{g_2,n_2}}\left(\mathcal{V}_1^{(2)},\dots,\mathcal{V}_{n_2-2}^{(2)},\phi_\beta(k')\right)$$

$$= \int_{\mathcal{S}_{g_2,n_2}} \bigwedge_{\lambda=1}^{\mathsf{M}_{g_2,n_2}} dt_\lambda^{(2)} \left\langle \Sigma_{n_2} | \phi_\beta(k') \right\rangle.$$

$$(14.17b)$$

The property (B.27) has been used to reverse the order of the BPZ product for the second Riemann surface, and this cancels the factor $(-1)^{|\phi_\alpha|}$.

Defining the second line of (14.16) as

$$\Delta_{\alpha\beta}(k,k') := \Delta\left(\phi_\alpha(k)^c, \phi_\beta(k')^c\right)$$

$$:= \frac{1}{2\pi i}\int \frac{dq}{q}\wedge\frac{d\bar{q}}{\bar{q}}\,\langle\phi_\alpha(k)^c|\,b_0\bar{b}_0\,q^{L_0}\bar{q}^{\bar{L}_0}\,|\phi_\beta(k')^c\rangle, \qquad (14.18)$$

one has

$$\mathcal{F}_{g,n}\left(\mathcal{V}_i^{(1)}|\mathcal{V}_j^{(2)}\right) = \int \frac{d^D k}{(2\pi)^D}\frac{d^D k'}{(2\pi)^D}\,A_{g_1,n_1}\left(\mathcal{V}_1^{(1)},\dots,\mathcal{V}_{n_1-1}^{(1)},\phi_\alpha(k)\right)\Delta_{\alpha\beta}(k,k')$$

$$\times A_{g_2,n_2}\left(\mathcal{V}_1^{(2)},\dots,\mathcal{V}_{n_2-1}^{(2)},\phi_\beta(k')\right).$$

$$(14.19)$$

We recover the expressions from Sect. 11.1.2, but for a more general amplitude. We had found that Δ corresponds to the propagator: its properties are studied further in Sect. 14.2.2. Hence, the object (14.16) corresponds to the product of two amplitudes connected by a propagator (Fig. 14.2).

There are several points to mention about this amplitude:

- We will find that the propagator depends only on one momentum because $\langle k|k'\rangle \sim \delta^{(D)}(k+k')$, which removes one of the integrals. Then, both amplitudes

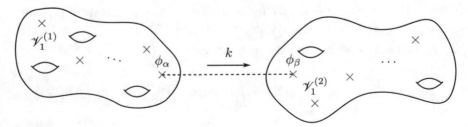

Fig. 14.2 Factorization of the amplitude into two sub-amplitudes connected by a propagator (dashed line)

A_{g_1,n_1} and A_{g_2,n_2} contain a delta function for the momenta:

$$A_{g_1,n_1} \sim \delta^{(D)}\big(k_1^{(1)} + \cdots + k_{n_1-1}^{(1)} + k\big), \ A_{g_2,n_2} \sim \delta^{(D)}\big(k_1^{(2)} + \cdots + k_{n_2-1}^{(2)} + k'\big).$$

$$(14.20)$$

As a consequence, the second momentum integral can be performed and yields a delta function:

$$\mathcal{F}_{g,n} \sim \delta^{(D)}\big(k_1^{(1)} + \cdots + k_{n_1-1}^{(1)} + k_1^{(2)} + \cdots + k_{n_2-1}^{(2)}\big). \qquad (14.21)$$

Hence, the momentum flowing in the internal line is fixed, and this ensures the overall momentum conservation as expected.

- The ghost numbers of the states ϕ_α and ϕ_β are also fixed (in terms of the external states). Indeed, because of the ghost number anomaly, the amplitudes on $\mathcal{M}_{g_1,n_1}$ and $\mathcal{M}_{g_2,n_2}$ are non-vanishing only if the ghost numbers of these states satisfy

$$N_{\text{gh}}(\phi_\alpha) = 2n_1 - \sum_{i=1}^{n_1-1} N_{\text{gh}}\big(\mathscr{V}_i^{(1)}\big), \qquad N_{\text{gh}}(\phi_\beta) = 2n_2 - \sum_{j=1}^{n_2-1} N_{\text{gh}}\big(\mathscr{V}_j^{(2)}\big).$$

$$(14.22)$$

The non-vanishing of $\mathcal{F}_{g,n}$ also gives another relation:

$$N_{\text{gh}}(\phi_\alpha) + N_{\text{gh}}(\phi_\beta) = 4. \qquad (14.23)$$

In particular, if the external states are on-shell with $N_{\text{gh}} = 2$, we find

$$N_{\text{gh}}(\phi_\alpha) = N_{\text{gh}}(\phi_\beta) = 2. \qquad (14.24)$$

As indicated in Chap. 10, such states are appropriate at the classical level since they do not contain spacetime ghosts.

- The sum over α and β is over an infinite number of states and could diverge. In fact, the sum can be made convergent by tuning the stub parameter (Sect. 14.2.4).

Properties of Feynman graphs and amplitudes in the momentum space will be discussed further in Chap. 18.

14.1.2 Non-separating Case

Next, we consider the non-separating plumbing fixture (Sect. 12.3.2). The computations are almost identical to the separating case; thus, we outline only the general steps.

Part of the moduli space $\mathcal{M}_{g,n}$ is covered by $\#\mathcal{M}_{g_1,n_1}$, with $g = g_1 + 1$ and $n = n_1 - 2$. The local coordinates are denoted as w_i for $i = 1, \ldots, n_1$ and the plumbing fixture reads

$$w_{n_1-1} w_{n_1} = q. \tag{14.25}$$

The g-loop n-point amplitude with external states $\{\mathcal{V}_1^{(1)}, \ldots, \mathcal{V}_{n_1-2}^{(1)}\}$ is denoted as

$$A_{g,n} = \int_{\mathcal{S}_{g,n}} \omega_{M_{g,n}}^{g,n} (\mathcal{V}_1^{(1)}, \ldots, \mathcal{V}_{n_1-2}^{(1)}). \tag{14.26}$$

When the $n_1 - 2$ punctures and $g_1 = g - 1$ holes move lose to each other, the form can be written as

$$\omega_{M_{g,n}} \left(V_1^{(1)}, \ldots, V_{M_{g_1,n_1}}^{(1)}, \partial_q, \partial_{\bar{q}} \right)$$

$$= (2\pi i)^{-M_{g,n}^c} \left\langle \prod_{\lambda=1}^{M_{g_1,n_1}} B\left(V_\lambda^{(1)} \right) B(\partial_q) B(\partial_{\bar{q}}) \prod_{i=1}^{n_1-2} \mathcal{V}_i^{(1)} \right\rangle_{\Sigma_{g,n}}. \tag{14.27}$$

To proceed, one needs to introduce the surface state Σ_{n_1-1,n_1}:

$$\langle \Sigma_{n_1-1,n_1} | \mathcal{V}_{n_1-1}^{(1)} \otimes \mathcal{V}_{n_1}^{(1)} \rangle := \omega_{M_{g_1,n_1}} (\mathcal{V}_1^{(1)}, \ldots, \mathcal{V}_{n_1}^{(1)}). \tag{14.28}$$

Following the same step as in the previous section leads to

$$\mathcal{F}_{g,n}(\mathcal{V}_i^{(1)}|)$$

$$= \int \frac{\mathrm{d}^D k}{(2\pi)^D} \frac{\mathrm{d}^D k'}{(2\pi)^D} A_{g_1,n_1} (\mathcal{V}_1^{(1)}, \ldots, \mathcal{V}_{n_1-2}^{(1)}, \phi_\alpha(k), \phi_\beta(k')) \Delta_{\alpha\beta}(k, k'), \tag{14.29}$$

where the propagator is given in (14.18). This is equivalent to an amplitude for which two external legs are glued together with a propagator, giving a loop (Fig. 14.3).

Since both states ϕ_α and ϕ_β are inserted on the same surface, their ghost numbers are not fixed, even if the external states are physical. The non-vanishing of $\mathcal{F}_{g,n}$ only leads to the constraint:

$$N_{\mathrm{gh}}(\phi_\alpha) + N_{\mathrm{gh}}(\phi_\beta) = 2n_1 - \sum_{i=1}^{n_1-2} N_{\mathrm{gh}}(\mathcal{V}_i^{(1)}) = 4. \tag{14.30}$$

As a consequence, loop diagrams force to introduce states of every ghost number. Internal states with $N_{\mathrm{gh}} \neq 2$ correspond to spacetime ghosts.

Fig. 14.3 Factorization of the amplitude into two sub-amplitudes connected by a propagator (dashed line). The propagator connects two punctures of the same surface, which is equivalent to a loop

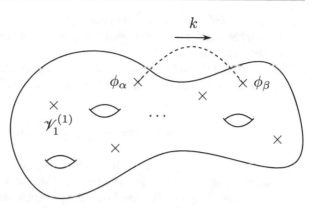

Since the propagator contains a delta function $\delta^{(D)}(k - k')$, the integral over k' can be removed by setting $k' = -k$. However, the integral over k remains since

$$A_{g_1, n_1}\left(\mathscr{V}_1^{(1)}, \ldots, \mathscr{V}_{n_1-2}^{(1)}, \phi_\alpha(k), \psi_\beta(-k)\right) \sim \delta^{(D)}\left(k_1^{(1)} + \cdots + k_{n_1-2}^{(1)}\right). \quad (14.31)$$

Hence, the loop momentum k is not fixed, as expected in QFT.

Remark 14.1 Not all values of the moduli associated to the holes can be associated to loops in Feynman diagrams. Only the values close to the degeneration limit can be interpreted in this way, the other being just standard (quantum) vertices.

14.2 Feynman Diagrams and Feynman Rules

In the standard QFT approach, Feynman graphs compute Green functions, and scattering amplitudes are obtained by amputating the external propagators through the LSZ prescription. For connected tree-level processes, this requires $n \geq 3$ (corresponding to $\chi_{0,n} < 0$).

Given a theory, there is a minimal set of Feynman diagrams—the Feynman rules—from which every other diagram can be constructed. These rules include the definitions of the fundamental vertices—the fundamental interactions—and of the propagator—how states propagate between two interactions (or, how to glue vertices together). In this section, we describe these different elements.

14.2.1 Feynman Graphs

The amplitude factorization described in Sect. 14.1 gives a natural separation of amplitudes into several contributions. Considering all the possible degeneration limits leads to a set of diagrams with amplitudes of lower order connected by propagators (Figs. 14.2 and 14.3). This corresponds exactly to the idea behind

Feynman graphs. Then, the goal is to find the Feynman rules of the theory: since the propagator has already been identified (further studied in Sect. 14.2.2), it is sufficient to find the interaction vertices.

Let us make this more precise by considering an amplitude $A_{g,n}(\mathcal{V}_1, \ldots, \mathcal{V}_n)$. The index of an amplitude is defined to be the index (12.50) of the corresponding Riemann surfaces

$$r(A_{g,n}) := r(\Sigma_{g,n}) = 3g + n - 2. \tag{14.32}$$

Contributions to an amplitude with a given $r(A_{g,n})$ can be described in terms of amplitudes $A_{g',n'}$ with $r(A_{g',n'}) < r(A_{g,n})$. But, the moduli space $\mathcal{M}_{g,n}$ cannot (generically) be completely covered with the plumbing fixture of lower-dimensional moduli spaces, i.e. with $r(\mathcal{M}_{g',n'}) < r(\mathcal{M}_{g,n})$ (Sect. 12.3.3). Then, the same must be true for the amplitudes, such that $A_{g,n}$ cannot be uniquely expressed in terms of amplitudes $A_{g',n'}$.

The g-loop n-point fundamental vertex is defined by

$$\mathcal{V}_{g,n}(\mathcal{V}_1, \ldots, \mathcal{V}_n) := \quad \mathcal{V}_1 \mathrel{\underrel{g}{\rule{1.5cm}{0.4pt}}}\!\!\!\Big\langle \begin{array}{c} \mathcal{V}_2 \\ \vdots \\ \mathcal{V}_n \end{array} \quad := \int_{\mathcal{R}_{g,n}} \omega_{\mathsf{M}_{g,n}}^{g,n}(\mathcal{V}_1, \ldots, \mathcal{V}_n),$$

$$\tag{14.33}$$

The form defined in (13.16) is integrated over a subsection $\mathcal{R}_{g,n} \subset \mathcal{S}_{g,n}$ of $\hat{\mathcal{P}}_{g,n}$. Its projection on the base is the region $\mathcal{V}_{g,n} \subset \mathcal{M}_{g,n}$ that cannot be described by the plumbing fixture, see (12.42b). In general, we will keep the choice of local coordinates implicit and always write $\mathcal{V}_{g,n}$ to avoid surcharging the notations.

It corresponds to the remaining contribution of the amplitude once all graphs containing propagators have been taken into account:

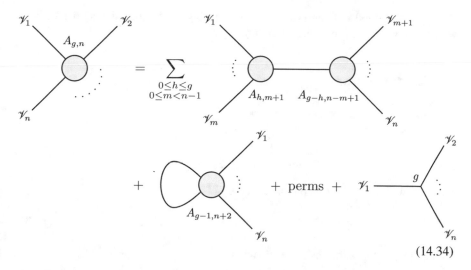

$$(14.34)$$

where the permutations are taken over all legs exiting the amplitudes in the first two terms (this includes the two legs glued together in the second term), including if necessary a weight to avoid overcounting. In the RHS, the amplitudes $A_{g,1}$ are tadpoles and have no external vertices $\mathcal{V}_i$ (from $A_{g,n}$); this corresponds to the terms for $m = 0$ and $m = n - 2$.

In general, the fundamental vertex is non-vanishing for every value $g, n \in \mathbb{N}$ such that $\chi_{g,n} < 0$. For this reason, the index g helps to distinguish between graphs with identical values of n. It may look strange that one needs vertices at every loop: the interpretation will be made clearer when translating this into the language of string field theory (Chap. 15). We stress again that the definition of the fundamental vertex (and the region covered) depends on the choice of local coordinates for *all* lower-order vertices $V_{g',n'}$ such that $r(V_{g',n'}) < r(V_{g,n})$.

Remark 14.2 There are different alternative notations for (14.33):

$$V_{g,n}(\mathcal{V}_1, \ldots, \mathcal{V}_n) := V_{g,n}(\otimes_i \mathcal{V}_i) := \{\mathcal{V}_1, \ldots, \mathcal{V}_n\}_g. \qquad (14.35)$$

Example 14.1: Scalar QFT

Consider a scalar field theory with a cubic and a quartic interaction. The 4-point amplitude contains four contributions, three from gluing 3-point vertices with a propagator, and one from the fundamental quartic vertex. The mismatch between the amplitude and the three graphs with a propagator hints at the existence of the quartic interactions. This example gives an idea of how one can identify the fundamental interactions recursively. ◄

The definition (14.34) of the vertex shows that it can also be interpreted as an amputated Green function without internal propagator (i.e. there is no propagator at all). This is expected from the definition of the interaction vertices from an action, as will be exemplified in Chap. 15. Before describing (and generalizing) the vertices, we describe first the properties of the propagator.

14.2.2 Propagator

The propagator has been defined in (14.18):

$$\Delta = \frac{1}{2\pi i} \int \frac{dq}{q} \wedge \frac{d\bar{q}}{\bar{q}} \, b_0 \bar{b}_0 \, q^{L_0} \bar{q}^{\bar{L}_0}. \tag{14.36}$$

The plumbing modulus q is parametrized by (12.31)

$$q = e^{-s+i\theta}, \qquad s \in \mathbb{R}_+, \qquad \theta \in [0, 2\pi), \tag{14.37}$$

such that the integration measure becomes

$$\frac{dq}{q} \wedge \frac{d\bar{q}}{\bar{q}} = -2i \, ds \wedge d\theta. \tag{14.38}$$

Using the variables $L_0^\pm = L_0 \pm \bar{L}_0$ and $b_0^\pm = b_0 \pm \bar{b}_0$, the propagator can be recast as

$$\Delta = \frac{1}{2\pi} b_0^+ b_0^- \int_0^\infty ds \, e^{-sL_0^+} \int_0^{2\pi} d\theta \, e^{i\theta L_0^-}. \tag{14.39}$$

The form of the first integral is recognized as the Schwinger parametrization of the propagator, while the second is the Fourier transformation of the discrete delta function:

$$\int_0^\infty ds \, e^{-sL_0^+} = \frac{1}{L_0^+}, \qquad \int_0^{2\pi} d\theta \, e^{i\theta L_0^-} = 2\pi \, \delta_{L_0^-, 0}. \tag{14.40}$$

In fact, the first integral converges only if $L_0^+ > 0$. As argued in the introduction, divergences for $L_0^+ \leq 0$ are either non-physical or IR divergences that can be cured by renormalization. For this reason, we take the RHS as a definition of the integral, which would be the correct result if one starts with a field theory action instead of a first-quantized formalism.

In this case, the propagator becomes

$$\Delta = \frac{b_0^+}{L_0^+} \, b_0^- \, \delta_{L_0^-, 0}. \tag{14.41}$$

This is the standard expression for the propagator. For completeness, the form in terms of the holomorphic and anti-holomorphic components is

$$\Delta = -2b_0 \bar{b}_0 \frac{1}{L_0 + \bar{L}_0} \delta_{L_0, \bar{L}_0}. \tag{14.42}$$

The delta function restricts the amplitude to states satisfying the level-matching condition, that is, annihilated by L_0^-.

Considering a basis $\{\phi_\alpha(k)\}$ of eigenstates of both L_0 and $\bar{L}_0$:

$$L_0^+ |\phi_\alpha(k)\rangle = \frac{\alpha'}{2}(k^2 + m_\alpha^2) |\phi_\alpha(k)\rangle, \qquad L_0^- |\phi_\alpha(k)\rangle = 0 \tag{14.43}$$

leads to the following momentum-space kernel for the propagator:

$$\Delta_{\alpha\beta}(k, k') := \langle \phi_\alpha(k)^c | \Delta | \phi_\beta(k')^c \rangle := (2\pi)^D \delta^{(D)}(k + k') \Delta_{\alpha\beta}(k), \tag{14.44a}$$

$$\Delta_{\alpha\beta}(k) := \frac{M_{\alpha\beta}(k)}{k^2 + m_\alpha^2}, \qquad M_{\alpha\beta}(k) := \frac{2}{\alpha'} \langle \phi_\alpha^c(k) | b_0^+ b_0^- | \phi_\beta^c(-k) \rangle, \tag{14.44b}$$

with $M_{\alpha\beta}$ a finite-dimensional matrix giving the overlap of states of identical masses (because the number of states at a given level is finite).

For the propagator to be well-defined, it must be invertible (in particular, to define a kinetic term). The propagator (14.41) is non-vanishing if the states it acts on satisfy

$$b_0^+ |\phi_\alpha^c\rangle \neq 0, \qquad b_0^- |\phi_\alpha^c\rangle \neq 0. \tag{14.45}$$

Necessary and sufficient conditions for this to be true are

$$c_0^+ |\phi_\alpha^c\rangle = 0, \qquad c_0^- |\phi_\alpha^c\rangle = 0. \tag{14.46}$$

Indeed, decomposing the state on the ghost zero-modes

$$|\phi_\alpha^c\rangle = |\phi_1\rangle + b_0^\pm |\phi_2\rangle, \qquad c_0^\pm |\phi_1\rangle = c_0^\pm |\phi_2\rangle = 0 \tag{14.47}$$

gives

$$c_0^\pm |\phi_\alpha^c\rangle = 0 \implies |\phi_2\rangle = 0, \tag{14.48}$$

and one correctly has $b_0^\pm |\phi_1\rangle \neq 0$.

These conditions are given for the dual states: translating them on the normal states reverses the roles of b_0 and c_0. Hence, the states must satisfy the conditions:

$$b_0^+ |\phi_\alpha\rangle = 0, \qquad b_0^- |\phi_\alpha\rangle = 0. \tag{14.49}$$

The second condition is satisfied automatically because the Hilbert space is $\mathcal{H}^-$ when working with $\hat{\mathcal{P}}_{g,n}$ (Sect. 13.3.1). However, the first condition further restricts the states that propagate in internal lines. This leads to postulate that the external states should also be taken to satisfy this condition

$$b_0^+ |\mathcal{V}_i\rangle = 0, \tag{14.50}$$

since external states are usually a subset of the internal states. This provides another motivation of the statement in Sect. 3.2.2 that scattering amplitudes for the states not annihilated by b_0^+ must be trivial. A field interpretation of this condition is given in Chaps. 10 and 15.

Under these constraints on the states, the propagator can be inverted:

$$\Delta^{-1} = c_0^+ c_0^- L_0^+ \delta_{L_0^-, 0}. \tag{14.51}$$

14.2.3 Fundamental Vertices

The vertices (14.33) can be constructed recursively assuming that all amplitudes are known. The starting point is the tree-level cubic amplitude $A_{0,3}$: since it does not contain any internal propagator, it is equal to the fundamental vertex $\mathcal{V}_{0,3}$.

The first thing to extract from the recursion relations are the background independent data. This amounts to find local coordinates and a characterization of the subspaces $\mathcal{V}_{g,n} \subset \mathcal{M}_{g,n}$, starting with $\mathcal{P}_{0,3}$ and iterating.

In the rest of this section, we show how this works schematically.

Recursive Definition: Tree-Level Vertices

The description of tree-level amplitudes $A_{0,n}$ is the simplest since only the separating plumbing fixture is used and Feynman graphs are trees. The possible factorizations of the amplitude correspond basically to all the partitions of the set $\{\mathcal{V}_i\}$ into subsets.

Tree-Level Cubic Vertex Since $M_{0,3} = 0$, the moduli space of the 3-punctured sphere $\Sigma_{0,3}$ reduces to a point, and so does the section $\mathcal{S}_{0,3}$ of $\mathcal{P}_{0,3}$ (Fig. 14.4a):

$$\mathcal{V}_{0,3}(\mathcal{V}_1, \mathcal{V}_2, \mathcal{V}_3) := A_{0,3}(\mathcal{V}_1, \mathcal{V}_2, \mathcal{V}_3) = \omega_0^{0,3}(\mathcal{V}_1, \mathcal{V}_2, \mathcal{V}_3). \tag{14.52}$$

The corresponding graph is indicated in Fig. 14.4b.

Tree-Level Quartic Vertex Part of the contributions to the 4-point amplitude $A_{0,4}$ with external states $\mathcal{V}_i$ ($i = 1, \ldots, 4$) comes from gluing two cubic vertices. Because there are four external states, there are three different partitions $2|2$ that are described in Fig. 14.5 (see also Fig. 12.7). The sum of these three diagrams does

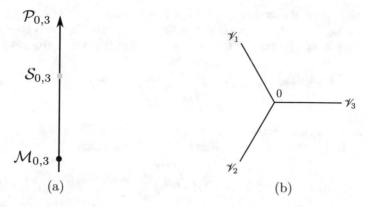

Fig. 14.4 Section of $\mathcal{P}_{0,3}$ and cubic vertex. (a) A section $\mathcal{S}_{0,3}$ over $\mathcal{P}_{0,3}$ reduces to a point. (b) Fundamental cubic vertex

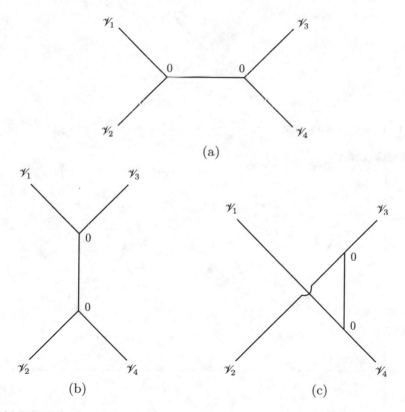

Fig. 14.5 Factorization of the quartic amplitude $A_{0,4}$ in the s-, t- and u-channels. (a) s-channel. (b) t-channel. (c) u-channel

not reproduce $A_{0,4}$: the moduli space $\mathcal{M}_{0,4}$ is not completely covered by the three amplitudes. Equivalently, the projection of the section over $\mathcal{P}_{0,4}$ does not cover all of $\mathcal{M}_{0,4}$. The missing contribution is defined by the quartic vertex (Fig. 14.6)

$$\mathcal{V}_{0,4}(\mathscr{V}_1, \mathscr{V}_2, \mathscr{V}_3, \mathscr{V}_4) := \int_{\mathcal{R}_{0,4}} \omega_2^{0,4}(\mathscr{V}_1, \dots, \mathscr{V}_4), \qquad (14.53)$$

and the corresponding section is denoted by $\mathcal{R}_{0,4}$ (Fig. 14.7). Denoting by $\mathcal{F}_{0,4}^{(s,t,u)}$ the graphs 14.5 in the s-, t- and u-channels, one has the relation

$$A_{0,4} = \mathcal{F}_{0,4}^{(s)} + \mathcal{F}_{0,4}^{(t)} + \mathcal{F}_{0,4}^{(u)} + \mathcal{V}_{0,4}. \qquad (14.54)$$

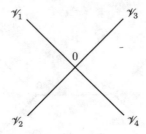

Fig. 14.6 Fundamental quartic vertex

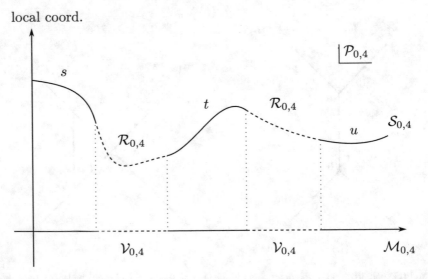

Fig. 14.7 A section $\mathcal{S}_{0,4}$ over $\mathcal{P}_{0,4}$ and the contribution from the s-, t- and u-channels (Fig. 14.5) are indicated by the corresponding indices. The fundamental vertex is defined by the section $\mathcal{V}_{0,4}$

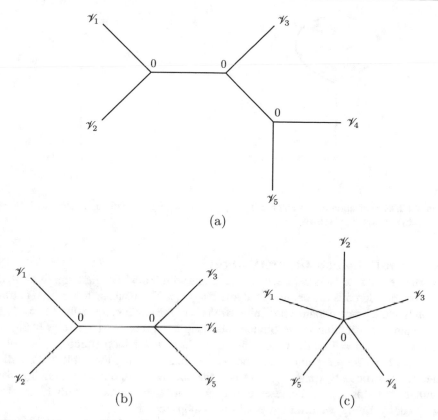

Fig. 14.8 Factorization of the amplitude $G_{0,5}$ in channels and fundamental quintic vertex. Only the cases where $\mathcal{V}_1$ and $\mathcal{V}_2$ factorize on one side are indicated, and the other cases follow by permutations of the external states. (**a**) Factorization 12|3|45. (**b**) Factorization 12|345. (**c**) Fundamental vertex

Tree-Level Quintic Vertex The amplitude $A_{0,5}$ can be factorized in a greater number of channels, the two types being 2|2|1 and 2|3. The possible Feynman graphs are built either from three cubic vertices and two propagators (Fig. 14.8a and permutations) or from one cubic and one quartic vertices together with one propagator (Fig. 14.8b and permutations). The remaining contribution is the fundamental vertex (Fig. 14.8c):

$$\mathcal{V}_{0,5}(\mathcal{V}_1, \ldots, \mathcal{V}_5) := \int_{\mathcal{R}_{0,5}} \omega_4^{0,5}(\mathcal{V}_1, \ldots, \mathcal{V}_5). \tag{14.55}$$

The construction to higher order follows exactly this scheme.

Fig. 14.9 Factorization of the amplitude $G_{1,1}$ and fundamental tadpole at 1-loop. (**a**) Internal loop. (**b**) Fundamental vertex

Recursive Definition: General Vertices

Next, one needs to consider Feynman diagrams with loops. The first amplitude that can be considered is the one-loop tadpole $A_{1,1}(\mathcal{V}_1)$. The factorization region corresponds to the graph obtained by gluing two legs of the cubic vertex (Fig. 14.9a). The remaining contribution is the fundamental tadpole vertex $\mathcal{V}_{1,1}(\mathcal{V}_1)$ (Fig. 14.9b)—note the index $g = 1$ on the vertex, indicating that it is a 1-loop effect.

Next, the 1-loop 2-point amplitude can be obtained using the cubic and quartic tree-level vertices $\mathcal{V}_{0,3}$ and $\mathcal{V}_{0,4}$, but also the one-loop tadpole $\mathcal{V}_{1,1}$. Iterating, the number of loops can be increased either by gluing together two external legs of a graph or by gluing two different graphs with loops together.

For $g \geq 2$, the recursion implies the existence of vertices with no external states $\mathcal{V}_{g,0}$: they should be interpreted as loop corrections to the vacuum energy density.

It is important to realize that, in this language, a handle in the Riemann surface is not necessarily mapped to a loop in the Feynman graph: only handles described by the region $\mathcal{F}_{g,n} = \mathcal{M}_{g,n} - \mathcal{V}_{g,n}$ do. The higher-order vertices—corresponding to surfaces with small handles only and described by $\mathcal{V}_{g,n}$—should be regarded as quantum fundamental interactions. In Chap. 15, it will be explained that they really correspond to (finite) counter-terms: the measure is not invariant under the gauge symmetry of the theory, and these terms must be introduced to restore it.

Other Vertices

The definition given at the end of (14.2.1) suggests to introduce additional vertices. The previous recursive definition gives only vertices with $\chi_{g,n} = 2 - 2g - n < 0$, but, in fact, it makes sense to consider the additional cases: $g = 0$ and $n = 0, 1, 2$, and $g = 1, n = 0$.

The definition of the vertices as amputated Green function without internal propagators provides a hint for the tree-level quadratic vertex $\mathcal{V}_{0,2}$. We define the latter as the amputated tree-level 2-point Green function:

$$\mathcal{V}_{0,2} := \Delta^{-1} \Delta \Delta^{-1} = \Delta^{-1}. \tag{14.56}$$

Hence, we have

$$\mathcal{V}_{0,2}(\mathcal{V}_1, \mathcal{V}_2) := \langle \mathcal{V}_1 | c_0^+ c_0^- L_0^+ \delta_{L_0^-,0} | \mathcal{V}_2 \rangle. \tag{14.57}$$

Note that $\mathcal{V}_{0,2}$ is *not* the 2-point scattering amplitude.

We denote the tree-level 1-point and 0-point vertices as $\mathcal{V}_{0,1}(\mathcal{V}_1)$ and $\mathcal{V}_{0,0}$. The first can be interpreted as a classical source in the action, while the second is a classical vacuum energy. They are set to zero in most applications and can be safely ignored. However, they appear when formulating the theory on a background that does not solve the equation of motion [4].

Finally, the 1-loop vacuum energy $\mathcal{V}_{1,0}$ can also be defined as the partition function of the worldsheet CFT integrated over the torus modulus.

This allows to define the vertices $\mathcal{V}_{g,n}$ for all $g, n \in \mathbb{N}$. We define the sum of all loop contributions for a fixed n as

$$\mathcal{V}_n(\mathcal{V}_1, \ldots, \mathcal{V}_n) := \sum_{g \geq 0} (\hbar g_s^2)^g \, \mathcal{V}_{g,n}(\mathcal{V}_1, \ldots, \mathcal{V}_n). \tag{14.58}$$

14.2.4 Stubs

In Sect. 12.3.4, we have indicated that the plumbing fixture can be modified by adding stubs or, equivalently, by rescaling the local coordinates. This amounts to introducing a cut-off (12.53) on the variable s such that

$$q = e^{-s+i\theta}, \qquad s \in [s_0, \infty), \qquad \theta \in [0, 2\pi). \tag{14.59}$$

instead of (14.37). In this case, the s-integral in the propagator (14.36) is modified to

$$\int_{s_0}^{\infty} ds \, e^{-sL_0^+} = \frac{e^{-s_0 L_0^+}}{L_0^+}. \tag{14.60}$$

This leads to a new expression for the propagator:

$$\Delta(s_0) = b_0^+ \frac{e^{-s_0 L_0^+}}{L_0^+} b_0^- \delta_{L_0^-,0}. \tag{14.61}$$

In momentum space, this reads

$$\Delta_{\alpha\beta}(k) := \frac{e^{-\frac{\alpha' s_0}{2}(k^2+m_\alpha^2)}}{k^2 + m_\alpha^2} M_{\alpha\beta}(k). \tag{14.62}$$

It is more convenient to work with the canonical propagator (14.41). This can be achieved by absorbing $e^{-\frac{s_0}{2}L_0^+}$ in the interaction vertex: a n-point interaction will get n such factors.[1]

Since s_0 changes the local coordinates, this means that it also changes the region $\mathcal{V}_{g,n}$ (Fig. 12.10). The freedom in the choice of s_0 translates into a freedom to choose which part of the amplitude is described by propagator graphs $\mathcal{F}_{g,n}(s_0)$, and which part is described by a fundamental vertex $\mathcal{V}_{g,n}(s_0)$. The amplitude $A_{g,n}$ is independent of s_0 since it is described in terms of the complete moduli space $\mathcal{M}_{g,n}$. This also means that the parameter s_0 must disappear when summing over the contributions from $\mathcal{V}_{g,n}(s_0)$ and $\mathcal{F}_{g,n}(s_0)$. This indicates that the value of s_0 is not relevant, even off-shell: it can be taken to any convenient value.

The possibility of adding stubs solves the problem that the sum over all states could diverge (see Sect. 14.1.1). Indeed, the expression (14.62) in momentum space shows that the propagator includes an exponential suppression for very massive particle propagating as intermediate states. Since the mass of a particle increases with the level, this shows that the sum converges for a sufficiently large value of s_0, thanks to the factor $e^{-\alpha' s_0 m^2}$. A second interesting aspect is the exponential momentum suppression $e^{-\alpha' s_0 k^2}$: this is responsible for the nice UV behaviour of string theory. Since the value of s_0 is not physical, this means that all Feynman graphs must share these properties. These two points will be made more precise in Chap. 18.

14.2.5 1PI Vertices

We can follow the same procedure as before, but considering only the separating plumbing fixture. In this case, the Feynman diagrams are all 1PR (1-particle reducible): if the propagator line is cut, then the graphs are split into two disconnected components. The region of the moduli space covered by these graphs is written as $\mathcal{F}_{g,n}^{\text{1PR}}$ (12.43a). The complement defines the 1PI region $\mathcal{V}_{g,n}^{\text{1PI}}$ (12.43b). Then, the 1PI g-loop n-point fundamental vertices are defined as

$$\mathcal{V}_{g,n}^{\text{1PI}}(\mathscr{V}_1,\ldots,\mathscr{V}_n) := \quad \mathscr{V}_1 \!-\!\!\!\bigcirc\!\!\!\text{1PI} \;\vdots\; := \int_{\mathcal{R}_{g,n}^{\text{1PI}}} \omega_{\mathsf{M}_{g,n}}^{g,n}(\mathscr{V}_1,\ldots,\mathscr{V}_n),$$

where $\mathcal{R}_{g,n}^{\text{1PI}}$ is a section of $\mathcal{P}_{g,n}$ which projection on the base is $\mathcal{V}_{g,n}^{\text{1PI}}$.

$$(14.63)$$

---

[1]To make this identification precise for vertices involving external states, one has to consider the non-amputated Green functions.

where $\mathcal{R}_{g,n}^{1PI}$ is a section of $\mathcal{P}_{g,n}$ whose projection on the base is $V_{g,n}^{1PI}$.

14.3 Properties of Fundamental Vertices

14.3.1 String Product

Following the definition of surfaces states (Sect. 13.1.2), the vertex state is defined as

$$\langle V^{g,n}| \otimes_i \mathcal{V}_i \rangle := V_{g,n}(\otimes_i \mathcal{V}_i). \tag{14.64}$$

The vertex is a map $V_{g,n} : \mathcal{H}^{\otimes n} \to \mathbb{C}$, where $\mathbb{C} \simeq \mathcal{H}^{\otimes 0}$. We will find very useful to introduce the string products $\ell_{g,n} : \mathcal{H}^{\otimes n} \to \mathcal{H}$ through the closed string inner product:

$$V_{g,n+1}(\mathcal{V}_0, \mathcal{V}_1, \ldots, \mathcal{V}_n) := \langle \mathcal{V}_0| c_0^- |\ell_{g,n}(\mathcal{V}_1, \ldots, \mathcal{V}_n)\rangle. \tag{14.65}$$

An alternative notation is

$$\ell_{g,n}(\mathcal{V}_1, \ldots, \mathcal{V}_n) := [\mathcal{V}_1, \ldots, \mathcal{V}_n]_g. \tag{14.66}$$

The advantage of the second notation is to show that the products with $n \geq 3$ are direct generalization of the 2-product, which is very similar to a super-Lie bracket. These products play a central role in SFT—in fact, the description of SFT is more natural using $\ell_{g,n}$ rather than $V_{g,n}$.

Note that the products with $n = 0$ are maps $\mathbb{C} \to \mathcal{H}$, which means that they correspond to a particular fixed state.

$$\ell_{g,0} := [\cdot]_g \in \mathcal{H}. \tag{14.67}$$

The ghost number of the product (14.65) is

$$N_{gh}\big(\ell_{g,n}(\mathcal{V}_1, \ldots, \mathcal{V}_n)\big) = 3-2n+\sum_{i=1}^n N_{gh}(\mathcal{V}_i) = 3+\sum_{i=1}^n \big(N_{gh}(\mathcal{V}_i)-2\big), \tag{14.68}$$

and it is independent of the genus g. As a consequence, the parity of the product is

$$\big|\ell_{g,n}(\mathcal{V}_1, \ldots, \mathcal{V}_n)\big| = 1 + \sum_{i=1}^n |\mathcal{V}_i| \mod 2, \tag{14.69}$$

and the string product itself is always odd.

The vertices satisfy the following identity for $g \geq 0$ and $n \geq 1$ [3, pp. 41–42]:

$$0 = \sum_{\substack{g_1,g_2 \geq 0 \\ g_1+g_2=g}} \sum_{\substack{n_1,n_2 \geq 0 \\ n_1+n_2=n}} \frac{n!}{n_1! \, n_2!} V_{g_1,n_1+1}\left(\Psi^{n_1}, \ell_{g_2,n_2}(\Psi^{n_2})\right)$$

$$+ (-1)^{|\phi_s|} V_{g-1,n+2}\left(\phi_s, b_0^- \phi_s^c, \Psi^n\right). \tag{14.70}$$

The last term is absent for $g = 0$. It is a consequence of the definition of the vertices as the missing region from gluing lower-order vertices.

14.3.2 Feynman Graph Interpretation

The vertices must satisfy a certain number of conditions to be interpreted as Feynman diagrams. The first is that they must be symmetric under permutations of the states. Not every choice of local coordinates satisfies this requirement: this can be solved by defining the vertex over a generalized section. In this case, the vertex is defined as the average of the integrals over N sections $S_{g,n}^{(a)}$ of $\mathcal{P}_{g,n}$:

$$V_{g,n}(\mathscr{V}_1, \ldots, \mathscr{V}_n) = \frac{1}{N} \sum_{a=1}^{N} \int_{\mathcal{R}_{g,n}^{(a)}} \omega_{M_{g,n}}^{g,n}(\mathscr{V}_1, \ldots, \mathscr{V}_n). \tag{14.71}$$

Example 14.2: 3-Point Vertex

The cubic vertex must be symmetric under permutations

$$V_{0,3}(\mathscr{V}_1, \mathscr{V}_2, \mathscr{V}_3) = V_{0,3}(\mathscr{V}_3, \mathscr{V}_1, \mathscr{V}_2) + \cdots \tag{14.72}$$

Taking the vertex to be given by a section $S_{0,3}$ with local coordinates f_i

$$V_{0,3}(\mathscr{V}_1, \mathscr{V}_2, \mathscr{V}_3) = \omega_0^{0,3}(\mathscr{V}_1, \mathscr{V}_2, \mathscr{V}_3)|_{S_{0,3}} = \langle f_1 \circ \mathscr{V}_1(0) \, f_2 \circ \mathscr{V}_2(0) \, f_3 \circ \mathscr{V}_3(0)\rangle, \tag{14.73}$$

one finds that a permutation looks different

$$V_{0,3}(\mathscr{V}_3, \mathscr{V}_1, \mathscr{V}_2) = \langle f_1 \circ \mathscr{V}_3(0) \, f_2 \circ \mathscr{V}_1(0) \, f_3 \circ \mathscr{V}_2(0)\rangle \neq V_{0,3}(\mathscr{V}_1, \mathscr{V}_2, \mathscr{V}_3), \tag{14.74}$$

unless the local coordinates satisfy special properties (remember that the local coordinates are specified by the vertex state V and not by the external states $\mathscr{V}_i$, so a permutation of them does not permute the local maps). Obviously, both amplitudes agree on-shell since the dependence in the local coordinates gets cancelled (equivalently one can rotate the punctures using $SL(2, \mathbb{C})$).

Writing $z_i = f_i(0)$, there is a SL(2, $\mathbb{C}$) transformation $g(z)$ such that

$$g(z_1) = z_2, \qquad g(z_2) = z_3, \qquad g(z_3) = z_1 \tag{14.75}$$

such that

$$\mathcal{V}_{0,3}(\mathcal{V}_3, \mathcal{V}_1, \mathcal{V}_2) = \langle g \circ f_1 \circ \mathcal{V}_3(0) g \circ f_2 \circ \mathcal{V}_1(0) g \circ f_3 \circ \mathcal{V}_2(0) \rangle. \tag{14.76}$$

While the state $\mathcal{V}_i$ is correctly inserted at the puncture z_i in this expression, this is not sufficient to guarantee the equality of the amplitudes. Indeed the fibre is defined by the complete functions $f_i(w)$ and not only by their values at $w = 0$. For this reason, the amplitudes can be equal only if

$$g \circ f_1 = f_2, \qquad g \circ f_2 = f_3, \qquad g \circ f_3 = f_1. \tag{14.77}$$

This provides constraints on the functions f_i, but it is often not possible to solve them.

If the constraints cannot be solved, then one must introduce a general section. In this case a generalized section will be made of 6 sections $S^{(a)}$ ($a = 1, \ldots, 6$) because there are 6 permutations (Fig. 14.10). Then the amplitude reads

$$\mathcal{V}_{0,3}(\mathcal{V}_1, \mathcal{V}_2, \mathcal{V}_3) = \frac{1}{6} \sum_{a=1}^{6} \omega_0^{0,3}(\mathcal{V}_1, \mathcal{V}_2, \mathcal{V}_3)|_{S_{0,3}^{(a)}}. \tag{14.78}$$

◀

When computing the Feynman graphs by gluing lower-dimensional amplitudes, it is possible that parts of the section overlap, meaning that several graphs cover the same part of the moduli space. In this case, the fundamental vertex should be defined as a negative contribution in the overlap region. This procedure is perfectly well-defined since all graphs are finite and there is no ambiguity. In practice, it is always simpler to work with non-overlapping sections (i.e. a single covering of the

Fig. 14.10 A generalized section $\{S_{0,3}^{(a)}\}$ ($a = 1, \ldots, 6$) of $\mathcal{P}_{0,3}$ for the 3-point vertex. This is to be compared with Fig. 14.4a

moduli space). A simple way to prevent overlaps is to tune the stub parameter s_0 to a large value.

By construction, the integral over $\mathcal{V}_{g,n}$ should be finite. If this is not the case, it means that the propagator graphs also diverge and that the parametrization is not good. This can also be solved by considering a sufficiently large value of the stub parameter s_0.

14.4   Suggested Readings

* Plumbing fixture and amplitude factorization [1, sec. 9.3, 9.4, 2, sec. 6].

References

1. J. Polchinski, *String Theory: Volume 1, An Introduction to the Bosonic String* (Cambridge University Press, Cambridge, 2005)
2. E. Witten, Superstring perturbation theory revisited (2012). arXiv:1209. 5461
3. B. Zwiebach, Closed string field theory: quantum action and the BV master equation. Nucl. Phys. B **390**(1), 33–152 (1993). https://doi.org/10.1016/0550-3213(93)90388-6. arXiv: hep-th/9206084
4. B. Zwiebach, Building string field theory around non-conformal backgrounds. Nucl. Phys. B **480**(3), 541–572 (1996). https://doi.org/10.1016/S0550-3213(96)00502-0. arXiv: hep-th/9606153

Closed String Field Theory

15

Abstract

We bring together the elements from the previous chapters in order to write the closed string field action. We first study the gauge fixed theory before reintroducing the gauge invariance. We then prove that the action satisfies the BV master equation meaning that closed SFT is completely consistent at the quantum level. Finally, we describe the 1PI effective action.

15.1 Closed String Field Expansion

In Chaps. 11, 13 and 14, constraints on the external and internal states were found to be necessary. But, to provide another perspective and decouple the properties of the field from the ones of the state, we assume that the string field does not obey any constraint. They will be derived later in order to reproduce the scattering amplitudes from the action and to make the latter well-defined.

The string field is expanded on a basis $\{\phi_r\}$ of the CFT Hilbert space $\mathcal{H}$ (see Sect. 11.2 for more details)

$$|\Psi\rangle = \sum_r \psi_r \, |\phi_r\rangle. \tag{15.1}$$

Using the decomposition (11.40) of the Hilbert space according to the ghost zero-modes, the string field can also be expanded as

$$|\Psi\rangle = \sum_r \Big(\psi_{\downarrow\downarrow,r} \, |\phi_{\downarrow\downarrow,r}\rangle + \psi_{\downarrow\uparrow,r} \, |\phi_{\downarrow\uparrow,r}\rangle + \psi_{\uparrow\downarrow,r} \, |\phi_{\uparrow\downarrow,r}\rangle + \psi_{\uparrow\uparrow,r} \, |\phi_{\uparrow\uparrow,r}\rangle \Big),$$

$$\tag{15.2}$$

© Springer Nature Switzerland AG 2021
H. Erbin, *String Field Theory*, Lecture Notes in Physics 980,
https://doi.org/10.1007/978-3-030-65321-7_15

where we recall that the basis states satisfy

$$b_0 |\phi_{\downarrow\downarrow,r}\rangle = \bar{b}_0 |\phi_{\downarrow\downarrow,r}\rangle = 0, \qquad b_0 |\phi_{\downarrow\uparrow,r}\rangle = \bar{c}_0 |\phi_{\downarrow\uparrow,r}\rangle = 0,$$
$$c_0 |\phi_{\uparrow\downarrow,r}\rangle = \bar{b}_0 |\phi_{\uparrow\downarrow,r}\rangle = 0, \qquad c_0 |\phi_{\uparrow\uparrow,r}\rangle = \bar{c}_0 |\phi_{\uparrow\uparrow,r}\rangle = 0. \tag{15.3}$$

We recall the definition of the dual basis $\{\phi_r^c\}$ through the BPZ inner product

$$\langle \phi_r^c | \phi_s \rangle = \delta_{rs}. \tag{15.4}$$

In terms of the ghost decomposition, the components of the dual states satisfy

$$\langle \phi_{\downarrow\downarrow,r}^c | c_0 = \langle \phi_{\downarrow\downarrow,r}^c | \bar{c}_0 = 0, \qquad \langle \phi_{\downarrow\uparrow,r}^c | c_0 = \langle \phi_{\downarrow\uparrow,r}^c | \bar{b}_0 = 0,$$
$$\langle \phi_{\uparrow\downarrow,r}^c | b_0 = \langle \phi_{\uparrow\downarrow,r}^c | \bar{c}_0 = 0, \qquad \langle \phi_{\uparrow\uparrow,r}^c | b_0 = \langle \phi_{\uparrow\uparrow,r}^c | \bar{b}_0 = 0, \tag{15.5}$$
$$\langle \phi_{x,r}^c | \phi_{y,s} \rangle = \delta_{xy}\delta_{rs},$$

where $x, y =\downarrow\downarrow, \uparrow\downarrow, \downarrow\uparrow, \uparrow\uparrow$. The spacetime ghost number of the fields ψ_r is defined by

$$G(\psi_r) = 2 - n_r. \tag{15.6}$$

Remember that the ghost number of the basis states is denoted by

$$n_r = N_{\mathrm{gh}}(\phi_r), \qquad n_r^c = N_{\mathrm{gh}}(\phi_r^c) = 6 - n_r. \tag{15.7}$$

15.2 Gauge Fixed Theory

Having built the kinetic term (Chap. 9), one needs to construct the interactions. For the same reason—our ignorance of SFT first principles—that forced us to start with the free equation of motion to derive the quadratic action (Chap. 10), we also need to infer the interactions from the scattering amplitudes. Preparing the stage for this analysis was the goal of Chap. 14, where we introduced the factorization of amplitudes to derive the fundamental interactions.

Scattering amplitudes are expressed in terms of gauge fixed states since only them are physical. This allows to give an alternative derivation of the kinetic term by defining it as the inverse of the propagator, which is well-defined for gauge fixed states.[1] The price to pay by constructing interactions in this way is that the SFT

[1]This step is not necessary because the propagator corresponding to the plumbing fixture (Sect. 14.2.2) matches the one found in Sect. 10.5 by considering the simplest gauge fixing. However, this would have been necessary if the factorization had given another propagator, or if the structure of the theory was more complicated, for example for the superstring.

action itself is gauge fixed. To undercover its deeper structure, it is necessary to release the gauge fixing condition. In view of the analysis of the quadratic action in Chap. 10, we can expect that the BV formalism is required. Another possibility is to consider directly the 1PI action.

In this section, we first derive the kinetic term by inverting the propagator. For this to be possible, the string field must obey some constraints: we will find that they correspond to the level-matching and Siegel gauge conditions. Then, we introduce the interactions into the action.

15.2.1 Kinetic Term and Propagator

In Chap. 14, it was found that the propagator reads (14.41)

$$\Delta = b_0^+ b_0^- \frac{1}{L_0^+} \delta_{L_0^-,0}, \qquad \Delta_{rs} = \langle \phi_r^c | b_0^+ b_0^- \frac{1}{L_0^+} \delta_{L_0^-,0} | \phi_s^c \rangle. \tag{15.8}$$

The most natural guess for the kinetic term is

$$S_{0,2} = \frac{1}{2} \langle \Psi | K | \Psi \rangle = \frac{1}{2} \psi_r K_{rs} \psi_s, \tag{15.9}$$

where

$$K = c_0^- c_0^+ L_0^+ \delta_{L_0^-,0} \qquad K_{rs} = \langle \phi_r | c_0^- c_0^+ L_0^+ \delta_{L_0^-,0} | \psi_s \rangle. \tag{15.10}$$

Indeed, it looks like $K\Delta = 1$ using the identities $c_0^\pm b_0^\pm \sim 1$, and it matches (10.115). In terms of the holomorphic and anti-holomorphic modes, we have

$$K = \frac{1}{2} c_0 \bar{c}_0 L_0^+ \delta_{L_0^-,0}. \tag{15.11}$$

But, when writing $c_0^\pm b_0^\pm \sim 1$, the second part of the anti-commutator $\{b_0^\pm, c_0^\pm\} = 1$ is missing. The relation $c_0^\pm b_0^\pm \sim 1$ is correct only when acting on basis dual states annihilated by $c_0^\pm$. The problem stems from the fact that Ψ is not yet subject to any constraint. Moreover, some of the string field components will not appear in the expression since they are annihilated by the ghost zero-mode. As a consequence, the kinetic operator in (15.10) (or equivalently the propagator) is not invertible in the Hilbert space $\mathcal{H}$ because its kernel is not empty:

$$\ker K |_{\mathcal{H}} \neq \emptyset. \tag{15.12}$$

This can be seen by writing ϕ_r as a 4-vector and K_{rs} as a 4×4-matrix:

$$K_{rs} = \frac{1}{2} \begin{pmatrix} \langle \phi_{\downarrow\downarrow,r}| \\ \langle \phi_{\downarrow\uparrow,r}| \\ \langle \phi_{\uparrow\downarrow,r}| \\ \langle \phi_{\uparrow\uparrow,r}| \end{pmatrix}^t \begin{pmatrix} c_0\bar{c}_0 L_0^+ & 0 & 0 & 0 \\ 0 & 0 & 0 & 0 \\ 0 & 0 & 0 & 0 \\ 0 & 0 & 0 & 0 \end{pmatrix} \begin{pmatrix} |\phi_{\downarrow\downarrow,s}\rangle \\ |\phi_{\downarrow\uparrow,s}\rangle \\ |\phi_{\uparrow\downarrow,s}\rangle \\ |\phi_{\uparrow\uparrow,s}\rangle \end{pmatrix}. \tag{15.13}$$

The matrix is mostly empty because the states $\phi_{x,r}$ with different $x = \downarrow\downarrow, \uparrow\downarrow$, $\downarrow\uparrow, \uparrow\uparrow$ are orthogonal (no non-diagonal terms) and the states with $x \neq \downarrow\downarrow$ are annihilated by c_0 or $\bar{c}_0$. The same consideration applies for the delta function: if the field does not satisfy $L_0^- = 0$, then the kinetic operator is non-invertible.

To summarize, the string field must satisfy three conditions in order to have an invertible kinetic term

$$L_0^- |\Psi\rangle = 0, \qquad b_0^- |\Psi\rangle = 0, \qquad b_0^+ |\Psi\rangle = 0. \tag{15.14}$$

This means that the string field is expanded on the $\mathcal{H}_0 \cap \ker L_0^-$ Hilbert space:

$$|\Psi\rangle = \sum_r \psi_{\downarrow\downarrow,r} |\phi_{\downarrow\downarrow,r}\rangle. \tag{15.15}$$

Ill-defined kinetic terms are expected in the presence of a gauge symmetry: this was already discussed in Sects. 10.1.4 and 10.5 for the free theory, and this will be discussed further later in this chapter for the interacting case.

Computation
Let us check that K_{rs} is correctly the inverse of Δ_{rs} when Ψ is restricted to $\mathcal{H}_0$:

$$K_{rs}\Delta_{st} = \langle \phi_r| c_0^- c_0^+ L_0^+ \delta_{L_0^-,0} |\phi_s\rangle \langle \phi_s^c| b_0^+ b_0^- \frac{1}{L_0^+} \delta_{L_0^-,0} |\phi_t^c\rangle$$

$$= \langle \phi_r| c_0^- c_0^+ L_0^+ \delta_{L_0^-,0} b_0^+ b_0^- \frac{1}{L_0^+} \delta_{L_0^-,0} |\phi_t^c\rangle$$

$$= \langle \phi_r| \{c_0^-, b_0^-\}\{c_0^+, b_0^+\} |\phi_t^c\rangle$$

$$= \langle \phi_r|\phi_t^c\rangle = \delta_{rt}.$$

The second equality follows from the resolution of the identity (11.36): due to the zero-mode insertions, the resolution of the identity collapses to a sum over the $\downarrow\downarrow$ states

$$1 = \sum_r |\phi_r\rangle\langle \phi_r^c| = \sum_r |\phi_{\downarrow\downarrow,r}\rangle\langle \phi_{\downarrow\downarrow,r}^c|. \tag{15.16}$$

The third equality uses that L_0^+ commutes with the ghost modes, that ϕ_r is annihilated by $b_0^{\pm}$, and that $(\delta_{L_0^-,0})^2 = \delta_{L_0^-,0} = 1$ on states with $L_0^- = 0$.

Finally, we find that the kinetic term matches the classical quadratic vertex $\mathcal{V}_{0,2}$ defined in (14.56) such that

$$S_{0,2} = \frac{1}{2} \mathcal{V}_{0,2}(\Psi^2) = \frac{1}{2} \langle \Psi | c_0^- c_0^+ L_0^+ \delta_{L_0^-,0} | \Psi \rangle . \tag{15.17}$$

15.2.2 Interactions

The second step to build the action is to write the interaction terms from the Feynman rules. Before proceeding to SFT, it is useful to remember how this works for a standard QFT.

Example 15.1: Feynman Rules for a Scalar Field

Consider a scalar field with a standard kinetic term and a n-point interaction:

$$S = \int \mathrm{d}^D x \left(\frac{1}{2} \phi(x)(-\partial^2 + m^2)\phi(x) + \frac{\lambda}{n!} \phi(x)^n \right). \tag{15.18}$$

First, one needs to find the physical states that correspond to solutions of the linearized equation of motion. In the current case, they are plane-waves (in momentum representation):

$$\phi_k(x) = e^{ik \cdot x}. \tag{15.19}$$

Then, the vertex (in momentum representation) $V_n(k_1, \ldots, k_n)$ is found by replacing in the interaction each occurrence of the field by a different state, and summing over all the different contributions. Here, this means that one considers states $\phi_{k_i}(x)$ with different momenta:

$$V_n(k_1, \ldots, k_n) = \frac{\lambda}{n!} \int \mathrm{d}^D x \, n! \prod_{i=1}^{n} \phi_{k_i}(x) = \lambda \int \mathrm{d}^D x \, e^{i(k_1 + \cdots + k_n)x}$$

$$= \lambda (2\pi)^D \delta^{(D)}(k_1 + \cdots + k_n). \tag{15.20}$$

The factor $n!$ comes from all the permutations of the n states in the monomial of order n. Reversing the argument, one sees how to move from the vertex $V_n(k_1, \ldots, k_n)$ written in terms of states to the interaction in the action in terms of the field.

Obviously, if the field has more states (e.g. if it has a spin or if it is in a representation of a group), then one needs to consider all the different possibilities. The above prescription also yields directly the insertion of the momentum necessary if the interaction contains derivatives. ◄

In Sect. 14.2, the Feynman rule for a g-loop n-point fundamental vertex of states $(\mathcal{V}_1, \ldots, \mathcal{V}_n)$ was found to be given by (14.33):

$$
\mathcal{V}_{g,n}(\mathcal{V}_1, \ldots, \mathcal{V}_n) = \int_{\mathcal{R}_{g,n}} \omega^{g,n}_{\mathsf{M}_{g,n}}(\mathcal{V}_1, \ldots, \mathcal{V}_n) = \mathcal{V}_1 \underset{g}{\diagdown} \vdots \quad , \qquad (15.21)
$$

where $\mathcal{R}_{g,n}$ is a section over the fundamental region $\mathcal{V}_{g,n} \subset \mathcal{M}_{g,n}$ (12.42b) that cannot be covered from the plumbing fixture of lower-dimensional surfaces.

From Example 15.1, it should be clear that the g-loop n-point contribution to the action can be obtained simply by replacing every state with a string field in $\mathcal{V}_{g,n}$:

$$
S_{g,n} = \hbar^g \frac{g_s^{2g-2+n}}{n!} \mathcal{V}_{g,n}(\Psi^n), \qquad (15.22)
$$

where $\Psi^n := \Psi^{\otimes n}$. The power of the coupling constant has been reinstated: it can be motivated by the fact that it should have the same power as the corresponding amplitude (Sect. 3.1.1). Note that the interactions are defined only when the power of g_s is positive: $\chi_{g,n} = 2 - 2g - n < 0$. We have also written explicitly the power of $\hbar$, which counts the number of loops.

Before closing this section, we need to comment on the effect of the constraints (15.14) on the interactions. Building a Feynman graph by gluing two m- and n-point interactions with a propagator, one finds that the states proportional to $\phi_{x,r}$ for $x \neq \downarrow\downarrow$ do not propagate inside internal legs

$$
\mathcal{V}_{g,m}(\mathcal{V}_1, \ldots, \mathcal{V}_{m-1}, \phi_r) \langle \phi_r^c | b_0^+ b_0^- \frac{1}{L_0^+} | \phi_s^c \rangle \mathcal{V}_{g',n}(\mathcal{W}_1, \ldots, \mathcal{W}_{n-1}, \phi_s)
$$

$$
= \mathcal{V}_{g,m}(\mathcal{V}_1, \ldots, \mathcal{V}_{m-1}, \phi_{\downarrow\downarrow,r}) \langle \phi_{\downarrow\downarrow,r}^c | b_0^+ b_0^- \frac{1}{L_0^+} | \phi_{\downarrow\downarrow,s}^c \rangle
$$

$$
\times \mathcal{V}_{g',n}(\mathcal{W}_1, \ldots, \mathcal{W}_{n-1}, \phi_{\downarrow\downarrow,s}). \qquad (15.23)
$$

Thus, they do not contribute to the final result even if the interactions contain them. While the conditions $L_0^- = b_0^- = 0$ were found to be necessary for defining off-shell amplitudes, the condition $b_0^+ = 0$ does not arise from any consistency requirement. But, it is also consistent with the interactions, since only fundamental vertices have a chance to give a non-vanishing result for states that do not satisfy (15.14). Hence, the interactions (15.22) are compatible with the definition of the kinetic term and the restriction of the string field.

15.2.3 Action

The interacting gauge fixed action is built from the kinetic term $\mathcal{V}_{0,2}$ (15.17) and from the interactions $\mathcal{V}_{g,n}$ (15.22) with $\chi_{g,n} < 0$. However, this is not sufficient: we have seen in Sect. 14.3 that it makes sense to consider the vertices with $\chi_{g,n} \geq 0$. First, we should consider the 1-loop cosmological constant $\mathcal{V}_{1,0}$. Then, we can also add the classical source $\mathcal{V}_{0,1}$ and the tree-level cosmological constant $\mathcal{V}_{0,0}$. With all the terms together, the action reads

$$
\begin{aligned}
S &= \sum_{g,n \geq 0} \hbar^g \frac{g_s^{2g-2+n}}{n!} \, \mathcal{V}_{g,n}(\Psi^n) \\
&:= \frac{1}{2} \langle \Psi | \, c_0^- c_0^+ L_0^+ \delta_{L_0^-,0} \, | \Psi \rangle + \sum_{g,n \geq 0}' \hbar^g \frac{g_s^{2g-2+n}}{n!} \, \mathcal{V}_{g,n}(\Psi^n),
\end{aligned}
\tag{15.24}
$$

where $\mathcal{V}_n$ was defined in (14.58). A prime on the sum indicates that the term $g = 0, n = 2$ is removed, such that one can single out the kinetic term. We will often drop the delta function imposing $L_0^- = 0$ because the field is taken to satisfy this constraint.

Rewriting the vertices in terms of the products $\ell_{g,n}$ defined in (14.65)

$$
\mathcal{V}_{g,n}(\Psi^n) := \langle \Psi | \, c_0^- \, \big| \ell_{g,n-1}(\Psi^{n-1}) \big\rangle
\tag{15.25}
$$

leads to the alternative form

$$
S = \sum_{g,n \geq 0} \hbar^g \frac{g_s^{2g-2+n}}{n!} \, \langle \Psi | \, c_0^- \, \big| \ell_{g,n-1}(\Psi^{n-1}) \big\rangle.
\tag{15.26}
$$

The definition (14.56) leads to the following explicit expression for $\ell_{0,1}$:

$$
\ell_{0,1}(\Psi_{cl}) = c_0^+ L_0^+ \, | \Psi_{cl} \rangle.
\tag{15.27}
$$

In most cases, the terms $g = 0, n = 0, 1$ vanish such that the action reads

$$
S = \sum_{\substack{g,n \geq 0 \\ \chi_{g,n} \leq 0}} \hbar^g \frac{g_s^{2g-2+n}}{n!} \, \mathcal{V}_{g,n}(\Psi^n).
\tag{15.28}
$$

However, we will often omit the condition $\chi_{g,n} \leq 0$ to simplify the notation, except when the distinction is important, and the reader can safely assume $V_{0,0} = V_{0,1} = 0$ if not otherwise stated. The classical action is obtained by setting $\hbar = 0$

$$S_{\text{cl}} = \frac{1}{2} \langle \Psi_{\text{cl}}| c_0^- c_0^+ L_0^+ |\Psi_{\text{cl}}\rangle + \sum_{n \geq 3} \frac{g_s^n}{n!} V_{0,n}(\Psi_{\text{cl}}^n). \tag{15.29}$$

Rescaling the string field by g_s^{-1} gives the more canonical form of the action (using the same symbol):

$$\begin{aligned}
S &= \sum_{g,n \geq 0} \hbar^g g_s^{2g-2} \frac{1}{n!} V_{g,n}(\Psi^n) \\
&:= \frac{1}{2g_s^2} \langle \Psi| c_0^- c_0^+ L_0^+ \delta_{L_0^-,0} |\Psi\rangle + \frac{1}{g_s^2} \sum_{g,n \geq 0}' \frac{(\hbar g_s^2)^g}{n!} V_{g,n}(\Psi^n).
\end{aligned} \tag{15.30}$$

In the path integral, the action is divided by $\hbar$ such that

$$\frac{S}{\hbar} = \sum_{g,n \geq 0} (\hbar g_s^2)^{g-1} \frac{1}{n!} V_{g,n}(\Psi^n). \tag{15.31}$$

This shows that there is a single coupling constant $\hbar g_s^2$, instead of two ($\hbar$ and g_s separately) as it looks at the first sight. This makes sense because g_s is in fact the expectation value of the dilaton field (2.166) and its value can be changed by deforming the background with dilatons [1, 2, 12].

The previous remark also allows to easily change the normalization of the action, for example, to perform a Wick rotation, to normalize canonically the action in terms of spacetime fields or to reintroduce $\hbar$. Rescaling the action by α is equivalent to rescaling g_s^2 by α^{-1}:

$$S \to \alpha S \quad \Longrightarrow \quad g_s^2 \to \frac{g_s^2}{\alpha}. \tag{15.32}$$

The linearized equation of motion is

$$L_0^+ |\Psi\rangle = 0, \tag{15.33}$$

which corresponds to the Siegel gauge equation of motion of the free theory (10.116). Hence, this equation is not sufficient to determine the physical states (cohomology of the BRST operator, Chap. 8), as discussed in Chap. 10, and additional constraints must be imposed. One can interpret this by saying that the action (15.24) provides only the Feynman rules, not the physical states. Removing the gauge fixing will be done in Sects. 15.3 and 15.4.

The action (15.24) looks overly more complicated than a typical QFT theory: instead of few interaction terms for low n ($n \leq 4$ in $d = 4$ renormalizable theories), it has contact interactions of all orders $n \in \mathbb{N}$. The terms with $g \geq 1$ are associated to quantum corrections as they indicate the power of $\hbar$, which means that they can be interpreted as counter-terms. But, how is it that one needs counter-terms despite the claim that every Feynman graphs (including the fundamental vertices) in SFT are finite? The role of renormalization is not only to cure UV divergences, but also IR divergences (due to vacuum shift and mass renormalization). Equivalently, this can be understood by the necessity to correct the asymptotic states of the theory, or to consider renormalized instead of bare quantities. Indeed, the asymptotic states obtained from the linearized classical equations of motion are idealization: turning on interactions modifies the states. In typical QFTs, these corrections are infinite, and renormalization is crucial to extract a number; however, even if the effect is finite, it is needed to describe correctly the physical quantities [16, p. 411]. There is a second reason for these additional terms: when relaxing the gauge fixing condition, the path integral is anomalous under the gauge symmetry, and the terms with $g > 0$ are necessary to cancel the anomaly (this will be discussed more precisely in Sect. 15.4). It may thus seem that SFT cannot be predictive because of the infinite number of counter-terms. Fortunately, this is not the case: the main reason for the loss of predictability in non-renormalizable theory is that the renormalization procedure introduces an infinite number[2] of arbitrary parameters (and thus making a prediction would require to have already made an infinite number of observations to determine all the parameters). These parameters come from the subtraction of two infinities: there is no unique way to perform it, and thus one needs to introduce a new parameter. The case of SFT is different: since every quantity is finite, the renormalization has no ambiguity because one subtracts two finite numbers, and the result is unambiguous. As a consequence, renormalization does not introduce any new parameter, and there is a unique coupling constant g_s in the theory, which is determined by the tree-level cubic interaction. The coupling constants of higher-order and higher-loop interactions are all determined by powers of g_s, and thus a unique measurement is sufficient to make predictions.

Another important point is that the action (15.24) is not uniquely defined. The definition of the vertices depends on the choice of the local coordinates and of the stub parameter s_0. Changing them modifies the vertices, and thus the action. But, one can show that the different theories are related by field redefinitions and are thus equivalent.

---

[2] In practice, this number does not need to be infinite to wreck predictability, and it is sufficient that it is very large.

15.3   Classical Gauge Invariant Theory

In the previous section, we have found the gauge fixed action (15.24). Since the complete gauge invariant quantum action has a complicated structure, it is instructive to first focus on the classical action (15.29). The full action is discussed in Sect. 15.4.

The gauge fixing is removed by relaxing the $b_0^+ = 0$ constraint on the field (the other constraints must be kept in order to have well-defined the interactions). The classical field Ψ_{cl} is then defined by

$$\Psi_{\text{cl}} \in \mathcal{H}^- \cap \ker L_0^-, \qquad N_{\text{gh}}(\Psi_{\text{cl}}) = 2. \tag{15.34}$$

The restriction on the ghost number translates the condition that the field is classical, i.e. that there are no spacetime ghosts at the classical level. The relation (15.6) implies that all components have vanishing spacetime ghost number.

In the free limit, the gauge invariant action should match (10.105)

$$S_{0,2} = \frac{1}{2} \langle \Psi | c_0^- Q_B | \Psi \rangle, \tag{15.35}$$

and lead to the results from Sect. 10.5. A natural guess is that the form of the interactions is not affected by the gauge fixing (the latter usually modifies the propagator but not the interactions). This leads to the gauge invariant classical action:

$$S_{\text{cl}} = \frac{1}{2} \langle \Psi_{\text{cl}} | c_0^- Q_B | \Psi_{\text{cl}} \rangle + \frac{1}{g_s^2} \sum_{n \geq 3} \frac{g_s^n}{n!} V_{0,n}(\Psi_{\text{cl}}^n), \tag{15.36}$$

where the vertices $V_{0,n}$ with $n \geq 3$ are the ones defined in (14.33) (we consider the case where $V_{0,0} = V_{0,1} = 0$). It is natural to generalize the definition of $V_{0,2}$ as

$$V_{0,2}(\Psi_{\text{cl}}^2) := \langle \Psi_{\text{cl}} | c_0^- Q_B | \Psi_{\text{cl}} \rangle \tag{15.37}$$

such that

$$S_{\text{cl}} = \frac{1}{g_s^2} \sum_{n \geq 2} \frac{g_s^n}{n!} V_{0,n}(\Psi_{\text{cl}}^n) = \frac{1}{g_s^2} \sum_{n \geq 2} \frac{g_s^n}{n!} \langle \Psi_{\text{cl}} | c_0^- \left| \ell_{0,n-1}(\Psi_{\text{cl}}^{n-1}) \right\rangle, \tag{15.38}$$

where (15.37) implies

$$\ell_{0,1}(\Psi_{\text{cl}}) = Q_B | \Psi_{\text{cl}} \rangle. \tag{15.39}$$

The equation of motion is

$$\mathcal{F}_{cl}(\Psi_{cl}) := \sum_{n \geq 1} \frac{g_s^{n-1}}{n!} \ell_{0,n}(\Psi_{cl}^n) = Q_B |\Psi_{cl}\rangle + \sum_{n \geq 2} \frac{g_s^{n-1}}{n!} \ell_{0,n}(\Psi_{cl}^n) = 0. \quad (15.40)$$

Computation: Equation (15.40)

$$\delta S_{cl} = \frac{1}{g_s^2} \sum_{n \geq 2} \frac{g_s^n}{n!} n \{\delta\Psi_{cl}, \Psi_{cl}^{n-1}\}_0 = \frac{1}{g_s^2} \sum_{n \geq 2} \frac{g_s^n}{(n-1)!} \langle \delta\Psi_{cl}| c_0^- \Big| \ell_{0,n-1}(\Psi_{cl}^{n-1}) \Big\rangle.$$

$$(15.41)$$

The first equality follows because the vertex is completely symmetric. Simplifying and shifting n, one obtains $c_0^- |\mathcal{F}_{cl}\rangle$. The factor c_0^- is invertible because of the constraint $b_0^- = 0$ imposed on the field.

The action is invariant

$$\delta_\Lambda S_{cl} = 0 \quad (15.42)$$

under the gauge transformation

$$\delta_\Lambda \Psi_{cl} = \sum_{n \geq 0} \frac{g_s^n}{n!} \ell_{0,n+1}(\Psi_{cl}^n, \Lambda) = Q_B |\Lambda\rangle + \sum_{n \geq 1} \frac{g_s^n}{n!} \ell_{0,n+1}(\Psi_{cl}^n, \Lambda). \quad (15.43)$$

The gauge algebra is [17, sec. 4]:

$$[\delta_{\Lambda_2}, \delta_{\Lambda_1}]\Psi_{cl} = \delta_{\Lambda(\Lambda_1, \Lambda_2, \Psi_{cl})} |\Psi_{cl}\rangle + \sum_{n \geq 0} \frac{g_s^{n+2}}{n!} \ell_{0,n+3}(\Psi_{cl}^n, \Lambda_2, \Lambda_1, \mathcal{F}_{cl}(\Psi_{cl})),$$

$$(15.44a)$$

where $\mathcal{F}_{cl}$ is the equation of motion (15.40), and $\Lambda(\Lambda_1, \Lambda_2, \Psi_{cl})$ is a field-dependent gauge parameter:

$$\Lambda(\Lambda_1, \Lambda_2, \Psi_{cl}) = \sum_{n \geq 0} \frac{g_s^{n+1}}{n!} \ell_{0,n+2}(\Lambda_1, \Lambda_2, \Psi_{cl}^n)$$

$$(15.44b)$$

$$= g_s \ell_{0,2}(\Lambda_1, \Lambda_2) + \sum_{n \geq 1} \frac{g_s^{n+1}}{n!} \ell_{0,n+2}(\Lambda_1, \Lambda_2, \Psi_{cl}^n).$$

The classical gauge algebra is complicated, which explains why a direct quantization (e.g. through the Faddeev–Popov procedure) cannot work: the second term in (15.44a) indicates that the algebra is open (it closes only on-shell), while the first term is a gauge transformation with a field-dependent parameter. As reviewed in Appendix C.3, both properties require using the BV formalism for the quantization, and the latter is performed in Sect. 15.4. An important point is that if the theory had only cubic interactions, i.e. if

$$\forall n \geq 4: \quad \mathcal{V}_{0,4}(\mathcal{V}_1, \ldots, \mathcal{V}_n) = 0, \quad \ell_{g,n-1}(\mathcal{V}_1, \ldots, \mathcal{V}_{n-1}) = 0, \quad \text{(cubic theory)},$$
$$(15.45)$$

then the algebra closes off-shell and $\Lambda(\Lambda_1, \Lambda_2, \Psi_{\text{cl}})$ becomes field independent.

Computation: Equation (15.42)

$$\delta_\Lambda S_{\text{cl}} = \sum_{n \geq 2} \frac{g_s^{n-2}}{n!} \, n \mathcal{V}_{0,n}(\delta \Psi_{\text{cl}}, \Psi_{\text{cl}}^{n-1})$$

$$= \sum_{m,n \geq 0} \frac{g_s^{m+n-1}}{m! \, n!} \, \mathcal{V}_{0,n+1}\big(\ell_{0,m+1}(\Psi_{\text{cl}}^m, \Lambda), \Psi_{\text{cl}}^n\big)$$

$$= \sum_{m \geq 0} \sum_{n=0}^m \frac{g_s^{m-1}}{(m-n)! \, n!} \, \big\langle \ell_{m-n+1}(\Psi_{\text{cl}}^{m-n}, \Lambda) \big| c_0^- \big| \ell_{0,n}(\Psi_{\text{cl}}^n) \big\rangle.$$

For simplicity, we have extended the sum up to $n = 0$ and $m = 0$ by using the fact that lower-order vertices vanish. The bracket can be rewritten as

$$= \big\langle \ell_{0,n}(\Psi_{\text{cl}}^n) \big| c_0^- \big| \ell_{0,m-n+1}(\Psi_{\text{cl}}^{m-n}, \Lambda) \big\rangle$$

$$= \mathcal{V}_{0,m-n+2}\big(\ell_{0,n}(\Psi_{\text{cl}}^n), \Psi_{\text{cl}}^{m-n}, \Lambda\big)$$

$$= -\mathcal{V}_{0,m-n+2}\big(\Lambda, \ell_{0,n}(\Psi_{\text{cl}}^n), \Psi_{\text{cl}}^{m-n}\big)$$

$$= \big\langle \Lambda \big| c_0^- \big| \ell_{0,m-n+1}\big(\ell_{0,n}(\Psi_{\text{cl}}^n), \Psi_{\text{cl}}^{m-n}\big) \big\rangle.$$

Then, one needs to use the identity (defined for all $m \geq 0$)

$$0 = \sum_{n=0}^m \frac{m!}{(m-n)! \, n!} \, \ell_{0,m-n+1}\big(\ell_{0,n}(\Psi_{\text{cl}}^n), \Psi_{\text{cl}}^{m-n}\big), \qquad (15.46)$$

which comes from (14.70). Multiplying this by $g_s^{m-1}/m!$ and summing over $m \geq 0$ proves (15.42).

Remark 15.1 (L_∞ Algebra) The identities satisfied by the products $\ell_{0,n}$ from the gauge invariance of the action imply that they form a L_∞ homotopy algebra [4, 11, 17] (for more general references, see [5,6,9,10]). The latter can also be mapped to a

BV structure, which explains why the BV quantization Sect. 15.4 is straightforward. This interplay between gauge invariance, covering of the moduli space, BV and homotopy algebra is particularly beautiful. It has also been fruitful in constructing super-SFT.

15.4 BV Theory

As indicated in the previous section (Sect. 15.3), the classical gauge algebra is open and has field-dependent structure constants. The BV formalism (Appendix C.3) is necessary to define the theory.

In the BV formalism, the classical action for the physical fields is extended to the quantum master action by solving the quantum master equation (C.114). It is generically difficult to build this action exactly, but the discussion of Sect. 10.3 can serve as a guide: it was found that the free quantum action (with the tower of ghosts) has exactly the same form as the free classical action (without ghosts). Hence, this motivates the ansatz that it should be of the same form as the classical action (15.36) to which are added the counter-terms from (15.24):

$$S = \frac{1}{g_s^2} \sum_{g \geq 0} \hbar^g g_s^{2g} \sum_{n \geq 0} \frac{g_s^n}{n!} V_{g,n}(\Psi^n) \tag{15.47a}$$

$$= \frac{1}{2} \langle \Psi | c_0^- Q_B | \Psi \rangle + \sideset{}{'}\sum_{g,n \geq 0} \frac{\hbar^g g_s^{2g-2+n}}{n!} V_{g,n}(\Psi^n) \tag{15.47b}$$

$$= \frac{1}{g_s^2} \sum_{g,n \geq 0} \frac{\hbar^g g_s^{2g-2+n}}{n!} \langle \Psi | c_0^- \big| \ell_{g,n-1}(\Psi^{n-1}) \big\rangle, \tag{15.47c}$$

but without any constraint on the ghost number of Ψ:

$$\Psi \in \mathcal{H}^- \cap \ker L_0^-. \tag{15.48}$$

In order to show that (15.47) is a consistent quantum master action, it is necessary to show that it solves the master BV equation (C.114):

$$(S, S) - 2\hbar \Delta S = 0. \tag{15.49}$$

The first step is to introduce the fields and antifields. In fact, because the CFT ghost number induces a spacetime ghost number, there is a natural candidate set.

The string field is expanded as (15.1)

$$|\Psi\rangle = \sum_r \psi_r |\phi_r\rangle, \tag{15.50}$$

where the $\{\phi_r\}$ forms a basis of $\mathcal{H}^-$. The string field can be further separated as

$$\Psi = \Psi_+ + \Psi_-, \tag{15.51}$$

where Ψ_- (Ψ_+) contains only states that have negative (positive) cylinder ghost numbers (this gives an offset of 3 when using the plane ghost number):

$$\Psi_- = \sum_r \sum_{n_r \le 2} |\phi_r\rangle \psi^r, \qquad \Psi_+ = \sum_r \sum_{n_r^c > 2} b_0^- |\phi_r^c\rangle \psi_r^*. \tag{15.52}$$

The order of the basis states and coefficients matters if they anti-commute. The sum in Ψ_+ can be rewritten as a sum over $n_r \le 2$ like the first term since $n_r + n_r^c = 6$. Correspondingly, the spacetime ghost numbers (15.6) for the coefficients in Ψ_- (Ψ_+) are positive (negative)

$$G(\psi^r) \ge 0, \qquad G(\psi_r^*) < 0. \tag{15.53}$$

Moreover, one finds that the ghost numbers of ψ^r and ψ_r^* are related as

$$G(\psi_r^*) = -1 - G(\psi^r), \tag{15.54}$$

which also implies that they have opposite parity. Comparing with Appendix C.3, this shows that the ψ^r (ψ_r^*) contained in Ψ_- (Ψ_+) can be identified with the fields (antifields).

> **Computation: Equation (15.54)**
>
> $$G(\psi_r^*) = 2 - N_{\mathrm{gh}}(b_0^- \phi_r^c) = 2 + 1 - n_r^c$$
> $$= 3 - (6 - n_r) = -3 + (2 - G(\psi^r)) = -1 - G(\psi^r).$$

In terms of fields and antifields, the master action is

$$\frac{\partial_R S}{\partial \psi^r} \frac{\partial_L S}{\partial \psi_r^*} + \hbar \frac{\partial_R \partial_L S}{\partial \psi^r \partial \psi_r^*} = 0. \tag{15.55}$$

Plugging the expression (15.47) of S inside and requiring that the expression vanishes order by order in g and n give the set of equations:

$$\sum_{\substack{g_1, g_2 \ge 0 \\ g_1 + g_2 = g}} \sum_{\substack{n_1, n_2 \ge 0 \\ n_1 + n_2 = n}} \frac{\partial_R S_{g_1, n_1}}{\partial \psi^r} \frac{\partial_L S_{g_2, n_2}}{\partial \psi_r^*} + \hbar \frac{\partial_R \partial_L S_{g-1, n}}{\partial \psi^r \partial \psi_r^*} = 0, \tag{15.56}$$

where $S_{g,n}$ was defined in (15.22). This holds true due to the identity (14.70) (the complete proof can be found in [17, pp. 42–45]). The fact that the second term is not identically zero means that the measure is not invariant under the classical gauge symmetry (anomalous symmetry): corrections need to be introduced to cancel the anomaly. It is a remarkable fact that one can construct directly the quantum master action in SFT and that it takes the same form as the classical action.

15.5 1PI Theory

The BV action is complicated: instead, it is often simpler and sufficient to work with the 1PI effective action. The latter incorporates all the quantum corrections in 1PI vertices such that scattering amplitudes are expressed only in terms of tree Feynman graphs (there are no loops in diagrams since they correspond to quantum effects, already included in the definitions of the vertices).

A 1PI graph is a Feynman graph that stays connected if one cuts any single internal line. On the other hand, a 1PR graph splits into two disconnected by cutting one of the lines. The scattering amplitudes $A_{g,n}$ are built by summing all the different ways to connect two 1PI vertices with a propagator: diagrams connecting two legs of the same 1PI vertex are forbidden by definition.

The g-loop n point 1PR and 1PI Feynman diagrams are associated to some regions of the moduli space $\mathcal{M}_{g,n}$. Comparing the previous definitions with the gluing of Riemann surfaces (Sect. 12.3), 1PR diagrams are obtained by gluing surfaces with the separating plumbing fixture (Sect. 14.1.1). Thus, the 1PR and 1PI regions $\mathcal{F}_{g,n}^{1PR}$ and $\mathcal{V}_{g,n}^{1PI}$ can be identified with the regions defined in (12.43a) and (12.43b). In particular, the n-point 1PI interaction is the sum over g of the g-loop n-point 1PI interactions (14.63):

$$\mathcal{V}_n^{1PI}(\mathcal{V}_1,\ldots,\mathcal{V}_n) := \mathcal{V}_1 \!\!-\!\!\!\!\boxed{1PI}\!\!\!\!\begin{array}{c}\mathcal{V}_2\\ \vdots \\ \mathcal{V}_n\end{array} := \sum_{g\geq 0}(\hbar g_s^2)^g\, \mathcal{V}_{g,n}^{1PI}(\mathcal{V}_1,\ldots,\mathcal{V}_n),$$

$$\mathcal{V}_{g,n}^{1PI}(V_1,\ldots,V_n) := \int_{\mathcal{R}_{g,n}^{1PI}} \omega_{\mathcal{M}_{g,n}}^{g,n}(\mathcal{V}_1,\ldots,\mathcal{V}_n),$$

$$(15.57)$$

where $\mathcal{R}_{g,n}^{1PI}$ is a section of $\mathcal{P}_{g,n}$ over $\mathcal{V}_{g,n}^{1PI}$.

Given the interaction vertices, it is possible to follow the same reasoning as in Sects. 15.2 and 15.3.

The gauge fixed 1PI effective action reads

$$S_{1PI} = \frac{1}{g_s^2} \sum_{n\geq 0} \frac{g_s^n}{n!} \mathcal{V}_n^{1PI}(\Psi^n) := \frac{1}{2} \langle \Psi | c_0^- c_0^+ L_0^+ | \Psi \rangle + \frac{1}{g_s^2} \sum_{n\geq 0}' \frac{g_s^n}{n!} \mathcal{V}_n^{1PI}(\Psi^n).$$

(15.58)

Here, the prime means again that the terms $g = 0, n = 2$ are excluded from the definition of $\mathcal{V}_2^{1PI}$. The action has the same form as the classical gauge fixed action (15.29), which is logical since it generates only tree-level Feynman graphs. For this reason, the vertices $\mathcal{V}_n^{1PI}$ have exactly the same properties as the brackets $\mathcal{V}_{0,n}$. This fact can be used to write the 1PI gauge invariant action:

$$S_{1PI} = \frac{1}{2} \langle \Psi | c_0^- Q_B | \Psi \rangle + \frac{1}{g_s^2} \sum_{n\geq 0}' \frac{g_s^n}{n!} \mathcal{V}_n^{1PI}(\Psi^n),$$

(15.59)

which mirrors the classical gauge invariant action (15.36). Then, it is straightforward to see that it enjoys the same gauge symmetry upon replacing the tree-level vertices by the 1PI vertices. But, since this action incorporates all quantum corrections, this also proves that the quantum theory is correctly invariant under a quantum gauge symmetry.

Remark 15.2 The 1PI action (15.59) can also be directly constructed from the BV action (15.47).

15.6   Suggested Readings

- Gauge fixed and classical gauge invariant closed SFT [17] (see also [7, 8]).
- BV closed SFT [17] (see also [15]).
- Construction of the open–closed BV SFT [18].
- 1PI SFT [3, 13, 14, sec. 4.1, 5.2].

References

1. A. Belopolsky, B. Zwiebach, Who changes the string coupling? Nucl. Phys. B **472**(1–2), 109–138 (1996). https://doi.org/10.1016/0550-3213(96)00203-9. arXiv: hep-th/9511077
2. O. Bergman, B. Zwiebach, The Dilaton theorem and closed string backgrounds. Nucl. Phys. B **441**(1–2), 76–118 (1995). https://doi.org/10.1016/0550-3213(95)00022-K. arXiv: hep-th/9411047
3. C. de Lacroix, H. Erbin, S.P. Kashyap, A. Sen, M. Verma, Closed super-string field theory and its applications. Int. J. Mod. Phys. A **32**(28–29), 1730021 (2017). https://doi.org/10.1142/S0217751X17300216. arXiv: 1703.06410
4. T. Erler, S. Konopka, I. Sachs, NS-NS sector of closed superstring field theory. J. High Energy Phys. **2014**(8) (2014). https://doi.org/10.1007/JHEP08(2014)158. arXiv: 1403.0940
5. O. Hohm, B. Zwiebach, L_∞ algebras and field theory. Fortschr. Phys. **65**(3–4), 1700014 (2017). https://doi.org/10.1002/prop.201700014. arXiv: 1701.08824

6. O. Hohm, V. Kupriyanov, D. Lüst, M. Traube, General constructions of L_∞ algebras (2017). arXiv: 1709.10004
7. T. Kugo, H. Kunitomo, K. Suehiro, Non-polynomial closed string field theory. Phys. Lett. B **226**(1), 48–54 (1989). https://doi.org/10.1016/0370-2693(89)90287-6
8. T. Kugo, K. Suehiro, Nonpolynomial closed string field theory: action and its gauge invariance. Nucl. Phys. B **337**(2), 434–466 (1990). https://doi.org/10.1016/0550-3213(90)90277-K
9. T. Lada, M. Markl, Strongly homotopy lie algebras (1994). arXiv: hep-th/9406095
10. T. Lada, J. Stasheff, Introduction to SH Lie algebras for physicists. Int. J. Theor. Phys. **32**(7), 1087–1103 (1993). https://doi.org/10.1007/BF00671791. arXiv: hep-th/9209099
11. K. Muenster, I. Sachs, Homotopy classification of bosonic string field theory. Commun. Math. Phys. **330**, 1227–1262 (2014). https://doi.org/10.1007/s00220-014-2027-8. arXiv: 1208.5626
12. S. Rahman, B. Zwiebach, Vacuum vertices and the ghost-dilaton. Nucl. Phys. B **471**(1–2), 233–245 (1996). https://doi.org/10.1016/0550-3213(96)00179-4. arXiv: hep-th/9507038
13. A. Sen, Gauge invariant 1PI effective action for superstring field theory. J. High Energy Phys. **1506**, 022 (2015). https://doi.org/10.1007/JHEP06(2015)022. arXiv: 1411.7478
14. A. Sen, Gauge invariant 1PI effective superstring field theory: inclusion of the Ramond sector. J. High Energy Phys. **2015**(8) (2015). https://doi.org/10.1007/JHEP08(2015)025. arXiv: 1501.00988
15. C.B. Thorn, String field theory. Phys. Rep. **175**(1–2), 1–101 (1989). https://doi.org/10.1016/0370-1573(89)90015-X
16. S. Weinberg, *The Quantum Theory of Fields, Volume 1: Foundations* (Cambridge University Press, Cambridge, 2005)
17. B. Zwiebach, Closed string field theory: quantum action and the BV master equation. Nucl. Phys. B **390**(1), 33–152 (1993). https://doi.org/10.1016/0550-3213(93)90388-6. arXiv: hep-th/9206084
18. B. Zwiebach, Oriented open-closed string theory revisited. Annal. Phys. **267**(2), 193–248 (1998). https://doi.org/10.1006/aphy.1998.5803. arXiv: hep-th/9705241

Background Independence

16

Abstract

Spacetime background independence is a fundamental property of any candidate quantum gravity theory. In this chapter, we outline the proof of background independence for the closed SFT by proving that the equations of motion of two backgrounds related by a marginal deformation are equivalent after a field redefinition.

16.1 The Concept of Background Independence

Background independence means that the formalism does not depend on the background—if any—used to write the theory. A dependence in the background would imply that there is a distinguished background among all possibilities, which seems in tension with the dynamics of spacetime and the superposition principle from quantum mechanics. Moreover, one would expect a fundamental theory to tell which backgrounds are consistent and that they could be derived instead of postulated. Background independence allows spacetime to emerge as a consequence of the dynamics of the theory and of its defining fundamental laws.

Background independence can be manifest or not. In the second case, one needs to fix a background to define the theory, but the dynamics on different backgrounds are physically equivalent.[1] This implies that two theories with different backgrounds can be related, for example, by a field redefinition.

While fields other than the metric can also be expanded around a background, no difficulty is expected in this case. Indeed, the topic of background independence is particularly sensible only for the metric because it provides the frame for all other

---

[1] This does not mean that the physics in all backgrounds are identical, but that the laws are. Hence, a computation made in one specific background can be translated into another background.

© Springer Nature Switzerland AG 2021
H. Erbin, *String Field Theory*, Lecture Notes in Physics 980,
https://doi.org/10.1007/978-3-030-65321-7_16

computations—and in particular for the questions of dynamics and quantization. Generally, these questions are subsumed into the problem of the emergence of time in a generally covariant theory. In the previous language, QFTs without gravity are (generically) manifestly background independent after minimal coupling.[2] For example, a classical field theory is defined on a fixed Minkowski background, and a well-defined time is necessary to perform its quantization and to obtain a QFT, but it is not needed to choose a background for the other fields. For this reason, the extension of a QFT on a curved background is generally possible if the spacetime is hyperbolic, implying that there is a distinguished time direction. But the coupling to gravity is difficult and restricted to a (semi-)classical description.

What is the status of background independence in string theory? The worldsheet formulation requires to fix a background (usually Minkowski) to quantize the theory and to compute scattering amplitudes. Thus, the quantum theory is at least not manifestly background independent. On the other hand, the worldsheet action can be modified to a generic CFT including a generic non-linear sigma model describing an arbitrary target spacetime. Conformal invariance reproduces (at leading order) Einstein equations coupled to various matters and gauge field equations of motion. From this point of view, the classical theory can be written as a manifestly background independent theory, and this provides hopes that the quantum theory may also be background independent, even if non-manifestly. This idea is supported by other definitions of string theory (e.g. through the AdS/CFT conjecture—and other holographic realizations—or through matrix models) that provide, at least partially, background independent formulations.

Ultimately, the greatest avenue to establish the background independence is string field theory. Indeed, the form of the SFT action and of its properties (gauge invariance, equation of motion...) are identical irrespective of the background [13]. This provides a good starting point. The background dependence enters in the precise definition of the string products (BRST operator and vertices). The origin of this dependence lies in the derivation of the action (Chaps. 14 and 15): one begins with a particular CFT describing a given background (spacetime compactifications, fluxes, etc.) and defines the vertices from correlation functions of vertex operators, and the Hilbert space from the CFT operators. As a consequence, even though it is clear that no specific property of the background has been used in the derivation—and that the final action describes SFT for any background—this is not sufficient to establish background independence. Since the theory assumes implicitly a background choice, one cannot guarantee that the physical quantities have no residual dependence in the background, even if the action looks superficially background independent. Background independence in SFT is thus the statement that theories characterized by different CFTs can be related by a field redefinition.

In this chapter, we will sketch the proof of background independence for backgrounds related by marginal deformations.[3] It is possible to prove it at the level

---

[2]However, non-minimal coupling terms may be necessary to make the theory physical.

[3]An alternative approach based on morphism of L_∞ algebra is followed in [4].

of the action [11, 12], or at the level of the equations of motion [10]. The advantage of the second approach is that one can use the 1PI theory, which simplifies vastly the analysis. It also generalizes directly to the super-SFT.

Remark 16.1 (Field Theory on the CFT Space) As mentioned earlier, the string field is defined as a functional on the state space of a given CFT and not as a functional on the field theory space (off-shell states would correspond to general QFTs, and only on-shell states are CFTs). In this case, background independence would amount to reparametrization invariance of the action in the theory space and would thus almost automatically hold. A complete formulation of SFT following this line is currently out of reach, but some ideas can be found in [15].

16.2 Problem Setup

Given a SFT on a background, there are two ways to describe it on another background:

- deform the worldsheet CFT and express the SFT on the new background;
- expand the original action around the infinitesimal classical solution (to the linearized equations of motion) corresponding to the deformation.

Background independence amounts to the equivalence of both theories up to a field redefinition. The derivation can be performed at the level of the action or of the equations of motion. To prove the background independence at the quantum level, one needs to take into account the changes in the path integral measure or to work with the 1PI action.

The simplest case is when the two CFTs are related by an infinitesimal marginal deformation

$$\delta S_{\text{cft}} = \frac{\lambda}{2\pi} \int d^2z \, \varphi(z, \bar{z}), \qquad (16.1)$$

with φ a $(1, 1)$ primary operator and λ infinitesimal. The two CFTs are denoted by CFT_1 and CFT_2, and quantities associated to each CFTs are indexed with the appropriate number.

Establishing background independence in this case also implies it for finite marginal deformation since they can be built from a series of successive deformations. In the latter case, the field redefinition may be singular, which reflects that the parametrization of one CFT is not adapted for the other (equivalently, the coordinate systems for the string field break down), which is expected if both CFTs are far in the field theory space.

Remember the form of the 1PI action (15.59):

$$S_1[\Psi_1] = \frac{1}{g_s^2} \left(\frac{1}{2} \langle \Psi_1 | c_0^- Q_B | \Psi_1 \rangle + \sum_{n \geq 0}' \frac{1}{n!} V_n^{1PI}(\Psi_1^n) \right), \tag{16.2}$$

where the prime indicates that vertices with $n < 3$ do not include contributions from the sphere. In all this chapter, we remove the index 1PI to lighten the notations. The equation of motion is

$$\mathcal{F}_1(\Psi_1) = Q_B | \Psi_1 \rangle + \sum_n \frac{1}{n!} \ell_n(\Psi_1^n) = 0. \tag{16.3}$$

16.3 Deformation of the CFT

Consider the case where the theory CFT_1 is described by an action $S_{\mathrm{cft},1}[\psi_1]$ given in terms of fields ψ_1. Then, the deformation of this action by (16.1) gives an action for CFT_2:

$$S_{\mathrm{cft},2}[\psi_1] = S_{\mathrm{cft},1}[\psi_1] + \frac{\lambda}{2\pi} \int d^2z \, \varphi(z, \bar{z}). \tag{16.4}$$

Correlation functions on a Riemann surface Σ in both theories can be related by expanding the action to first order in λ in the path integral:

$$\left\langle \prod_i \mathcal{O}_i(z_i, \bar{z}_i) \right\rangle_2 = \left\langle \exp\left(-\frac{\lambda}{2\pi} \int d^2z \, \varphi(z, \bar{z}) \right) \prod_i \mathcal{O}_i(z_i, \bar{z}_i) \right\rangle_1 \tag{16.5a}$$

$$\approx \left\langle \prod_i \mathcal{O}_i(z_i, \bar{z}_i) \right\rangle_1 - \frac{\lambda}{2\pi} \int_\Sigma d^2z \left\langle \varphi(z, \bar{z}) \prod_i \mathcal{O}_i(z_i, \bar{z}_i) \right\rangle_1, \tag{16.5b}$$

where the $\mathcal{O}_i$ are operators built from the matter fields ψ_1. This expression presents two obvious problems. First, the correlation function may diverge when φ collides with one of the insertions, i.e. when $z = z_i$ in the integration. Second, there is an inherent ambiguity: the correlation functions are written in terms of operators in the Hilbert space of CFT_1, which are different from the CFT_2 Hilbert space, and there is no canonical isomorphism between both spaces.

Seeing the Hilbert space as a vector bundle over the CFT theory space, the second problem can be solved by introducing a connection on this bundle. This allows to relate Hilbert spaces of neighbouring CFTs. In fact, the choice of a non-singular connection also regularizes the divergences.

The simplest definition of a connection corresponds to cut unit disks around each operator insertions [1, 5, 6, 8, 14]. This amounts to define the variation between the two correlation functions as

$$
\delta \left\langle \prod_i \mathcal{O}_i(z_i, \bar{z}_i) \right\rangle_1 = -\frac{\lambda}{2\pi} \int_{\Sigma - \cup_i D_i} d^2 z \left\langle \varphi(z, \bar{z}) \prod_i \mathcal{O}_i(z_i, \bar{z}_i) \right\rangle_1. \tag{16.6}
$$

The integration is over Σ minus the disks $D_i = \{|w_i| \leq 1\}$, where w_i is the local coordinate for the insertion $\mathcal{O}_i$. The divergences are cured because φ never approaches another operator since the corresponding regions have been removed. The changes in the correlation functions induce a change in the string vertices denoted by $\delta V_n(\mathcal{V}_1, \ldots, \mathcal{V}_n)$.

The next step consists in computing the deformations of the operator modes. Since it involves only a matter operator, the modes in the ghost sector are left unchanged. The Virasoro generators change as

$$
\delta L_n = \lambda \oint_{|z|=1} \frac{d\bar{z}}{2\pi i} z^{n+1} \varphi(z, \bar{z}), \qquad \delta \bar{L}_n = \lambda \oint_{|z|=1} \frac{dz}{2\pi i} \bar{z}^{n+1} \varphi(z, \bar{z}). \tag{16.7}
$$

As a consequence, the BRST operator changes as

$$
\delta Q_B = \lambda \oint_{|z|=1} \frac{d\bar{z}}{2\pi i} c(z)\varphi(z, \bar{z}) + \lambda \oint_{|z|=1} \frac{d\bar{z}}{2\pi i} \bar{c}(\bar{z})\varphi(z, \bar{z}). \tag{16.8}
$$

One can prove that

$$
\{Q_B, \delta Q_B\} = O(\lambda^2), \tag{16.9}
$$

such that the BRST charge $Q_B + \delta Q_B$ in CFT_2 is correctly nilpotent if Q_B is nilpotent in CFT_1.

For the deformation to provide a consistent SFT, the conditions $b_0^- = 0$ and $L_0^- = 0$ must be preserved. The first is automatically satisfied since the ghost modes are not modified. Considering a weight-(h, h) operator $\mathcal{O}$, one finds

$$
\delta L_0^- |\mathcal{O}\rangle = \lambda \oint_{|z|=1} \frac{d\bar{z}}{2\pi i} z \sum_{p,q} z^{p-1} \bar{z}^{q-1} |\mathcal{O}_{p,q}\rangle - \lambda
$$

$$
\times \oint_{|z|=1} \frac{d\bar{z}}{2\pi i} \sum_{p,q} z^{p-1} \bar{z}^{q-1} |\mathcal{O}_{p,q}\rangle, \tag{16.10}
$$

where $\mathcal{O}_{p,q}$ are the fields appearing in the OPE with φ:

$$
\varphi(z, \bar{z})\mathcal{O}(0, 0) = \sum_{p,q} z^{p-1} \bar{z}^{q-1} \mathcal{O}_{p,q}(0, 0). \tag{16.11}
$$

The terms with $p \neq q$ vanish because the contour integrals are performed around circles of unit radius centred at the origin. Moreover, the terms $p = q$ are identical and cancel with each other, showing that $\delta L_0^- = 0$ when acting on states satisfying $L_0^- = 0$.

The SFT action $S_2[\Psi_1]$ in the new background reads

$$S_2[\Psi_1] = S_1[\Psi_1] + \delta S_1[\Psi_1], \tag{16.12}$$

where the change δS_1 in the action is induced by the changes in the string vertices:

$$\delta S_1[\Psi_1] = \frac{1}{g_s^2} \left(\frac{1}{2} \langle \Psi_1 | c_0^- \delta Q_B | \Psi_1 \rangle + \sum_{n \geq 0} \frac{1}{n!} \delta \mathcal{V}_n(\Psi_1^n) \right). \tag{16.13}$$

The equation of motion is

$$\mathcal{F}_2(\Psi_1) = \mathcal{F}_1(\Psi_1) + \lambda \, \delta \mathcal{F}_1(\Psi_1) = 0, \tag{16.14}$$

where $\mathcal{F}_1$ is given in (16.3) and

$$\lambda \, \delta \mathcal{F}_1(\Psi_1) = \delta Q_B | \Psi_1 \rangle + \sum_n \frac{1}{n!} \delta \ell_n(\Psi_1^n). \tag{16.15}$$

16.4 Expansion of the Action

Given a $(1, 1)$ primary φ, a BRST invariant operator is $c\bar{c}\varphi$. Hence, the field

$$|\Psi_1\rangle = \lambda \, |\Psi_0\rangle, \qquad |\Psi_0\rangle = c_1 \bar{c}_1(0) \, |\varphi\rangle \tag{16.16}$$

is a classical solution to first order in λ since the interactions on the sphere are at least cubic.

Separating the string field as the contribution from the (fixed) background and a fluctuation Ψ'

$$|\Psi_1\rangle = \lambda \, |\Psi_0\rangle + |\Psi'\rangle, \tag{16.17}$$

the action expanded to first order in λ reads

$$S_1[\Psi_1] = S_1[\Psi_0] + S'[\Psi'], \tag{16.18}$$

where

$$S'[\Psi'] = \frac{1}{g_s^2}\left(\frac{1}{2}\,\langle\Psi'|\,c_0^-\,Q_B\,|\Psi'\rangle + \sum_n \frac{1}{n!}\left(\mathcal{V}_n(\Psi'^n) + \lambda\,\mathcal{V}_{n+1}(\Psi_0,\Psi'^n)\right)\right).$$

(16.19)

The equation of motion is

$$\mathcal{F}'(\Psi') := \mathcal{F}_1(\Psi') + \lambda\,\delta\mathcal{F}'(\Psi') = 0,$$

(16.20)

where $\mathcal{F}_1$ is given in (16.3) and

$$\delta\mathcal{F}'(\Psi') = \sum_n \frac{1}{n!}\,\ell_{n+1}(\Psi_0,\Psi'^n).$$

(16.21)

16.5 Relating the Equations of Motion

In the previous section, we have derived the equations of motion for two different descriptions of a SFT obtained after shifting the background: (16.14) arises by deforming the CFT and computing the changes in the BRST operator and string products, while (16.20) arises by expanding the SFT action around the new background. The theory is background independent if both sets of Eqs. (16.14) and (16.20) are related by a (possibly field-dependent) linear transformation $\mathcal{M}(\Psi')$ after a field redefinition of $\Psi_1 = \Psi_1(\Psi')$:

$$\mathcal{F}_1(\Psi_1) + \lambda\,\delta\mathcal{F}_1(\Psi_1) = \left(1 + \lambda\mathcal{M}(\Psi')\right)\left(\mathcal{F}_1(\Psi') + \lambda\,\delta\mathcal{F}'(\Psi')\right),$$

(16.22a)

$$|\Psi_1\rangle = |\Psi'\rangle + \lambda\,|\delta\Psi'\rangle.$$

(16.22b)

The zero-order equation is automatically satisfied. To first order, this becomes

$$\frac{d}{d\lambda}\mathcal{F}_1(\Psi' + \lambda\delta\Psi')\Big|_{\lambda=0} + \delta\mathcal{F}_1(\Psi_1) - \delta\mathcal{F}'(\Psi') = \mathcal{M}(\Psi')\mathcal{F}_1(\Psi').$$

(16.23)

Taking Ψ' to be a solution of the original action removes the RHS, such that

$$\lambda\,Q_B\,|\delta\Psi'\rangle + \lambda\sum_n \frac{1}{n!}\,\ell_{n+1}(\delta\Psi',\Psi'^n) + \delta Q_B\,|\Psi'\rangle$$

$$+ \sum_n \frac{1}{n!}\,\delta\ell_n(\Psi'^n) - \lambda\sum_n \frac{1}{n!}\,\ell_{n+1}(\Psi_0,\Psi'^n) = 0.$$

(16.24)

To simplify the computations, it is simpler to consider the inner product of this quantity with an arbitrary state A (assumed to be even):

$$\Delta := \lambda \langle A | c_0^- Q_B | \delta \Psi' \rangle + \lambda \sum_n \frac{1}{n!} \mathcal{V}_{n+2}(A, \delta \Psi', \Psi'^n) + \langle A | c_0^- \delta Q_B | \Psi' \rangle$$

$$+ \sum_n \frac{1}{n!} \delta \mathcal{V}_{n+1}(A, \Psi'^n) - \lambda \sum_n \frac{1}{n!} \mathcal{V}_{n+2}(A, \Psi_0, \Psi'^n). \qquad (16.25)$$

The goal is to prove the existence of $\delta \Psi'$ such that $\Delta = 0$ up to the zero-order equation of motion $\mathcal{F}_1(\Psi') = 0$.

16.6 Idea of the Proof

In this section, we give an idea of how the proof ends, referring to [10] for the details.

The first step is to introduce new vertices $\mathcal{V}'_{0,3}$ and $\mathcal{V}'_n$ parametrizing the variations of the string vertices:

$$\langle A | c_0^- \delta Q_B | B \rangle = \lambda \, \mathcal{V}'_{0,3}(\Psi_0, B, A), \qquad \delta \mathcal{V}_n(\Psi'^n) = \lambda \, \mathcal{V}'_{n+1}(\Psi_0, \Psi'^n),$$
$$(16.26)$$

where the notation (13.19) has been used. Each subspace $\mathcal{V}'_{g,n}$ is defined such that the LHS is recovered upon integrating the appropriate $\omega_{g,n}$ over this section segment. Next, the field redefinition $\delta \Psi'$ is parametrized as

$$\langle A | c_0^- | \delta \Psi' \rangle = \sum_n \frac{1}{n!} \mathcal{B}_{n+2}(\Psi_0, \Psi'^n, A). \qquad (16.27)$$

The objective is to prove the existence (and if possible the form) of the subspaces $\mathcal{B}_{n+2}$. Both the vertices $\mathcal{V}'_n$ and $\mathcal{B}_n$ admit a genus expansion:

$$\mathcal{V}'_n = \sum_{g \geq 0} \mathcal{V}'_{g,n}, \qquad \mathcal{B}_n = \sum_{g \geq 0} \mathcal{B}_{g,n}. \qquad (16.28)$$

Plugging the new expressions in (16.25) gives

$$\Delta = - \sum_n \frac{1}{n!} \mathcal{B}_{n+2}(\Psi_0, \Psi'^n, Q_B A) + \sum_{m,n} \frac{1}{m!n!} \mathcal{B}_{n+2}(\Psi_0, \Psi'^m, \ell_{n+1}(A, \Psi'^n))$$

$$+ \sum_n \frac{1}{n!} \mathcal{V}'_{n+2}(A, \Psi_0, \Psi'^n) - \sum_n \frac{1}{n!} \mathcal{V}_{n+2}(A, \Psi_0, \Psi'^n). \qquad (16.29)$$

Next, the BRST identity (13.46) and the equation of motion $\mathcal{F}_1(\Psi') = 0$ allow to rewrite the first term as

$$\mathcal{B}_{n+2}(\Psi_0, \Psi'^n, Q_B A) = \partial \mathcal{B}_{n+2}(\Psi_0, \Psi'^n, A) + n\, \mathcal{B}_{n+2}(\Psi_0, \Psi'^{n-1}, Q_B \Psi', A)$$

$$(16.30a)$$

$$= \partial \mathcal{B}_{n+2}(\Psi_0, \Psi'^n, A)$$
$$- \sum_m \frac{n}{m!} \mathcal{B}_{n+2}(\Psi_0, \Psi'^{n-1}, \ell_m(\Psi'^m), A). \qquad (16.30b)$$

In the second term, the sum over n is shifted. Combining everything together gives

$$\Delta = \sum_n \frac{1}{n!} \partial \mathcal{B}_{n+2}(\Psi_0, \Psi'^n, A) - \sum_{m,n} \frac{1}{m!n!} \mathcal{B}_{n+3}(\Psi_0, \Psi'^n, \ell_m(\Psi'^m), A)$$
$$+ \sum_{m,n} \frac{1}{m!n!} \mathcal{B}_{n+2}(\Psi_0, \Psi'^m, \ell_{n+1}(A, \Psi'^n)) + \sum_n \frac{1}{n!} \mathcal{V}'_{n+2}(A, \Psi_0, \Psi'^n)$$
$$- \sum_n \frac{1}{n!} \mathcal{V}_{n+2}(A, \Psi_0, \Psi'^n).$$

$$(16.31)$$

Solving for $\Delta = 0$ requires that each term with a different power of Ψ' vanishes independently:

$$\partial \mathcal{B}_{n+2}(\Psi_0, \Psi'^n, A) = -\mathcal{V}'_{n+2}(A, \Psi_0, \Psi'^n) + \mathcal{V}_{n+2}(A, \Psi_0, \Psi'^n)$$
$$+ \sum_{\substack{m_1, m_2 \\ m_1 + m_2 = n}} \frac{n!}{m_1! m_2!} \mathcal{B}_{m_1+3}(\Psi_0, \Psi'^{m_1}, \ell_{m_2}(\Psi'^{m_2}), A)$$
$$- \sum_{\substack{m_1, m_2 \\ m_1 + m_2 = n}} \frac{n!}{m_1! m_2!} \mathcal{B}_{m_1+2}\big(\Psi_0, \Psi'^{m_1}, \ell_{m_2+1}(A, \Psi'^{m_2})\big).$$

$$(16.32)$$

In order to proceed, one needs to perform a genus expansion of the various spaces: this allows to solve recursively for all $\mathcal{B}_{g,n}$ starting from $\mathcal{B}_{0,3}$. One can then build $|\delta\Psi'\rangle$ recursively, which provides the field redefinition. Indeed, the RHS of this equation contains only $\mathcal{B}_{g',n'}$ for $g' < g$ or $n' < n$, and the equation for $\mathcal{B}_{0,3}$ contains no $\mathcal{B}_{g,n}$ in the RHS. It should be noted that the field redefinition is not unique, but there is the freedom of performing (infinite-dimensional) gauge transformations. Finding an obstruction to solve these equations means that the field redefinition does not exist, and thus that the theory is not background independent.

The form of the equation

$$\partial \mathcal{B}_{0,3} = \mathcal{V}_{0,3} - \mathcal{V}'_{0,3} \tag{16.33}$$

suggests to use homology theory. The interpretation of $\mathcal{B}_{0,3}$ is that it is a space interpolating between $\mathcal{V}_{0,3}$ and $\mathcal{V}'_{0,3}$. A preliminary step is to check that there is no obstruction: since the LHS is already a boundary, one has $\partial^2 \mathcal{B}_{0,3} = 0$ and one should check that $\partial(\text{RHS}) = 0$ as well. It can be shown that it is indeed true. It was proved in [10] that this equation admits a solution and that the equations for higher g and n can all be solved. Hence, there exists a field redefinition and SFT is background independent.

16.7   Suggested Readings

- Proof of the background independence under marginal deformations [10–12] (see also [7–9] for earlier results laying foundations for the complete proof).
- L_∞ perspective [4, sec. 4] (see also [2, 3, sec. III.B]).
- Connection on the space of CFTs [1, 5, 6, 8, 14].

References

1. M. Campbell, P. Nelson, E. Wong, Stress tensor perturbations in conformal field theory. Int. J. Mod. Phys. A **06**(27), 4909–4924 (1991). https://doi.org/10.1142/S0217751X9100232X
2. K. Muenster, I. Sachs, On homotopy algebras and quantum string field theory (2013). arXiv: 1303.3444
3. K. Muenster, I. Sachs, Quantum open-closed homotopy algebra and string field theory. Commun. Math. Phys. **321**(3), 769–801 (2013). https://doi.org/10.1007/s00220-012-1654-1. arXiv: 1109.4101
4. K. Muenster, I. Sachs, Homotopy classification of bosonic string field theory. Commun. Math. Phys. **330**, 1227–1262 (2014). https://doi.org/10.1007/s00220-014-2027-8. arXiv: 1208.5626
5. K. Ranganathan, Nearby CFT's in the operator formalism: the role of a connection. Nucl. Phys. B **408**(1), 180–206 (1993). https://doi.org/10.1016/0550-3213(93)90136-D. arXiv: hep-th/9210090
6. K. Ranganathan, H. Sonoda, B. Zwiebach, Connections on the state-space over conformal field theories. Nucl. Phys. B **414**(1–2), 405–460 (1994). https://doi.org/10.1016/0550-3213(94)90436-7. arXiv: hep-th/9304053
7. A. Sen, On the background independence of string field theory: II. Analysis of on-shell S-matrix elements. Nucl. Phys. B **347**(1), 270–318 (1990). https://doi.org/10.1016/0550-3213(90)90560-Z
8. A. Sen, On the background independence of string field theory. Nucl. Phys. B **345**(2–3), 551–583 (1990). https://doi.org/10.1016/0550-3213(90)90400-8
9. A. Sen, On the background independence of string field theory: III. Explicit field redefinitions. Nucl. Phys. B **391**(3), 550–590 (1993). https://doi.org/10.1016/0550-3213(93)90084-3. arXiv: hep-th/9201041
10. A. Sen, Background independence of closed superstring field theory. J. High Energy Phys. **2018**(2), 155 (2018). https://doi.org/10.1007/JHEP02(2018)155. arXiv: 1711.08468

11. A. Sen, B. Zwiebach, A proof of local background independence of classical closed string field theory. Nucl. Phys. B **414**(3), 649–711 (1994). https://doi.org/10.1016/0550-3213(94)90258-5. arXiv: hep-th/9307088
12. A. Sen, B. Zwiebach, Quantum background independence of closed string field theory. Nucl. Phys. B **423**(2–3), 580–630 (1994). https://doi.org/10.1016/0550-3213(94)90145-7. arXiv: hep-th/9311009
13. A. Sen, B. Zwiebach, Background independent algebraic structures in closed string field theory. Commun. Math. Phys. **177**(2), 305–326 (1996). https://doi.org/10.1007/BF02101895. arXiv: hep-th/9408053
14. H. Sonoda, Connection on the theory space (1993). arXiv:Hep-Th/9306119
15. E. Witten, On background independent open-string field theory. Phys. Rev. D **46**(12), 5467–5473 (1992). https://doi.org/10.1103/PhysRevD.46.5467. arXiv: hep-th/9208027

Superstring

17

Abstract

Superstring theory is generally the starting point for physical model building. It has indeed several advantages over the bosonic string, most importantly, the removal of the tachyon and the inclusion of fermions in the spectrum. The goal of this chapter is to introduce the most important concepts needed to generalize the bosonic string to the superstring, both for off-shell amplitudes and string field theory. We refer to the review [2] for more details.

17.1 Worldsheet Superstring Theory

There are five different superstring theories with spacetime supersymmetry: the types I, IIA and IIB, and the $E_8 \times E_8$ and $SO(32)$ heterotic models.

In the Ramond–Neveu–Schwarz formalism (RNS), the left- and right-moving sectors of the superstring worldsheet are described by a two-dimensional superconformal field theory (SCFT), possibly with different numbers of supersymmetries. The prototypical example is the heterotic string with $N = (1, 0)$, and we will focus on this case: only the left-moving sector is supersymmetric, while the right-moving is given by the same bosonic theory as in the other chapters. Up to minor modifications, the type II theory follows by duplicating the formulas of the left-moving sector to the right-moving one.

17.1.1 Heterotic Worldsheet

The ghost super-CFT is characterized by anti-commuting ghosts (b, c) (left-moving) and $(\bar{b}, \bar{c})$ (right-moving) with central charge $c = (-26, -26)$, associated to diffeomorphisms, and by commuting ghosts (β, γ) with central charge $c = (11, 0)$, associated to local supersymmetry. As a consequence, the matter SCFT must have a

© Springer Nature Switzerland AG 2021   339
H. Erbin, *String Field Theory*, Lecture Notes in Physics 980,
https://doi.org/10.1007/978-3-030-65321-7_17

central charge $c = (15, 26)$. If spacetime has D non-compact dimensions, then the matter CFT is made of

- a free theory of D scalars X^μ and D left-moving fermions ψ^μ ($\mu = 0, \ldots, D-1$) such that $c_{\text{free}} = 3D/2$ and $\bar{c}_{\text{free}} = D$;
- an internal theory with $c_{\text{int}} = 15 - 3D/2$ and $\bar{c}_{\text{int}} = 26 - D$.

The critical dimension is reached when $c_{\text{int}} = 0$, which corresponds to $D = 10$.

The diffeomorphisms are generated by the energy–momentum tensor $T(z)$; correspondingly, supersymmetry is generated by its super-partner $G(z)$ (sometimes also denoted by T_F). The OPEs of the algebra formed by $T(z)$ and $G(z)$ are

$$T(z)T(w) \sim \frac{c/2}{(z-w)^4} + \frac{2T(w)}{(z-w)^2} + \frac{\partial T(w)}{z-w}, \tag{17.1a}$$

$$G(z)G(w) \sim \frac{2c/3}{(z-w)^3} + \frac{2T(w)}{(z-w)}, \tag{17.1b}$$

$$T(z)G(w) \sim \frac{3}{2}\frac{G(w)}{(z-w)^2} + \frac{\partial G(w)}{(z-w)}. \tag{17.1c}$$

The superconformal ghosts form a first-order system (see Sect. 7.2) with $\epsilon = -1$ and $\lambda = 3/2$. Hence, they have conformal weights

$$h(\beta) = \left(\frac{3}{2}, 0\right), \qquad h(\gamma) = \left(-\frac{1}{2}, 0\right) \tag{17.2}$$

and OPEs

$$\gamma(z)\beta(w) \sim \frac{1}{z-w}, \qquad \beta(z)\gamma(w) \sim -\frac{1}{z-w}. \tag{17.3}$$

The expressions of the ghost energy–momentum tensors are

$$T^{\text{gh}} = -2b\,\partial c + c\partial b, \qquad T^{\beta\gamma} = \frac{3}{2}\beta\partial\gamma + \frac{1}{2}\gamma\,\partial\beta. \tag{17.4}$$

The ghost numbers of the different fields are

$$N_{\text{gh}}(b) = N_{\text{gh}}(\beta) = -1, \qquad N_{\text{gh}}(c) = N_{\text{gh}}(\gamma) = 1. \tag{17.5}$$

The worldsheet scalars satisfy periodic boundary conditions. On the other hand, fermions can satisfy anti-periodic or periodic conditions: this leads to two different sectors, called Neveu–Schwarz (NS) and Ramond (R), respectively.

$\beta\gamma$ **System**

The $\beta\gamma$ system can be bosonized as

$$\gamma = \eta\,e^{\phi}, \qquad \beta = \partial\xi\,e^{-\phi}, \tag{17.6}$$

where (ξ, η) are fermions with conformal weights 0 and 1 (this is a first-order system with $\epsilon = 1$ and $\lambda = 1$), and ϕ is a scalar field with a background charge (Coulomb gas). This provides an alternative representation of the delta functions:

$$\delta(\gamma) = e^{-\phi}, \qquad \delta(\beta) = e^{\phi}. \tag{17.7}$$

Introducing these operators is necessary to properly define the path integral with bosonic zero-modes. They play the same role as the zero-modes insertions for fermionic fields needed to obtain a finite result (see also Appendix C.1.3):

$$\int dc_0 = 0 \implies \int dc_0\,c_0 = 1, \tag{17.8}$$

because $c_0 = \delta(c_0)$. For a bosonic path integral, one needs a delta function:

$$\int d\gamma_0 = \infty \implies \int d\gamma_0\,\delta(\gamma_0) = 1 \tag{17.9}$$

By definition of the bosonization, one has

$$T^{\beta\gamma} = T^{\eta\xi} + T^{\phi}, \tag{17.10}$$

where

$$T^{\eta\xi} = -\eta\,\partial\xi, \qquad T^{\phi} = -\frac{1}{2}(\partial\phi)^2 - \partial^2\phi. \tag{17.11}$$

The OPEs between the new fields are

$$\xi(z)\eta(w) \sim \frac{1}{z-w}, \quad e^{q_1\phi(z)}e^{q_2\phi(w)} \sim \frac{e^{(q_1+q_2)\phi(w)}}{(z-w)^{q_1 q_2}}, \quad \partial\phi(z)\partial\phi(w) \sim -\frac{1}{(z-w)^2}. \tag{17.12}$$

The simplest attribution of ghost numbers to the new fields is

$$N_{gh}(\eta) = 1, \qquad N_{gh}(\xi) = -1, \qquad N_{gh}(\phi) = 0. \tag{17.13}$$

To the scalar field, ϕ is associated to another U(1) symmetry whose quantum number is called the picture number N_{pic}. The picture numbers of η and ξ are assigned[1] such that β and γ have $N_{\text{pic}} = 0$:

$$N_{\text{pic}}(e^{q\phi}) = q, \qquad N_{\text{pic}}(\xi) = 1, \qquad N_{\text{pic}}(\eta) = -1. \tag{17.14}$$

Because of the background charge, this symmetry is anomalous and correlation functions are non-vanishing if the total picture number (equivalently, the number of ϕ zero-modes) is

$$N_{\text{pic}} = 2(g - 1) = -\chi_g. \tag{17.15}$$

For the same reason, the vertex operators $e^{q\phi}$ are the only primary operators:

$$h(e^{q\phi}) = -\frac{q}{2}(q + 2), \tag{17.16}$$

and the Grassmann parity of these operators is $(-1)^q$. Special values are

$$h(e^{\phi}) = \frac{3}{2}, \qquad h(e^{-\phi}) = \frac{1}{2}. \tag{17.17}$$

The superstring theory features a $\mathbb{Z}_2$ symmetry called the GSO symmetry. All fields are taken to be GSO even, except β and γ that are GSO odd and $e^{q\phi}$ whose parity is $(-1)^q$. Physical states in the NS sector are restricted to be GSO even: it is required to remove the tachyon of the spectrum and to get a spacetime with supersymmetry. In type II, the Ramond sector can be projected in two different ways, leading to the type IIA and type IIB theories.

The components of the BRST current are

$$j_B = c(T^{\text{m}} + T^{\beta\gamma}) + \gamma G + bc\partial c - \frac{1}{4}\gamma^2 b, \tag{17.18a}$$

$$\bar{j}_B = \bar{c}\bar{T}^{\text{m}} + \bar{b}\bar{c}\bar{\partial}\bar{c}. \tag{17.18b}$$

From there, it is useful to define the picture changing operator (PCO):

$$\mathcal{X}(z) = \{Q_B, \xi(z)\} = c\partial\xi + e^{\phi}G - \frac{1}{4}\partial\eta\,e^{2\phi}\,b - \frac{1}{4}\partial(\eta\,e^{2\phi}b), \tag{17.19}$$

which is a weight-$(0, 0)$ primary operator that carries a unit picture number. It is obviously BRST exact. This operator will be necessary to saturate the picture

---

[1] Any linear combination of both U(1) could have been used. The one given here is conventional, but also the most convenient.

number condition: the naive insertion of $e^{\phi} \sim \delta(\beta)$ breaks the BRST invariance. The PCO zero-mode is obtained from the contour integral:

$$\mathcal{X}_0 = \frac{1}{2\pi i} \oint \frac{dz}{z} \mathcal{X}(z). \tag{17.20}$$

It can be interpreted as delocalizing a PCO insertion from a point to a circle, which decreases the risk of divergence.

17.1.2 Hilbert Spaces

The description in terms of the (η, ξ, ϕ) fields leads to a subtlety: the bosonization involves only the derivative $\partial \xi$ and not the field ξ itself, meaning that the zero-mode ξ_0 is absent from the original Hilbert space defined from (β, γ). In the bosonized language, the Hilbert space without the ξ zero-mode is called the *small Hilbert space* and is made of state annihilated by η_0 (the η zero-mode)

$$\mathcal{H}_{small} = \{ |\psi\rangle \mid \eta_0 |\psi\rangle = 0 \}. \tag{17.21}$$

Removing this condition leads to the *large Hilbert space*:[2]

$$\mathcal{H}_{small} = \mathcal{H}_{large} \cap \ker \eta_0. \tag{17.22}$$

A state in $\mathcal{H}_{small}$ contains ξ with at least one derivative acting on it.

A correlation function defined in terms of the (η, ξ, ϕ) system is in the large Hilbert space and will vanish since there is no ξ factor to absorb the zero-mode of the path integral. As a consequence, correlation functions (and the inner product) are defined with a ξ_0 insertion (by convention at the extreme left) or, equivalently, $\xi(z)$. The position does not matter since only the zero-mode contribution survives, and the correlation function is independent of z. Sometimes it is more convenient to work in the large Hilbert space and to restrict later to the small Hilbert space.

The $SL(2, \mathbb{C})$ invariant vacuum is normalized as

$$\langle k | c_{-1}\bar{c}_{-1}c_0\bar{c}_0c_1\bar{c}_1 \, e^{-2\phi(z)} |k'\rangle = (2\pi)^D \delta^{(D)}(k + k'). \tag{17.23}$$

Remark 17.1 (Normalization in Type II) In type II theory, the $SL(2, \mathbb{C})$ is normalized as

$$\langle k | c_{-1}\bar{c}_{-1}c_0\bar{c}_0c_1\bar{c}_1 \, e^{-2\phi(z)}e^{-\bar{\phi}(\bar{w})} |k'\rangle = -(2\pi)^D \delta^{(D)}(k + k'). \tag{17.24}$$

---

[2]The relation between the small and large Hilbert spaces is similar to the one between the $\mathcal{H}$ and $\mathcal{H}_0 = b_0\mathcal{H}$ Hilbert space from the open string since the (b, c) and (η, ξ) are both fermionic first-order systems.

The sign difference allows to avoid sign differences between type II and heterotic string theories in most formulas [2].

The Hilbert space of GSO even states satisfying the $b_0^- = 0$ and $L_0^- = 0$ conditions is denoted by $\mathcal{H}_T$ (ghost and picture numbers are arbitrary). This Hilbert space is the direct sum of the NS and R Hilbert spaces:

$$\mathcal{H}_T = \mathcal{H}_{NS} \oplus \mathcal{H}_R. \tag{17.25}$$

The subspace of states with picture number $N_{pic} = n$ is written as $\mathcal{H}_n$. The picture numbers of the NS and R states are, respectively, integer and half-integer. Two special subspaces of $\mathcal{H}_T$ play a distinguished role:

$$\widehat{\mathcal{H}}_T = \mathcal{H}_{-1} \oplus \mathcal{H}_{-1/2}, \qquad \widetilde{\mathcal{H}}_T = \mathcal{H}_{-1} \oplus \mathcal{H}_{-3/2}. \tag{17.26}$$

To understand this, consider the vacuum $|p\rangle$ of the ϕ field with picture number p:

$$|p\rangle = e^{p\phi}(0) |0\rangle . \tag{17.27}$$

Then, acting on the vacuum with the β_n and γ_n modes implies

$$\forall n \geq -p - \frac{1}{2}: \qquad \beta_n |p\rangle = 0,$$
$$\forall n \geq p + \frac{3}{2}: \qquad \gamma_n |p\rangle = 0. \tag{17.28}$$

For $p = -1$, all positive modes (starting with $n = 1/2$) annihilate the vacuum in the NS sector. This is a positive asset because positive modes that do not annihilate the vacuum can create states with arbitrary negative energy (since it is bosonic).[3] For $p = -1/2$ or $p = -3/2$, the vacuum is annihilated by all positive modes, but not by one of the zero-modes γ_0 or β_0. Nonetheless, one can show that the propagator in the R sector allows to propagate only a finite number of states if one chooses $\mathcal{H}_{-1/2}$; the role of $\mathcal{H}_{-3/2}$ will become apparent when discussing how to build the superstring field theory.

Basis states are introduced as in the bosonic case:

$$\widehat{\mathcal{H}}_T = \text{Span}\{|\phi_r\rangle\}, \qquad \widetilde{\mathcal{H}}_T = \text{Span}\{|\phi_r^c\rangle\}, \tag{17.29}$$

such that

$$\langle \phi_r^c | \phi_s \rangle = \delta_{rs}. \tag{17.30}$$

---

[3]This is not a problem on-shell since the BRST cohomology is independent of the picture number. However, this matters off-shell since states would propagate in loops and make the theory inconsistent.

The completeness relations are

$$1 = \sum_r |\phi_r\rangle\langle\phi_r^c|, \tag{17.31}$$

$\widehat{\mathcal{H}}_T$ and

$$1 = \sum_r (-1)^{|\phi_r|} |\phi_r^c\rangle\langle\phi_r| \tag{17.32}$$

on $\widetilde{\mathcal{H}}_T$.

Finally, the operator $\mathcal{G}$ is defined as

$$\mathcal{G} = \begin{cases} 1 & \text{NS sector,} \\ \mathcal{X}_0 & \text{R sector.} \end{cases} \tag{17.33}$$

Note the following properties

$$[\mathcal{G}, L_0^{\pm}] = [\mathcal{G}, b_0^{\pm}] = [\mathcal{G}, Q_B] = 0. \tag{17.34}$$

It will appear in the propagator and kinetic terms of the superstring field theory.

17.2 Off-Shell Superstring Amplitudes

In this section, we are going to build the scattering amplitudes. The procedure is very similar to the bosonic case, except for the PCO insertions and of the Ramond sector. For this reason, we will simply state the result and motivate the modifications with respect to the bosonic case.

17.2.1 Amplitudes

External states can be either NS or R: the Riemann surface corresponding to the g-loop scattering of m external NS states and n external R states is denoted by $\Sigma_{g,m,n}$. R states must come in pairs because they correspond to fermions. As in the bosonic case, the amplitude is written as the integration of an appropriate p-form $\Omega_p^{(g,m,n)}$ over the moduli space $\mathcal{M}_{g,m,n}$ (or, more precisely, of a section of a fibre bundle with this moduli space as a basis). From the geometric point of view, nothing distinguishes the punctures, and thus,

$$M_{g,m,n} := \dim \mathcal{M}_{g,m,n} = 6g - 6 + 2m + 2n. \tag{17.35}$$

The form $\Omega_{M_{g,m,n}}$ is defined as a SCFT correlation function of the physical vertex operators together with ghost and PCO insertions.

Remark 17.2 A simple way to avoid making errors with signs is to multiply every Grassmann odd external state with a Grassmann odd number. These can be removed at the end to read the sign.

The two conditions from the U(1) anomalies on the scattering amplitude are

$$N_{gh} = 6 - 6g, \qquad N_{pic} = 2g - 2. \qquad (17.36)$$

Given an amplitude with m NS states $\mathscr{V}_i^{NS} \in \mathcal{H}_{-1}$ and n R states $\mathscr{V}_j^{R} \in \mathcal{H}_{-1/2}$, the above picture number can be reached by introducing a certain number of PCOs $\mathcal{X}(y_A)$

$$n_{pco} := 2g - 2 + m + \frac{n}{2}. \qquad (17.37)$$

These PCOs are inserted at various positions: while the amplitude does not depend on these locations on-shell, off-shell it will (because the vertex operators are not BRST invariant). The choices of PCO locations are arbitrary except for several consistency conditions:

1. avoid spurious poles (Sect. 17.2.3);
2. consistent with factorization (each component of the surface in the degeneration limits must saturate the picture number condition).

This parallels the discussion of the choices of local coordinates: as a consequence, the natural object is a fibre bundle $\widetilde{\mathcal{P}}_{g,m,n}$ with the local coordinate choices (up to global phase rotations) and the PCO locations as fibre, and the moduli space $\mathcal{M}_{g,m,n}$ as base. Forgetting about the PCO locations leads to a fibre bundle $\widehat{\mathcal{P}}_{g,m,n}$ that is a generalization of the one found in the bosonic case. The coordinate system of the fibre bundle presented in the bosonic case is extended by including the PCO locations $\{y_A\}$.

With these information, the amplitude can be written as

$$A_{g,m,n}(\mathscr{V}_i^{NS}, \mathscr{V}_j^{R}) = \int_{\mathcal{S}_{g,m,n}} \Omega_{M_{g,m,n}}(\mathscr{V}_i^{NS}, \mathscr{V}_j^{R}), \qquad (17.38a)$$

where

$$\Omega_{M_{g,m,n}} = (-2\pi i)^{-M_{g,m,n}^c} \left\langle \bigwedge_{\lambda=1}^{M_{g,m,n}} \mathcal{B}_\lambda \, dt_\lambda \prod_{A=1}^{n_{pco}} \mathcal{X}(y_A) \prod_{i=1}^{m} \mathscr{V}_i^{NS} \prod_{j=1}^{n} \mathscr{V}_j^{R} \right\rangle_{\Sigma_{g,n}}, \qquad (17.38b)$$

where $\mathcal{S}_{g,m,n}$ is a $\mathsf{M}_{g,m,n}$-dimensional section of $\widetilde{\mathcal{P}}_{g,m,n}$ parametrized by coordinates t_λ. The 1-form $\mathcal{B}$ corresponds to a generalization of the bosonic 1-form. It has ghost number 1 and includes a correction to compensate the variation of the PCO locations in terms of the moduli parameters:

$$
\mathcal{B}_\lambda = \sum_\alpha \oint_{C_\alpha} \frac{d\sigma_\alpha}{2\pi i} \, b(\sigma_\alpha) \, \frac{\partial F_\alpha}{\partial t_\lambda} \left(F_\alpha^{-1}(\sigma_\alpha) \right)
$$

$$
+ \sum_\alpha \oint_{C_\alpha} \frac{d\bar{\sigma}_\alpha}{2\pi i} \, \bar{b}(\bar{\sigma}_\alpha) \, \frac{\partial \bar{F}_\alpha}{\partial t_\lambda} \left(\bar{F}_\alpha^{-1}(\bar{\sigma}_\alpha) \right)
$$

$$
- \sum_A \frac{1}{\mathcal{X}(y_A)} \frac{\partial y_A}{\partial t_\lambda} \, \partial \xi(y_A). \tag{17.39}
$$

The last factor amounts to consider the combination

$$
\mathcal{X}(y_A) - \partial \xi(y_A) \, dy_A \tag{17.40}
$$

for each PCO insertion:[4] the correction is necessary to ensure that the BRST identity (13.46) holds. This can be understood as follows: the derivative acting on the PCO gives a term $d\mathcal{X}(z) = \partial\mathcal{X}(z)dz$ that must be cancelled. This is achieved by the second term since $\{Q_B, \partial\xi(z)\} = \partial\mathcal{X}(z)$.

Remark 17.3 While it is sufficient to work with $\mathcal{M}_{g,n}$ for on-shell bosonic amplitudes, on-shell superstring amplitudes are naturally expressed in $\widetilde{\mathcal{P}}_{g,m,n}$ (with the local coordinate removed) since the positions of the PCO must be specified even on-shell.

Remark 17.4 (Amplitudes on the Supermoduli Space) Following Polyakov's approach from Chaps. 2 and 3 to the superstring would lead to replace the moduli space by the supermoduli space. The latter includes Grassmann odd moduli parameters in addition to the moduli parameters from $\mathcal{M}_{g,m+n}$ (in the same way the superspace includes odd coordinates θ along with spacetime coordinates x). The natural question is whether it is possible to split the integration over the even and odd moduli and to integrate over the latter such that only an integral over $\mathcal{M}_{g,m+n}$ remains. In view of (17.38a), the answer seems positive. However, this is incorrect: it was proven in [3] that there is no global holomorphic projection of the supermoduli space to the moduli space. This is related to the problem of spurious poles described below. But, this does not prevent to do it locally: in that case, implementing the procedure carefully should give the rules of vertical integration [8, 28, 31].

---

[4]The sum is formal since it is composed of 0- and 1-forms.

17.2.2 Factorization

The plumbing fixture of two Riemann surfaces Σ_{g_1,m_1,n_1} and Σ_{g_2,m_2,n_2} can be performed in two different ways since two NS or two R punctures can be glued.

If two NS punctures are glued, the resulting Riemann surface is $\Sigma_{g_1+g_2,m_1+m_2-2,n_1+n_2}^{(\text{NS})}$. The number of PCOs inherited from the two original surfaces is

$$n_{\text{pco}}^{(1)} + n_{\text{pco}}^{(2)} = 2(g_1 + g_2) - 2 + (m_1 + m_2 - 2) + \frac{n_1 + n_2}{2} = n_{\text{pco}}^{(\text{NS})}, \qquad (17.41)$$

which is the required number for a non-vanishing amplitude. As a consequence, the propagator is the same as in the bosonic case:

$$\Delta_{\text{NS}} = b_0^+ b_0^- \frac{1}{L_0 + \bar{L}_0} \delta(L_0^-). \qquad (17.42)$$

If two R punctures are glued, the numbers of PCO do not match by one unit:

$$n_{\text{pco}}^{(1)} + n_{\text{pco}}^{(2)} = 2(g_1+g_2) - 2 + (m_1+m_2) + \frac{n_1 + n_2 - 2}{2} - 1 = n_{\text{pco}}^{(\text{R})} - 1. \qquad (17.43)$$

This means that an additional PCO must be inserted in the plumbing fixture procedure: the natural place for it is in the propagator since this is the only way to keep both vertices symmetric as required for a field theory interpretation. Another way to see the need of this modification is to study the propagator (17.42) for Ramond states: since Ramond states carry a picture number $-1/2$, the conjugate states have $N_{\text{pic}} = -3/2$, and thus the propagator has a total picture number -3 instead of -2 (the propagator graph is equivalent to a sphere). Then, to avoid localizing the PCO at a point of the propagator, one inserts the zero-mode that corresponds to smear the PCO:

$$\Delta_{\text{R}} = b_0^+ b_0^- \frac{\mathcal{X}_0}{L_0 + \bar{L}_0} \delta(L_0^-). \qquad (17.44)$$

Delocalizing the PCO amounts to average the amplitude over an infinite number of points (i.e. to consider a generalized section): this is necessary to preserve the L_0^- eigenvalue since $\mathcal{X}_0$ is rotationally invariant while $\mathcal{X}(z)$ is not. Note that the zero-mode can be written equivalently as a contour integral around one of the two glued punctures:

$$\mathcal{X}_0 = \frac{1}{2\pi i} \oint \frac{dw_n^{(1)}}{w_n^{(1)}} \mathcal{X}(w_n^{(1)}) = \frac{1}{2\pi i} \oint \frac{dw_n^{(2)}}{w_n^{(2)}} \mathcal{X}(w_n^{(2)}). \qquad (17.45)$$

The equality of both expressions holds because $\mathcal{X}(z)$ has conformal weight 0.

Using the operator $\mathcal{G}$ (17.33), the propagator can be written generically as

$$\Delta = b_0^+ b_0^- \frac{\mathcal{G}}{L_0 + \bar{L}_0} \delta(L_0^-). \qquad (17.46)$$

Remark 17.5 (Propagators) NS and R states correspond, respectively, to bosonic and fermionic fields: the operators L_0^+ and $\mathcal{X}_0$ can be interpreted as the (massive) Laplacian and Dirac operators, such that both propagators can be written as

$$\Delta_{NS} \sim \frac{1}{k^2 + m^2}, \qquad \Delta_R \sim \frac{i\partial\!\!\!/ + m}{k^2 + m^2}. \qquad (17.47)$$

To motivate the identification of $\mathcal{X}_0$ with the Dirac operator, remember that $\mathcal{X}(z)$ contains a term $e^{\phi(z)} G(z)$ (this is the only term that contributes on-shell), where $G(z)$ in turn contains $\psi_\mu \partial X^\mu$. But, the zero-modes of ψ_μ and ∂X^μ correspond, respectively, to the gamma matrix γ^μ and momentum k^μ when acting on a state.

The PCO zero-mode insertion inside the propagator has another virtue. It was noted previously that states with $N_{pic} = -3/2$ are infinitely degenerate since one can apply β_0 an arbitrary number of times. These states have large negative ghost numbers. Considering a loop amplitude, all these states would appear in the sum over the states and lead to a divergence. The problem is present only for loops because the ghost number is not fixed: in a tree propagator, the ghost number is fixed and only a finite number of β_0 can be applied. But, the PCO insertion turns these states into $N_{pic} = -1/2$ states. In this picture number, one cannot create an arbitrarily large negative ghost number since γ_0^n can only increase the ghost number.

17.2.3 Spurious Poles

A spurious pole corresponds to a singularity of the amplitude that cannot be interpreted as the degeneration limit of Riemann surfaces. As a consequence, they do not correspond to infrared divergences and do not have any physical meaning; they must be avoided in order to define a consistent theory. To achieve this, the section $\mathcal{S}_{g,m,n}$ must be chosen such that it avoids all spurious poles. However, while it is always possible to avoid these poles locally, it is not possible globally (this is related to the results from [3]). Poles can be avoided using vertical integration: two methods have been proposed, in the small (Sen–Witten) [28, 31] and large (Erler–Konopka) [8] Hilbert spaces, respectively. Before describing the essence of both approaches, we review the origin of spurious poles.

Origin

Spurious poles arise in three different ways:

- two PCOs collide;

- one PCO and one matter vertex collide;
- other singularities of the correlation functions.

The last source is the less intuitive one and we focus on it.

A general correlation function of (η, ξ, ϕ) on the torus[5] (satisfying the ghost number condition) reads

$$
C(x_i, y_j, z_q) = \left\langle \prod_{i=1}^{n+1} \xi(x_i) \prod_{j=1}^{n} \eta(y_j) \prod_{k=1}^{m} e^{q_k \phi(z_k)} \right\rangle
$$

$$
= \frac{\displaystyle\prod_{j'=1}^{n} \vartheta_\delta\left(-y_{j'} + \sum_i x_i - \sum_j y_j + \sum_k q_k z_k \right)}{\displaystyle\prod_{i'=1}^{n+1} \vartheta_\delta\left(-x_{i'} + \sum_i x_i - \sum_j y_j + \sum_k q_k z_k \right)}
$$

$$
\times \frac{\displaystyle\prod_{i<i'} E(x_i, x_{i'}) \prod_{j<j'} E(y_j, y_{j'})}{\displaystyle\prod_{i,j} E(x_i, y_j) \prod_{k,\ell} E(z_k, z_\ell)^{q_k q_\ell}} . \tag{17.48}
$$

The additional ξ insertion is necessary since it provides the ξ zero-mode, the correlation function being defined in the large Hilbert space. On the torus, the picture numbers must add to zero, and thus, the charges q_k satisfy

$$
\sum_k q_k = 0. \tag{17.49}
$$

The function $E(x, y)$ is called the prime form and is a generalization of the function $x - y$ on the torus:

$$
E(x, y) = \frac{\vartheta_1(x - y)}{\vartheta_1'(0)} \sim_{x \to y} x - y. \tag{17.50}
$$

Its presence ensures that C vanishes or diverges appropriately when the operators collide (i.e. that the zeros and poles of C are the expected ones from the OPE). The theta functions are used to make sure that the correlation function satisfies the appropriate boundary conditions (specified by the spin structure δ) for each cycle of the surface.

---

[5]The discussion generalizes directly to higher-genus Riemann surfaces.

However, theta functions can also vanish, and the ones in the denominator lead to additional singularities (not implied by any OPE) for the correlation function. Since x_1 can be chosen arbitrarily, the only theta function that can have poles is

$$
\vartheta_\delta \left(\sum_{i=2}^{n+1} x_i - \sum_{j=1}^{n} y_j + \sum_{k=1}^{m} q_k z_k \right) = 0. \tag{17.51}
$$

This defines a complex codimension 1 curve in $\widetilde{P}_{g,m,n}$, depending on the vertex and PCO locations, but also on the moduli parameters (appearing in the definition of the theta function). On the other hand, it does not depend on the local coordinate choice. If the section $\mathcal{S}_{g,m,n}$ intersects this curve, it will be ill-defined (even on-shell).

From this formula, several comments can be made. If an operator inserted at z contains n_ξ fields $\partial\xi$, n_η fields η and a factor $e^{p\phi}$, then the dependence in z of the theta function is of the form $(n_\xi - n_\eta + p)z = N_{\text{pic}} z$. Then, if the PCO locations are chosen as to avoid spurious poles for a given operator, this will also avoid them for any operator of the same picture number. This also implies that insertion of β and γ cannot lead to spurious poles since they have $N_{\text{pic}} = 0$; this is important since they appear in the BRST current, and thus, insertion of the latter cannot lead to new poles.

Since it is always possible to choose locally a distribution of PCO to avoid spurious poles, the idea is to discretize the moduli space in small pieces. But, since the PCO cannot be distributed continuously along the different components of the moduli space, correction terms are required. These can be generated in two different ways in both the small and large Hilbert spaces. The second is more general, while the first may be more adapted since it keeps the amplitude in the small Hilbert space.

Vertical Integration: Large Hilbert Space

Consider the n-point amplitude state $\langle A^{(p)}|$ that produces an amplitude with p PCOs when contracted with n external states are specified. The BRST identity implies that the amplitude state is closed (i.e. gauge invariant)

$$
\langle A^{(p)}| \, Q = 0, \tag{17.52}
$$

where

$$
Q = Q_B \otimes 1^{\otimes n-1} + \cdots + 1^{\otimes n-1} \otimes Q_B. \tag{17.53}
$$

Moreover, this state is in the small Hilbert space, which implies that it is in the kernel of η_0:

$$
\langle A^{(p)}| \, \eta = 0, \tag{17.54}
$$

where

$$\eta = \eta_0 \otimes 1^{\otimes n-1} + \cdots + 1^{\otimes n-1} \otimes \eta_0. \tag{17.55}$$

The BRST cohomology is trivial in the large Hilbert space: thus, if $\langle A^{(p)} |$ is closed, it must be exact in this space:

$$\langle A^{(p)} | = \langle \alpha^{(p)} | Q, \tag{17.56}$$

where the state $\alpha^{(p)}$ (called gauge amplitude) must be in the large Hilbert space. This is consistent with $\langle A^{(p)} | \eta = 0$ only if

$$\langle \alpha^{(p)} | \eta Q = 0 \tag{17.57}$$

(Q and η anti-commute). It is then natural to interpret the state on which Q acts as an amplitude with one less PCO

$$\langle A^{(p-1)} | = \langle \alpha^{(p)} | \eta \tag{17.58}$$

since $N_{\mathrm{pic}}(\eta) = -1$.

Continuing this procedure leads to an amplitude $\langle A^{(0)} |$ without any PCO insertion, and thus without spurious singularities. Consistency with the picture number anomaly requires the external state to have non-canonical picture numbers. But, this should not be a puzzle since the amplitude states should be viewed as intermediate object to obtain the final amplitude.

Hence, the amplitude $\langle A^{(p)} |$ can be constructed by starting with $\langle A^{(0)} |$: inserting $\xi(z)$ in the amplitude leads to the gauge amplitude $\langle \alpha^{(1)} |$, whose BRST variation yields $\langle A^{(2)} |$. Continuing recursively helps to construct the desired amplitude. Moreover, Q and ξ insertions automatically take care of the corrections at the interfaces of the components.

Showing that the amplitude is independent of the non-physical data (i.e. gauge invariance) is trivial since it is expressed as a BRST exact expression.

Vertical Integration: Small Hilbert Space

The section of $\widetilde{\mathcal{P}}_{g,m,n}$ is given by a series of discontinuous components linked by vertical segments. On the vertical segment, the PCO configuration interpolates continuously between the components, and the integrand can encounter a spurious pole. Since the integrand is not a total derivative in terms of the fibre coordinates, its integration over a segment depends on the path followed and not only on the end points. This implies that it diverges when it encounters the spurious pole. However, there is a specific prescription that avoids these problems. When only one PCO varies, the integrand is a total derivative of the PCO location and can thus be integrated directly, giving a difference in the two end points. In this case, the result is independent from the specific path and from the presence of the spurious pole.

To be more concrete, given a PCO insertion $\mathcal{X}(y_1)$, the variation of its location inserts a factor $-\partial\xi(y_1)$. Integrating this term between two components labelled by i and j—keeping everything else fixed—leads to a factor $\xi(y_1^{(j)}) - \xi(y_1^{(j)})$.

When several PCOs are involved, it is not sufficient to integrate the vertical segment along a path where only one PCO varies at a time. Indeed, because a hole is left in the process of the vertical integration. Additional segments must be added and integrated over. This avoids the spurious poles, and one can show that it yields a well-defined amplitude. Moreover, it agrees with the large Hilbert space approach.

Finally, it remains to address the question of the Feynman diagrams construction. In this case, every graph obtained by plumbing fixture inherits its PCO locations from the lower-dimensional surfaces, and there is no control on the resulting distribution. It can be shown that no spurious singularity is generated in the gluing process if the lower-dimensional graphs have no spurious poles. Hence, it is sufficient to ensure that the fundamental graphs have no spurious poles.

17.3 Superstring Field Theory

The construction of super-SFT has proceeded along different directions (for reviews, see [2,7,24]). There are two main strategies for constructing the superstring vertices:

1. brute-force construction: build the vertices recursively from amplitude factorization;
2. dress the bosonic products with superconformal ghosts.

While the second approach is simpler and preferred for explicit construction, the first allows to derive the general structure as was done for the bosonic string. There are two main strategies for dressing the vertices:

1. Munich construction (homotopy algebra bootstrap): Use the L_∞ and A_∞ structures to derive the superstring vertices from the bosonic vertices (small Hilbert space).
2. Berkovits' construction (WZW action): Generalize Witten's cubic bosonic open SFT (NS/R in large/small Hilbert space).

As indicated in parenthesis, a super-SFT can be written in the small or large Hilbert space (or a combination). The different approaches have been shown to be equivalent at the classical level.

The main difficulty in building a super-SFT is to properly describe the Ramond sector. This can be done following two different approaches:

- constraining the Ramond string field;
- using an auxiliary string field.

Berkovits' original SFT cannot describe the Ramond sector in the large Hilbert space, but it is possible to couple Berkovits' action for the NS field in the large Hilbert space to a Ramond field in the small Hilbert space. Another limitation of Berkovits' approach is that it works only for the open and heterotic superstrings (but not for type II).

We assume that the problems with PCO are absent (in Berkovits' and supermoduli constructions) or that they have been defined using vertical integration.

In the rest of this chapter, we will discuss the kinetic term for each of the first three approaches. At the level of the free action, the open and heterotic super-SFT differs only in the bosonic factors as in Chap. 10.

17.3.1 String Field and Propagator

As in the bosonic case, it is natural to consider a string field gathering all possible states

$$\Psi = \Psi_{-1} + \Psi_{-1/2}, \tag{17.59}$$

where Ψ_{-1} and $\Psi_{-1/2}$ are, respectively, the NS and R string fields. If the field is in the small Hilbert space, it satisfies

$$\eta_0 |\Psi\rangle = 0. \tag{17.60}$$

The propagator was found in (17.46) to be

$$\Delta = b_0^+ b_0^- \frac{\mathcal{G}}{L_0 + \bar{L}_0} \delta(L_0^-), \qquad \mathcal{G} = \begin{cases} 1 & \text{NS}, \\ \mathcal{X}_0 & \text{R}. \end{cases} \tag{17.61}$$

As for the bosonic case, the constraints

$$b_0^- |\Psi\rangle = L_0^- |\Psi\rangle = 0, \qquad b_0^+ |\Psi\rangle = 0 \tag{17.62}$$

must be imposed on the field to ensure that the propagator is invertible.

For similar reasons, the PCO insertion implies that the propagator is not invertible since $\mathcal{X}_0$ has zero-modes: this means equivalently that it has a non-empty kernel off-shell or that it contains derivatives. Two different solutions can be chosen to address this issue: imposing constraints as for the level-matching condition, or introducing auxiliary fields.

17.3.2 Constraint Approach

Two new PCO operators must be introduced

$$X = G_0\, \delta(\beta_0) + b_0\, \delta'(\beta_0), \qquad Y = -c_0\, \delta'(\gamma_0). \tag{17.63}$$

The first operator commutes with the BRST operator

$$[Q_B, X] = 0. \tag{17.64}$$

The product of these operators is a projector

$$XYX = X. \tag{17.65}$$

Then, the R string field is constrained to satisfy

$$XY\, |\Psi_{-1/2}\rangle = |\Psi_{-1/2}\rangle. \tag{17.66}$$

A state satisfying this condition is said to be in the restricted Hilbert space. It can be shown that it reproduces the cohomology of Q_B on-shell.

Remark 17.6 Since G_0 contains derivatives, the restriction is not purely algebraic as in the bosonic case. It prevents the degeneration due to γ_0^n.

Remark 17.7 (Comparison with Level-Matching) The conditions $b_0^- = L_0^- = 0$ can be rephrased as the statement that the string field Ψ is invariant under the action of the projector Bc_0^-

$$Bc_0^-\, |\Psi\rangle = |\Psi\rangle, \tag{17.67}$$

where

$$B = b_0^- \int_0^{2\pi} \frac{d\theta}{2\pi}\, e^{i\theta L_0^-} = \delta(b_0^-)\delta(L_0^-). \tag{17.68}$$

The kinetic term (after unfixing the gauge) reads

$$S_{0,2} = -\frac{1}{2}\, \langle\Psi_{-1}|\, c_0^-\, Q_B\, |\Psi_{-1}\rangle - \frac{1}{2}\, \langle\Psi_{-1/2}|\, c_0^-\, Y Q_B\, |\Psi_{-1/2}\rangle. \tag{17.69}$$

The action is invariant under the gauge transformation

$$\delta\, |\Psi\rangle = Q_B\, |\Lambda\rangle, \tag{17.70}$$

where

$$\Lambda = \Lambda_{-1} + \Lambda_{-1/2}. \tag{17.71}$$

Each gauge parameter satisfies the same conditions as the associated field (in particular, $\Lambda_{-1/2}$ is in the restricted Hilbert space).

17.3.3 Auxiliary Field Approach

The disadvantage of the constraint approach is two-fold. First, it treats both components of the field on a different footing. Second, the constraint must be imposed by hand and does not follow from any fundamental principles. Another possibility is to embed the propagator in a higher-dimensional field space by introducing additional fields: in this way, the propagator can be inverted without introducing the inverse of $\mathcal{X}_0$.

Let us introduce the new field

$$\widetilde{\Psi} = \widetilde{\Psi}_{-1} + \widetilde{\Psi}_{-3/2}, \tag{17.72}$$

which satisfies the same conditions as Ψ:

$$b_0^- |\widetilde{\Psi}\rangle = L_0^- |\widetilde{\Psi}\rangle = 0, \qquad b_0^+ |\widetilde{\Psi}\rangle = 0. \tag{17.73}$$

A tentative kinetic term is then

$$S_{0,2} = \frac{1}{2} \langle \widetilde{\Psi} | c_0^- c_0^+ L_0^+ \mathcal{G} | \widetilde{\Psi}\rangle - \langle \widetilde{\Psi} | c_0^- c_0^+ L_0^+ | \Psi\rangle. \tag{17.74}$$

The kinetic operator in matrix form for $(\widetilde{\Psi}, \Psi)$ reads

$$K = c_0^- c_0^+ L_0^+ \begin{pmatrix} -\mathcal{G} & 1 \\ 1 & 0, \end{pmatrix} \tag{17.75}$$

and its inverse is

$$\Delta = b_0^- b_0^+ \frac{1}{L_0^+} \begin{pmatrix} 0 & 1 \\ 1 & \mathcal{G} \end{pmatrix}. \tag{17.76}$$

This reproduces the expected propagator for (Ψ, Ψ) without needing to invert $\mathcal{X}_0$.

What is the interpretation of the additional fields? The gauge invariance of the action is

$$\delta |\Psi\rangle = Q_B |\Lambda\rangle, \qquad \delta |\widetilde{\Psi}\rangle = Q_B |\widetilde{\Lambda}\rangle, \tag{17.77}$$

where Λ satisfies the same constraints as Ψ (in particular, it contains more components than the Λ of the previous section). Then, the equations of motion are

$$Q_B |\Psi\rangle = 0, \qquad Q_B |\widetilde{\Psi}\rangle = 0. \tag{17.78}$$

This shows that both fields are free and decoupled and that the spectrum is doubled. To push the interpretation further, one needs to consider the interactions.

Amplitudes involve only the states contained in Ψ, and thus, the interactions are built solely in terms of Ψ. Then, the equations of motion have the form:

$$Q_B\left(|\Psi\rangle - \mathcal{G}|\widetilde{\Psi}\rangle\right) = 0, \qquad Q_B |\widetilde{\Psi}\rangle = |J(\Psi)\rangle, \tag{17.79}$$

where $J(\Psi)$ is a source term due to the interactions. An equation for Ψ only is obtained by multiplying the second with $\mathcal{G}$

$$Q_B |\Psi\rangle = \mathcal{G}|J(\Psi)\rangle. \tag{17.80}$$

Once Ψ is determined by solving this equation, the auxiliary field $\widetilde{\Psi}$ is completely fixed by the second equation up to free field solutions. This shows that $\widetilde{\Psi}$ describes only free fields even when Ψ is interacting. Note that this implies that the degrees of freedom contained in $\widetilde{\Psi}$ do not even couple to the gravitational field! This can also be shown at the level of Feynman diagrams.

Remark 17.8 The field $\widetilde{\Psi}$ is not an auxiliary field strictly speaking since it is propagating (its equation of motion is not algebraic).

17.3.4 Large Hilbert Space

The last formulation of the kinetic term considers the NS string field to be in the large Hilbert space, i.e. $\eta_0 \neq 0$. The Ramond field must be described with one of the two previous approaches.

Writing the action requires to use a NS field Ψ_0 with picture number 0. The kinetic term becomes

$$S_{0,2} = -\frac{1}{2} \langle\!\langle \Psi_0, \eta_0 Q_B, \Psi_0 \rangle\!\rangle, \tag{17.81}$$

where $\langle\!\langle \cdot, \cdot \rangle\!\rangle$ is the inner product in the large Hilbert space (contains a ξ_0 insertion). This action has an enlarged gauge invariance:

$$\delta |\Psi_0\rangle = Q_B |\Lambda_0\rangle + \eta_0 |\Omega_1\rangle, \tag{17.82}$$

and the equation of motion reads

$$Q_B \eta_0 |\Psi_0\rangle = 0. \tag{17.83}$$

The η_0 gauge invariance can be fixed with the condition

$$\xi_0 |\Psi_0\rangle = 0, \tag{17.84}$$

and one can introduce a new field Ψ_{-1} such that

$$|\Psi_0\rangle = \xi_0 |\Psi_{-1}\rangle \tag{17.85}$$

to satisfy automatically the condition. The equation of motion becomes

$$Q_B |\Psi_{-1}\rangle = 0, \tag{17.86}$$

and one recovers the small Hilbert space formulation.

17.4 Suggested Readings

- General reviews [2, 7, sec. 6].
- Spurious poles and vertical integration:
  - small Hilbert space [2, app. C, D, 28, 31];
  - large Hilbert space [8]
- Constructions of super-SFT:
  - "Sen's" amplitude factorization construction [2, 25–27, 29, 30];
  - "Munich" homotopy algebra bootstrap [9–11, 13, 17];
  - Berkovits' SFT [1, 6, 13, 20–22];
  - supermoduli space [23, 32];
  - democratic SFT [18, 19];
  - light-cone SFT [15, 16];
- Relations between different constructions [4, 5, 12–14].
- Ramond string field:
  - constrained field [6, 13, 21, 32];
  - auxiliary field [2, 13, 27, 30];

References

1. N. Berkovits, Super-poincare invariant superstring field theory. Nucl. Phys. B **450**(1–2), 90–102 (1995). https://doi.org/10.1016/0550-3213(95)00259-U. arXiv: hep-th/9503099
2. C. de Lacroix, H. Erbin, S.P. Kashyap, A. Sen, M. Verma, Closed super-string field theory and its applications. Int. J. Mod. Phys. A **32**(28–29), 1730021 (2017). https://doi.org/10.1142/S0217751X17300216. arXiv: 1703.06410
3. R. Donagi, E. Witten, Supermoduli space is not projected (2013). arXiv: 1304.7798
4. T. Erler, Relating Berkovits and A_∞ superstring field theories; large Hilbert space perspective. J. High Energy Phys. **1602**, 121 (2015). https://doi.org/10.1007/JHEP02(2016)121. arXiv: 1510.00364

5. T. Erler, Relating Berkovits and A_∞ superstring field theories; small Hilbert space perspective. J. High Energy Phys. **1510**, 157 (2015). https://doi.org/10.1007/JHEP10(2015)157. arXiv: 1505.02069

6. T. Erler, Superstring field theory and the Wess-Zumino-Witten Action. J. High Energy Phys. **2017**(10), 57 (2017). https://doi.org/10.1007/JHEP10(2017)057. arXiv: 1706.02629

7. T. Erler, Four lectures on closed string field theory. Phys. Rep. (2020), S0370157320300132. https://doi.org/10.1016/j.physrep.2020.01.003. arXiv: 1905.06785

8. T. Erler, S. Konopka, Vertical integration from the large Hilbert Space. J. High Energy Phys. **2017**(12) (2017). https://doi.org/10.1007/JHEP12(2017)112. arXiv: 1710.07232

9. T. Erler, S. Konopka, I. Sachs, NS-NS sector of closed superstring field theory. J. High Energy Phys. **2014**(8) (2014). https://doi.org/10.1007/JHEP08(2014)158. arXiv: 1403.0940

10. T. Erler, S. Konopka, I. Sachs, Resolving Witten's superstring field theory. J. High Energy Phys. **2014**(4) (2014). https://doi.org/10.1007/JHEP04(2014)150. arXiv: 1312.2948

11. T. Erler, S. Konopka, I. Sachs, Ramond equations of motion in superstring field theory. J. High Energy Phys. **2015**(11), 199 (2015). https://doi.org/10.1007/JHEP11(2015)199. arXiv: 1506.05774

12. T. Erler, Y. Okawa, T. Takezaki, A_∞ structure from the Berkovits formulation of open superstring field theory (2015). arXiv: 1505.01659

13. T. Erler, Y. Okawa, T. Takezaki, Complete action for open superstring field theory with cyclic A_∞ structure. J. High Energy Phys. **2016**(8), 12 (2016). https://doi.org/10.1007/JHEP08(2016)012. arXiv: 1602.02582

14. Y. Iimori, T. Noumi, Y. Okawa, S. Torii, From the Berkovits formulation to the Witten formulation in open superstring field theory. J. High Energy Phys. **2014**(3), 44 (2014). https://doi.org/10.1007/JHEP03(2014)044. arXiv: 1312.1677

15. N. Ishibashi, Multiloop amplitudes of light-cone gauge string field theory for type II superstrings (2018)

16. N. Ishibashi, K. Murakami, Multiloop amplitudes of light-cone gauge super-string field theory: odd spin structure contributions. J. High Energy Phys **2018**(3), 63 (2018). https://doi.org/10.1007/JHEP03(2018)063. arXiv: 1712.09049

17. S. Konopka, I. Sachs, Open superstring field theory on the restricted Hilbert space. J. High Energy Phys. **2016**(4), 1–12 (2016). https://doi.org/10.1007/JHEP04(2016)164. arXiv: 1602.02583

18. M. Kroyter, Superstring field theory in the democratic picture. Adv. Theor. Math. Phys. **15**, 741–781 (2009). MIT-CTP-4037. https://doi.org/10.4310/ATMP.2011.v15.n3.a3. arXiv: 0911.2962

19. M. Kroyter, Democratic superstring field theory: gauge fixing. J. High Energy Phys. **2011**(3) (2011). https://doi.org/10.1007/JHEP03(2011)081. arXiv: 1010.1662

20. M. Kroyter, Y. Okawa, M. Schnabl, S. Torii, B. Zwiebach, Open superstring field theory I: gauge fixing, ghost structure, and propagator. J. High Energy Phys. **2012**(3) (2012). https://doi.org/10.1007/JHEP03(2012)030. arXiv: 1201.1761

21. H. Kunitomo, Y. Okawa, Complete action for open superstring field theory. Prog. Theor. Exp. Phys. **2016**(2), 023B01 (2016). https://doi.org/10.1093/ptep/ptv189. arXiv: 1508.00366

22. H. Matsunaga, Comments on complete actions for open superstring field theory. J. High Energy Phys. **2016**(11) (2016). https://doi.org/10.1007/JHEP11(2016)115. arXiv: 1510.06023

23. K. Ohmori, Y. Okawa, Open superstring field theory based on the supermoduli space. J. High Energy Phys. **2018**(4), 35 (2018). https://doi.org/10.1007/JHEP04(2018)035. arXiv: 1703.08214

24. Y. Okawa, Construction of superstring field theories (2018). http://www.hri.res.in/~strings/okawa_school.pdf

25. R. Pius, Quantum closed superstring field theory and hyperbolic geometry I: construction of string vertices (2018). arXiv: 1808.09441

26. A. Sen, Gauge invariant 1PI effective action for superstring field theory. J. High Energy Phys. **1506**, 022 (2015). https://doi.org/10.1007/JHEP06(2015)022. arXiv: 1411.7478

27. A. Sen, Gauge invariant 1PI effective superstring field theory: inclusion of the Ramond sector. J. High Energy Phys. **2015**(8) (2015). https://doi.org/10.1007/JHEP08(2015)025. arXiv: 1501.00988

28. A. Sen, Off-shell amplitudes in superstring theory. Fortschr. Phys. **63**(3–4), 149–188 (2015). https://doi.org/10.1002/prop.201500002. arXiv: 1408.0571

29. A. Sen, Supersymmetry restoration in superstring perturbation theory. J. High Energy Phys. **2015**(12) (2015). https://doi.org/10.1007/JHEP12(2015)075. arXiv: 1508.02481

30. A. Sen, BV master action for heterotic and type II string field theories. J. High Energy Phys. **2016**(2) (2016). https://doi.org/10.1007/JHEP02(2016)087. arXiv: 1508.05387

31. A. Sen, E. Witten, Filling the gaps with PCO's. J. High Energy Phys. **1509**, 004 (2015). https://doi.org/10.1007/JHEP09(2015)004. arXiv: 1504.00609

32. T. Takezaki, Open superstring field theory including the Ramond sector based on the supermoduli space (2019). arXiv: 1901.02176

Momentum-Space SFT

18

Abstract

In this chapter, we describe the general properties of SFT actions in the momentum space. This allows to make SFT more intuitive, but also to use standard QFT methods to prove various properties of string theory. We explain how the Wick rotation is generalized for theories with vertices diverging at infinite real energies (Lorentzian signature). This allows to prove important properties of string theory, such as unitarity or crossing symmetry.

18.1 General Form

Since the explicit expressions of the string vertices are not known, it is not possible to write explicitly the SFT action. However, the general properties of the vertices are known: then, one can write a general QFT that contains SFT as a subcase. This is sufficient to already extract a lot of information. The other advantage is that the QFT language is more familiar and intuitive in many situations. Hence, one can use this general form to built intuition before translating the results in a more stringy language. In a nutshell, SFT is a QFT:

- with an infinite number of fields (of all spins);
- with an infinite number of interactions;
- with non-local interactions $\propto e^{-\#k^2}$;
- that reproduces the worldsheet amplitudes (if the latter are well-defined).

© Springer Nature Switzerland AG 2021
H. Erbin, *String Field Theory*, Lecture Notes in Physics 980,
https://doi.org/10.1007/978-3-030-65321-7_18

The non-locality of the interactions is the most salient property of SFT, beyond the infinite number of fields. This has a number of consequences:

- the Wick rotation is ill-defined;
- the position representation cannot be used, nor any property relying on it (micro-causality, largest time equation...);
- standard assumptions from local QFT (in particular, from the constructive S-matrix program, such as micro-causality) break down.

Together, these points imply that the usual arguments from QFTs must be improved. This has been an active topic in the recent years, and the results will be summarized in Sect. 18.2.

We expand the string field in Fourier space using a basis $\{\phi_\alpha(k)\}$ as (Chap. 9)

$$|\Psi\rangle = \sum_j \int \frac{d^D k}{(2\pi)^D} \, \psi_\alpha(k) \, |\phi_\alpha(k)\rangle , \qquad (18.1)$$

where k is the D-dimensional momentum and α the discrete indices (Lorentz indices, group representation, KK modes...) of the spacetime fields $\psi_\alpha(k)$. The action in momentum space takes the form (in Lorentzian signature):

$$S = - \int d^D k \, \psi_\alpha(k) K_{\alpha\beta}(k) \psi_\beta(-k)$$
$$- \sum_{n \geq 0} \int d^D k_1 \cdots d^D k_n \, V^{(n)}_{\alpha_1 \cdots \alpha_n}(k_1, \ldots, k_n) \, \psi_{\alpha_1}(k_1) \cdots \psi_{\alpha_n}(k_n). \qquad (18.2)$$

The kinetic matrix $K_{\alpha\beta}$ is usually quadratic in the momentum. In the direct Fourier expansion of the SFT action (15.24), it describes only the classical kinetic term: the quantum corrections are found in the vertex $V^{(2)}$.

From the action, we can write the Feynman rules (for the path integral weight e^{iS} and S-matrix $S = 1 + iT$). The propagator reads

$$\alpha \; \underset{k}{\xrightarrow{\hspace{1.5cm}}} \; \beta \;\; = K_{\alpha\beta}(k)^{-1} = \frac{-i \, M_{\alpha\beta}}{k^2 + m_\alpha^2} \, Q_\alpha(k), \qquad (18.3)$$

where $M_{\alpha\beta}$ is mixing matrix for states of equal mass and Q_α a polynomial in k (there is no sum over α). The interactions are obtained by plugging the basis states

$\{\phi_\alpha\}$ inside the vertices $\mathcal{V}_n$ (14.58):

$$= i\,V^{(n)}_{\alpha_1\cdots\alpha_n}(k_1,\ldots,k_n) := i\,\mathcal{V}_n\big(\phi_{\alpha_1}(k_1),\ldots,\phi_{\alpha_n}(k_n)\big)$$

$$= i \int dt\, e^{-g^{\{\alpha_k\}}_{ij}(t)\,k_i\cdot k_j - \lambda \sum_\alpha m_\alpha^2}\, P_{\alpha_1,\ldots,\alpha_n}\, k_1,\ldots,k_n; t\ ,$$

$$(18.4)$$

where t denotes collectively the moduli parameters, $P_{\{\alpha_i\}}$ is a polynomial in k, g_{ij} is a positive-definite matrix, and $\lambda > 0$ is a number. There is an implicit sum over the momentum indices.

The terms quadratic in the momenta inside the exponential arise from two sources:

- The correlation functions of the vertex operators $\langle \prod_i e^{ik_i\cdot X(z_i)} \rangle$ are proportional to $e^{-k_i\cdot k_j\,G(z_i,z_j)}$, where G is the Green function. Additional factors like ∂X contribute to the polynomial $P_{\alpha_1,\ldots,\alpha_n}$.
- It is possible to add stubs to the vertices. The effect is to multiply each leg by a factor $e^{-\lambda(k_i^2+m_i^2)}$ with $\lambda > 0$ (we take λ to be the same for all vertices for simplicity). The first term of the exponential contributes to the diagonal of the matrix g_{ij}. By taking λ sufficiently large, one can enforce that all eigenvalues are positive.

Finally, the exponential term with the masses m_α^2 ensures that the sum over all intermediate states converges despite an infinite number of states. Indeed, the number of states of mass m_α grows as e^{cm_α}, which is dominated by $e^{-\lambda m_\alpha^2}$ for sufficiently large λ. Hence, the addition of stubs makes explicit the absence of divergences in SFT.[1]

The vertices have no singularity for $k_i \in \mathbb{C}$ finite. As the energy becomes infinite $|k_i^0| \to \infty$, they behave as

$$\lim_{k^0\to\pm i\infty} V^{(n)} = 0, \qquad \lim_{k^0\to\pm\infty} V^{(n)} = \infty. \qquad (18.5)$$

---

[1] Remember that λ is not a physical parameter and disappears on-shell. This means that the cancellation of the divergences is independent of λ and must always happen on-shell.

The first property is responsible for the soft UV behaviour of string theory in Euclidean signature, while the second prevents from performing the Wick rotation (indeed, the pole at infinity implies that the arcs closing the contour contribute).

The g-loop n-point amputated Green functions are sums of Feynman diagrams, each of the form:

$$F_{g,n}(p_1, \ldots, p_n) \sim \int \mathrm{d}T \prod_s \mathrm{d}^D \ell_s \, e^{-G_{rs}(T)\,\ell_r \cdot \ell_s - 2H_{ri}(T)\,\ell_r \cdot p_i - F_{ij}(T)\,p_i \cdot p_j}$$

$$\times \prod_a \frac{1}{k_a^2 + m_a^2} \, \mathcal{P}(p_i, \ell_r; T), \tag{18.6}$$

where $\{p_i\}$ are the external momenta, $\{\ell_r\}$ the loop momenta and $\{k_i\}$ the internal momenta, with the latter given by a linear combination of the others. Moreover, T denotes the dependence in the moduli parameters of all the internal vertices, and $\mathcal{P}$ is a polynomial in (p_i, ℓ_r). The matrix G_{rs} is positive definite, which implies that:

- integrations over spatial loop momenta $\boldsymbol{\ell}_r$ converge;
- integrations over loop energies ℓ_r^0 diverge.

As a consequence, the Feynman diagrams in Lorentzian signature are ill-defined: we will explain in the next section how to fix this problem.

18.2 Generalized Wick Rotation

We have seen that loop integrals in Lorentzian signature are divergent because of the large energy behaviour of the interactions. But, this is not different from the usual QFT, where the loop integrals are also ill-defined in Lorentzian signature. Indeed, poles of the propagators sit on the real axis and also give divergent loop integrals (note that the same problem arises also here). In that case, the strategy is to define the Feynman diagrams in Euclidean space and to perform a Wick rotation: the latter matches the expressions in Lorentzian signature up to the $i\varepsilon$-prescription. The goal of the latter is to move slightly the poles away from the real axis.

Example 18.1: Scalar Field

Consider a scalar field of mass m with a quartic interaction. The 1-loop 4-point Feynman diagram is given in Fig. 18.1. The external momenta are p_i, $i = 1, \ldots, 4$. There are one-loop momentum ℓ and two internal momenta $k_1 = \ell$ and $k_2 = p - \ell$, where $p = p_1 + p_2$. The poles in the loop energy ℓ^0 are located at

$$p_{\pm} = \pm \sqrt{\boldsymbol{\ell}^2 + m^2}, \qquad q_{\pm} = p^0 \pm \sqrt{(\boldsymbol{p} - \boldsymbol{\ell})^2 + m^2}. \tag{18.7}$$

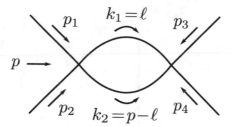

Fig. 18.1 1-loop 4-point function for a scalar field theory

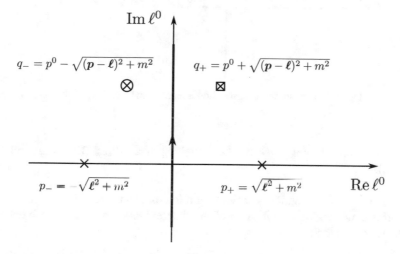

Fig. 18.2 Integration contour for external Euclidean momenta

The graph is first defined in Euclidean signature, where the external and loop energies are pure imaginary, $p_i^0, \ell^0 \in i\mathbb{R}$. The poles are shown in Fig. 18.2. Then, the external momenta are analytically continued to real values, $p_i^0 \in \mathbb{R}$. At the same time, the integration contour is also analytically continued thanks to the Wick rotation (Fig. 18.3). The contour is closed with arcs, but they do not contribute since there is no poles in the upper-right and lower-left quadrants, and no poles at infinity. However, one cannot continue the contour such that $\ell^0 \in \mathbb{R}$ because of the poles on the real axis. The Wick rotation is possible for ℓ^0 in the upper-right quadrant, $\mathrm{Re}\,\ell^0 \geq 0$, $\mathrm{Im}\,\ell^0 > 0$, which leads to the $i\varepsilon$-prescription $\ell^0 \in \mathbb{R} + i\varepsilon$. ◄

Since the Feynman diagram (18.6) is not defined in Lorentzian signature because of the poles at $\ell_r^0 \to \pm\infty$, it is also necessary to start with Euclidean momenta. However, the same behaviour at infinity prevents from using the Wick rotation since the contribution from the arcs does not vanish. It is then necessary to find another prescription for defining the Feynman diagrams in SFT starting from the Euclidean

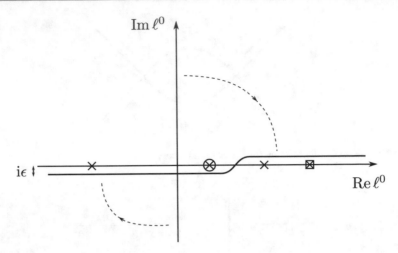

Fig. 18.3 Integration contour for external Lorentzian momenta after Wick rotation (regular vertices)

Green functions. This is given by the following *generalized Wick rotation* (Pius–Sen [6]):

1. Define the Green functions for Euclidean internal and external momenta.
2. Perform an analytic continuation of the external energies and of the integration contour such that:

- keep poles on the same side;
- keep the contour ends fixed at $\pm i\infty$.

One can show [6] that the Green functions are analytic in the upper-right quadrant $\mathrm{Im}\, p_a^0 > 0, \mathrm{Re}\, p_a^0 \geq 0$, for $p_a \in \mathbb{R}$, p_a^0. Moreover, the result is independent of the contour chosen as long as it satisfies the conditions described above. In fact, this generalized Wick rotation is valid even for normal QFT, which raises interesting questions. For example, it seems that the internal and external sets of states have no intersection, which can be puzzling when trying to interpret the Cutkosky rules. Nonetheless, everything works as expected. The generalized Wick rotation associated with the Feynman diagram from example 18.1 is shown in Fig. (18.4).

Remark 18.1 (Timelike Liouville Theory) It has been shown in [1] that this generalized Wick rotation is also the correct way for defining the timelike Liouville theory.

The fact that the amplitude is analytic only when the imaginary parts of the momenta are not zero, $\mathrm{Im}\, p_a^0 > 0$, is equivalent to the usual $i\varepsilon$-prescription for QFT. Moreover, it has been shown [11] to be equivalent to the moduli space $i\varepsilon$-prescription from [15]. Then, it has also been used to prove several important

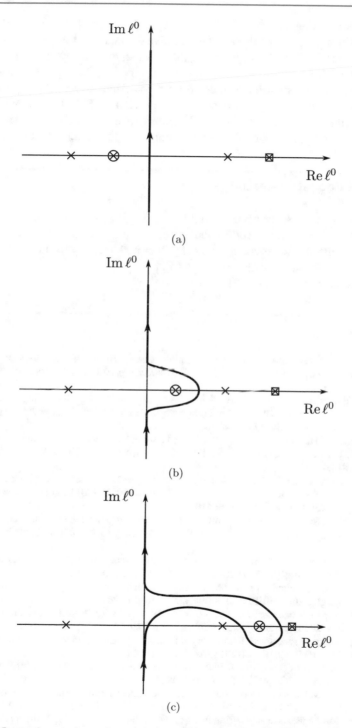

Fig. 18.4 Integration contour after analytic continuation to external Lorentzian momenta. Depending on the values of the external momenta, different cases can happen

properties of string theory shared by local QFTs: Cutkosky rules [6, 7], unitarity [9, 10], analyticity in a subset of the primitive domain and crossing symmetry [3]. Finally, general soft theorems for string theory (and, in fact, any theory of quantum gravity) have been proven in [2, 5, 12, 13]. All together, these properties establish string theory as a very strong candidate for a consistent theory of everything. The next main question is how to obtain an expression of SFT that is amenable to explicit computations. This will certainly require to understand even better the deep structure of SFT, a goal that this book will hopefully help the reader to achieve.

18.3 Suggested Readings

- SFT momentum-space action [4, 6, 14].
- Consistency properties of string theory [4]:
  - generalized Wick rotation, Cutkosky rules and unitarity [6–11].
  - analyticity and crossing symmetry [3].
  - soft theorems [2, 5, 12, 13].

References

1. T. Bautista, A. Dabholkar, H. Erbin, Quantum gravity from timelike Liouville theory. J. High Energy Phys. **2019**(10), 284 (2019). https://doi.org/10.1007/JHEP10(2019)284. arXiv: 1905.12689

2. S. Chakrabarti, S.P. Kashyap, B. Sahoo, A. Sen, M. Verma, Subleading soft theorem for multiple soft gravitons. J. High Energy Phys. **2017**(12) (2017). https://doi.org/10.1007/JHEP12(2017)150. arXiv: 1707.06803

3. C. de Lacroix, H. Erbin, A. Sen, Analyticity and crossing symmetry of super-string loop amplitudes. J. High Energy Phys. **2019**(5), 139 (2019). https://doi.org/10.1007/JHEP05(2019)139. arXiv: 1810.07197

4. C. de Lacroix, H. Erbin, S.P. Kashyap, A. Sen, M. Verma, Closed super-string field theory and its applications. Int. J. Mod. Phys. A **32**(28–29), 1730021 (2017). https://doi.org/10.1142/S0217751X17300216. arXiv: 1703.06410

5. A. Laddha, A. Sen, Sub-subleading soft graviton theorem in generic theories of quantum gravity. J. High Energy Phys. **2017**(10), 065 (2017). https://doi.org/10.1007/JHEP10(2017)065. arXiv: 1706.00759

6. R. Pius, A. Sen, Cutkosky rules for superstring field theory. J. High Energy Phys. **2016**(10) (2016). https://doi.org/10.1007/JHEP10(2016)024. arXiv: 1604.01783

7. R. Pius, A. Sen, Unitarity of the box diagram. J. High Energy Phys. **2018**(11), 094 (2018). https://doi.org/10.1007/JHEP11(2018)094. arXiv: 1805.00984

8. A. Sen, One loop mass renormalization of unstable particles in superstring theory. J. High Energy Phys. **2016**(11) (2016). https://doi.org/10.1007/JHEP11(2016)050. arXiv: 1607.06500

9. A. Sen, Reality of superstring field theory action. J. High Energy Phys. **2016**(11), 014 (2016). https://doi.org/10.1007/JHEP11(2016)014. arXiv: 1606.03455

10. A. Sen, Unitarity of superstring field theory. J. High Energy Phys. **2016**(12) (2016). https://doi.org/10.1007/JHEP12(2016)115. arXiv: 1607.08244

11. A. Sen, Equivalence of two contour prescriptions in superstring perturbation theory. J. High Energy Phys. **2017**(04) (2017). https://doi.org/10.1007/JHEP04(2017)025. arXiv: 1610.00443

12. A. Sen, Soft theorems in superstring theory. J. High Energy Phys. **2017**(06) (2017). https://doi.org/10.1007/JHEP06(2017)113. arXiv: 1702.03934

13. A. Sen, Subleading soft graviton theorem for loop amplitudes. J. High Energy Phys. **2017**(11) (2017). https://doi.org/10.1007/JHEP11(2017)123. arXiv: 1703.00024

14. A. Sen, Wilsonian effective action of superstring theory. J. High Energy Phys. **2017**(1), 108 (2017). https://doi.org/10.1007/JHEP01(2017)108. arXiv: 1609.00459

15. E. Witten, The Feynman $i\epsilon$ in string theory (2013). arXiv: 1307.5124

Conventions

This book uses natural units where $c = \hbar = 1$, but the string length ℓ_s (or Regge slope α') is kept.

A bar is used to denote both the complex conjugation and the anti-holomorphic operators. The symbol $:=$ (resp,. $=:$) means that the LHS (RHS) is defined by the expression in the RHS (LHS).

A.1   Coordinates

The number of spacetime (target-space) dimensions is denoted by $D = d+1$, where d is the number of spatial dimensions. The corresponding spacetime and spatial coordinates are written with Greek and Latin indices:

$$x^\mu = (x^0, x^i), \qquad \mu = 0, \ldots, D-1 = d \qquad i = 1, \ldots, d. \qquad \text{(A.1)}$$

When time is singled out, one writes $x^0 = t$ in Lorentzian signature and $x^0 = t_E$ in Euclidean signature (or $x^0 = \tau$ when there is no ambiguity with the worldsheet time).

A p-brane is a $(p+1)$-dimensional object whose worldvolume is parametrized by coordinates:

$$\sigma^a = (\sigma^0, \sigma^\alpha), \qquad a = 0, \ldots, p-1, \qquad \alpha = 1, \ldots, p. \qquad \text{(A.2)}$$

The time coordinate can also be singled out as $\sigma^0 = \tau_M$ in Lorentzian signature and $\sigma^0 = \tau$ in Euclidean signature. For the string, the index α is omitted since it takes only one value.

© Springer Nature Switzerland AG 2021
H. Erbin, *String Field Theory*, Lecture Notes in Physics 980,
https://doi.org/10.1007/978-3-030-65321-7

The Lorentzian signature is taken to be mostly plus, and the flat Minkowski metric reads

$$\eta_{\mu\nu} = \text{diag}(-1, \underbrace{1, \ldots, 1}_{d}). \tag{A.3}$$

The flat Euclidean metric is

$$\delta_{\mu\nu} = \text{diag}(\underbrace{1, \ldots, 1}_{D}). \tag{A.4}$$

Similar notations hold for the worldvolume metrics η_{ab} and δ_{ab}. The Levi-Civita (completely antisymmetric) tensor is normalized by

$$\epsilon_{01} = -\epsilon^{01} = 1. \tag{A.5}$$

Wick rotation from Lorentzian time t to Euclidean time τ (either worldsheet or target spacetime) is defined by

$$t = -i\tau. \tag{A.6}$$

Accordingly, contravariant (covariant) vector transforms with the same (opposite) factor:

$$V_M^0 = -iV_E^0, \qquad V_{M,0} = iV_{E,0}. \tag{A.7}$$

Most computations are performed with both spacetime and worldsheet Euclidean signatures. Expressions are Wick-rotated when needed.

Light-cone coordinates are defined by

$$x^\pm = x^0 \pm x^1. \tag{A.8}$$

A function depending only on x^+ (x^-) is said to be left-moving (right-moving) by analogy with the displacement of a wave. Under analytic continuation, the left-moving (right-moving) coordinate is mapped to the holomorphic[1] (anti-holomorphic) coordinate z ($\bar{z}$).

---

[1]The terms of *holomorphic* are simply used to indicate that the object depends only on z, but not on $\bar{z}$. Typically, the objects have singularities and are really *meromorphic* in z.

In chiral theories, the left-moving value is written first.

The worldsheet coordinates (τ, σ) on the cylinder are defined by

$$\tau \in \mathbb{R}, \qquad \sigma \in [0, L), \qquad \sigma \sim \sigma + L, \tag{A.9}$$

where typically $L = 2\pi$. The integration over the spatial coordinate is normalized such that the perimeter of spatial slice is normalized to 1 if $L = 2\pi$:

$$\mathcal{L} = \frac{1}{2\pi} \int_0^L d\sigma = \frac{L}{2\pi}. \tag{A.10}$$

This implies that $2d$ action, conserved charges, etc. are divided by an extra factor of 2π.

The coordinates can be written in terms of complex coordinates as

$$w = \tau + i\sigma, \qquad \bar{w} = \tau - i\sigma \tag{A.11}$$

such that the flat metric is

$$ds^2 = d\tau^2 + d\sigma^2 = dw d\bar{w}. \tag{A.12}$$

Under Wick rotation, the complex coordinates are mapped to light-cone coordinates as follows:

$$w = i\sigma^+, \qquad \bar{w} = i\sigma^-. \tag{A.13}$$

The cylinder can be mapped to the complex plane through

$$z = e^{2\pi w/L}, \qquad \bar{z} = e^{2\pi \bar{w}/L}. \tag{A.14}$$

The definition of the Levi-Civita tensor includes the $\sqrt{g}$ factor, such that

$$\epsilon_{z\bar{z}} = \frac{i}{2}, \qquad \epsilon^{z\bar{z}} = -2i \tag{A.15}$$

on the complex plane with flat metric conventions in the literature are compared in Table A.1 .

Table A.1 Conventions for the coordinates. The notations are the following (they can slightly vary depending on the references): the Euclidean time is obtained by the analytic continuation $\tau = it$ (denoted also by $\tau = \sigma^0 = \sigma^2$), the spatial direction is $\sigma = \sigma^1$ and the light-cone coordinates are $\sigma^{\pm} = t \pm \sigma$

Refs	Cylinder	Plane	Light cone	Left-moving
Here, Di Francesco et al. [5, 8, 14–16, 23, 28, 30]	$w = \tau + i\sigma$	$z = e^w$	$w = i\sigma^+, \bar{w} = i\sigma^-$	Holomorphic
Blumenhagen et al. [2, 4, 13, 22]	$w = \tau - i\sigma$	$z = e^w$	$w = i\sigma^-, \bar{w} = i\sigma^+$	Anti-holomorphic
Polchinski [18, 21, 25]	$w = \sigma + i\tau$	$z = e^{-iw}$	$w = -\sigma^-, \bar{w} = \sigma^+$	Anti-holomorphic[4]

A.2 Operators

Commutators and anti-commutators are denoted by

$$[A, B] := [A, B]_- = AB - BA, \qquad \{A, B\} := [A, B]_+ = AB + BA. \qquad (A.16)$$

The Grassmann parity of a field A is denoted by $|A|$

$$|A| = \begin{cases} +1 & \text{Grassmann odd,} \\ 0 & \text{Grassmann even.} \end{cases} \qquad (A.17)$$

Two (anti-)commuting operators satisfy

$$AB = (-1)^{|A||B|} BA. \qquad (A.18)$$

A.3 QFT

Energy is defined as the first component of the momentum vector

$$p^{\mu} := (E, p^i). \qquad (A.19)$$

The following notations are used to denote the number of supersymmetries:

$$(N_L, N_R), \qquad N = N_L + N_R, \qquad (A.20)$$

---

[4]In fact, the terms of "left"- and "right"-moving are interchanged in [21, p. 34] to get agreement with the literature. But, it means that the spatial axis orientation is reversed.

Moreover, concerning [8], the first definition agrees with (6.1) but not with (6.53) since the definition of ξ (our w) is modified in-between. This explains why the definitions of left- and right-moving [8, p. 161] do not agree with the one given in the table.

where N_L and N_R are the numbers of left- and right-chirality supersymmetries. The last form is used when it is not important to know the chirality of the supercharges.

The variation of a field $\phi(x)$ is defined by

$$\delta\phi(x) = \phi'(x) - \phi(x). \tag{A.21}$$

Given an internal symmetry with parameters α^a, the Noether current in Lorentzian signature is given by

$$J_a^\mu = \lambda \frac{\partial \mathcal{L}}{\partial(\partial_\mu \phi)} \frac{\delta\phi}{\delta\alpha^a}, \qquad \nabla_\mu J_a^\mu = 0, \tag{A.22}$$

where $\mathcal{L}$ is the Lagrangian, which does not include the factor $\sqrt{g}$ for curved spaces, and λ is some normalization.[5] The conserved charges Q_a associated to the currents J_a^μ for a fixed spatial slice $t = \text{cst}$ are

$$Q_a = \frac{1}{\lambda} \oint_\Sigma d^{D-1}x \sqrt{h} \, J_a^0, \tag{A.23}$$

where Σ is a spatial slice and h is the induced metric. One sets $\lambda = 2\pi$ in two dimensions, otherwise $\lambda = 1$. The variation of a field under a transformation generated by Q is

$$\delta_{\alpha^a}\phi(x) = i\alpha^a[Q_a, \phi(x)]. \tag{A.24}$$

In Euclidean signature, the current and variation are

$$J_a^\mu = i\lambda \frac{\partial \mathcal{L}}{\partial(\partial_\mu \phi)} \frac{\delta\phi}{\delta\alpha^a}, \tag{A.25a}$$

$$\delta_{\alpha^a}\phi(x) = -\alpha^a[Q_a, \phi(x)]. \tag{A.25b}$$

Note that the charge is still given by (A.23). The factor of i in (A.25a) can be understood as follows.[6] First, the time component J_a^0 of the current transforms like time such that $J_a^0 \to iJ_a^0$, which implies that the charge also gets a factor i, $Q_a \to iQ_a$. This explains the minus sign in (A.25b). Then, one needs to make this consistent with the formula (B.9) for the charge associated to a general surface. Given a spacelike n_μ, the integration measure includes the time which transforms with a factor of i: one can interpret it as coming from the spatial components of the current, $J_a^i \to iJ_a^i$, while working with a Euclidean region. Another way to

---

[5]Including the $\sqrt{g}$ would give the current density $\sqrt{g}J_a^\mu$. The simple derivative of the latter vanishes $\partial_\mu(\sqrt{g}J_a^\mu) = 0$ in view of the identity (B.4).

[6]We stress that these formulas and arguments do not apply to the energy–momentum tensor.

understand this factor for the spatial vector is by considering the electromagnetic case, where J contains a time derivative.

The term "zero-mode" has two (related) meanings:

1. given an operator D acting on a space of fields $\psi(z)$, zero-modes $\psi_{0,i}(z)$ of the operator are all fields with zero eigenvalue $D\psi_{0,i}(z) = 0$, $i = 1, \ldots, \dim\ker D$;
2. the zero-mode of a field expansion $\psi = \sum_n \psi_n z^{-n-h}$ is the mode ψ_0 for $n = 0$: on the cylinder, it corresponds to the constant term of the Fourier expansion on the cylinder (hence, a zero-mode of ∂_z according to the previous definition).

A prime indicates that the zero-modes are excluded. For example, $\det' D$ is the product of non-zero eigenvalues, ϕ' is a field without zero-mode and $d'\phi$ is the corresponding integration measure.

A.4   Curved Space and Gravity

The covariant derivative is defined by

$$\nabla_\mu = \partial_\mu + \Gamma_\mu, \tag{A.26}$$

where Γ_μ is the connection. For example, one has for a vector field

$$\nabla_\mu A^\nu = \partial_\mu A^\nu + \Gamma_{\mu\rho}{}^\nu A^\rho. \tag{A.27}$$

The negative-definite Laplacian (or Laplace–Beltrami operator) is defined by

$$\Delta = g^{\mu\nu}\nabla_\mu\nabla_\nu = \frac{1}{\sqrt{g}}\nabla_\mu\left(\sqrt{g}g^{\mu\nu}\nabla_\nu\right). \tag{A.28}$$

Note that ∇_μ does not contain the Christoffel symbol for the index ν because of the identity (B.4) (but it contains a connection for any other index of the field). For a scalar field, both derivatives become simple derivatives.

The energy–momentum tensor is defined by

$$T_{\mu\nu} = -\frac{2\lambda}{\sqrt{g}}\frac{\delta S}{\delta g^{\mu\nu}}, \tag{A.29}$$

where $\lambda = 2\pi$ for $D = 2$ and $\lambda = 1$ otherwise.

A.5   List of Symbols

General:

- D: number of non-compact spacetime dimensions
- g: loop order for a scattering amplitude
- n: number of external closed string states
- x^μ: spacetime non-compact coordinates
- $\sigma^a = (t, \sigma)$: worldsheet coordinates
- g_s: closed string coupling
- $Z_g = A_{g,0}$: genus-g vacuum amplitude
- $A_{g,n}(k_1, \ldots, k_n)_{\alpha_1,\ldots,\alpha_n} := A_{g,n}(\{k_i\})_{\{\alpha_i\}}$: g-loop n-point scattering amplitude for states with quantum numbers $\{k_i, \alpha_i\}$ (if connected, amputated Green functions for $n \geq 3$)
- $G_{g,n}(k_1, \ldots, k_n)_{\alpha_1,\ldots,\alpha_n}$: g-loop n-point Green function for states with quantum numbers $\{k_i, \alpha_i\}$
- $T_{ab}^\perp$: traceless symmetric tensor or traceless component of the tensor T_{ab}
- Ψ: generic (set of) matter field(s)

Hilbert spaces:

- $\mathcal{H}$: generic Hilbert space (in general, Hilbert space of the matter plus ghost CFT)
- $\mathcal{H}_\pm = \mathcal{H} \cap \ker b_0^\pm$
- $\mathcal{H}_0 = \mathcal{H} \cap \ker b_0^- \cap \ker b_0^+$
- $\mathcal{H}(Q_B)$: absolute cohomology of the operator Q_B inside the space $\mathcal{H}$
- $\mathcal{H}_-(Q_B) = \mathcal{H}(Q_B) \cap \mathcal{H}^-$: semi-relative cohomology of the operator Q_B inside the space $\mathcal{H}$
- $\mathcal{H}_0(Q_B) = \mathcal{H}(Q_B) \cap \mathcal{H}^0$: relative cohomology of the operator Q_B inside the space $\mathcal{H}$
- A: Grassmann parity of the operator or state A

Riemann surfaces:

- g: Riemann surface genus (number of holes/handles)
- n: number of bulk punctures/marked points
- $\Sigma_{g,n}$: genus-g Riemann surface with n punctures
- $\Sigma_g = \Sigma_{g,0}$: genus-g Riemann surface
- $\mathcal{M}_{g,n}$: moduli space of genus-g Riemann surfaces with n punctures
- $\mathcal{M}_g = \mathcal{M}_{g,0}$: moduli space of genus-$g$ Riemann surfaces
- $\mathsf{M}_{g,n} = \dim \mathcal{M}_{g,n}$
- $\mathsf{M}_{g,n}^c = \dim_{\mathbb{C}} \mathcal{M}_{g,n}$
- $\mathsf{M}_g = \mathsf{M}_{g,0} = \dim \mathcal{M}_g = \dim \ker P_1^\dagger$
- $\mathsf{M}_g^c = \mathsf{M}_{g,0}^c = \dim_{\mathbb{C}} \mathcal{M}_g$

- $\mathcal{K}_{g,n}$: conformal Killing vector group of genus-g Riemann surfaces with n punctures
- $\mathcal{K}_g = \mathcal{K}_{g,0} = \ker P_1$: conformal Killing vector group of genus-g Riemann surfaces
- $\mathsf{K}_{g,n} = \dim \mathcal{K}_{g,n}$
- $\mathsf{K}_{g,n}^c = \dim_{\mathbb{C}} \mathcal{K}_{g,n} = \dim_{\mathbb{C}} \ker P_1$
- $\mathsf{K}_g = \mathsf{K}_{g,0} = \dim \ker P_1$
- $\mathsf{K}_g^c \mathsf{K}_{g,0}^c = \dim_{\mathbb{C}} \ker P_1$
- ψ_i: real basis of $\ker P_1$, CKV
- ϕ_i: real basis of $\ker P_1^\dagger$, real quadratic differentials
- $(\psi_K, \bar{\psi}_K)$: complex basis of $\ker P_1$, (anti-)holomorphic CKV
- $(\phi_I, \bar{\phi}_I)$: complex basis of $\ker P_1^\dagger$, (anti-)holomorphic quadratic differentials
- $t_\lambda \in \mathcal{M}_{g,n}$: real moduli of $\mathcal{M}_{g,n}$
- $m_\Lambda \in \mathcal{M}_{g,n}$: complex moduli of $\mathcal{M}_{g,n}$
- $t_i \in \mathcal{M}_g$: real moduli of $\mathcal{M}_g$
- $m_I \in \mathcal{M}_g$: complex moduli of $\mathcal{M}_g$
- z: coordinate on the Riemann surface
- w_i: local coordinates around punctures
- z_a: local coordinates away from punctures
- $f_i(w_i)$: transition functions from w_i to z
- σ_α: coordinate system on the left of the contour C_α
- τ_α: coordinate system on the right of the contour C_α

CFT:

- $V_\alpha(k; \sigma^a) := V_{k,\alpha}(\sigma^a)$: matter vertex operator with[7] momentum k and quantum numbers α inserted at position $\sigma^a = (z, \bar{z})$
- $\mathcal{V}_\alpha(k; \sigma^a)$: unintegrated vertex operator with momentum k and quantum numbers α inserted at position σ^a
- $V_\alpha(k) = \int d^2\sigma \sqrt{g} \, V_\alpha(k; \sigma)$: integrated vertex operator
- on-shell (closed bosonic string): $\mathcal{V}_\alpha(k; \sigma^a) = c\bar{c} V_\alpha(k; \sigma^a)$ is a $(0, 0)$-primary, with $V_\alpha(k; \sigma^a)$ a $(1, 1)$-primary matter operator
- $\widehat{\mathcal{O}}$: operator $\mathcal{O}$ with zero-modes removed
- $\mathcal{O}^\dagger$: Hermitian adjoint
- $\mathcal{O}^\ddagger$: Euclidean adjoint
- $\mathcal{O}^t$: BPZ conjugation
- $\langle \mathcal{O}_1 | \mathcal{O}_2 \rangle$: BPZ inner product
- $\langle \mathcal{O}_1^\ddagger | \mathcal{O}_2 \rangle$: Hermitian inner product
- $|0\rangle$: SL$(2, \mathbb{C})$ (conformal) vacuum
- $|\Omega\rangle$: energy vacuum (lowest energy state)

---

[7]When the momentum and/or quantum numbers are not relevant, we remove them or simply index the operators by a number.

- $:\mathcal{O}::$ conformal normal ordering (with respect to $SL(2, \mathbb{C})$ vacuum $|0\rangle$)
- $^*_*\mathcal{O}^*_*:$ energy normal ordering (with respect to energy vacuum $|\Omega\rangle$)

SFT:

- Ψ: closed string field
- $\{\phi_r\} = \{\phi_\alpha(k)\}$: basis of $\mathcal{H}$ (or some subspace)

Indices:

- $\mu = 0, \ldots, D - 1$: non-compact spacetime dimensions
- $a = 0, \ldots, p$: worldvolume coordinates ($p = 1$: worldsheet)
- $i = 1, \ldots, n$: external states, local coordinates
- $\lambda = 1, \ldots, \mathsf{M}_{g,n}$: real moduli of $\mathcal{M}_{g,n}$
- $\Lambda = 1, \ldots, \mathsf{M}^c_{g,n}$: complex moduli of $\mathcal{M}_{g,n}$
- $i = 1, \ldots, \mathsf{M}_g$: real moduli of $\mathcal{M}_g$
- $I = 1, \ldots, \mathsf{M}^c_g$: complex moduli of $\mathcal{M}_g$
- $i = 1, \ldots, \mathsf{K}_g$: real CKV of $\mathcal{K}_g$
- $K = 1, \ldots, \mathsf{K}^c_g$: complex CKV of $\mathcal{K}_g$
- $r = (k, \alpha)$: index for basis state of $\mathcal{H}$ (or some subspaces) and α: non-momentum indices

Summary of Important Formulas

This appendix summarizes formulas that appear in the book or are needed but assumed to be known to the reader (such as formulas from QFT and general relativity).

B.1 Complex Analysis

The Cauchy–Riemann formula is

$$\oint_{C_z} \frac{dw}{2\pi i} \frac{f(w)}{(w-z)^n} = \frac{f^{(n-1)}(z)}{(n-1)!},$$

(B.1)

where $f(z)$ is a holomorphic function.

One has

$$\bar{\partial} \frac{1}{z} = 2\pi\,\delta^{(2)}(z).$$

(B.2)

B.2 QFT, Curved Spaces and Gravity

The Green function G of a differential operator D is defined by

$$D_x G(x, y) = \frac{\delta(x - y)}{\sqrt{g}} - P(x, y),$$

(B.3)

where P is the projector on the zero-modes of D.

© Springer Nature Switzerland AG 2021
H. Erbin, *String Field Theory*, Lecture Notes in Physics 980,
https://doi.org/10.1007/978-3-030-65321-7

The covariant divergence of a vector can be rewritten in terms of a simple derivative:

$$\nabla_\mu v^\mu = \frac{1}{\sqrt{g}} \partial_\mu(\sqrt{g} v^\mu).$$ (B.4)

Under an infinitesimal change of coordinates

$$\delta x^\mu = \xi^\mu,$$ (B.5)

the metric transforms as

$$\delta g_{\mu\nu} = \mathcal{L}_\xi g_{\mu\nu} = \nabla_\mu \xi_\nu + \nabla_\nu \xi_\mu.$$ (B.6)

Stokes' theorem reads

$$\int_V d^D x\, \nabla_\mu v^\mu = \oint_{\partial V} d\Sigma_\mu v^\mu, \qquad d\Sigma_\mu := \epsilon\, n_\mu\, d^{D-1}\Sigma,$$ (B.7)

where V is a spacetime region, $S = \partial V$ its boundary and $d^{D-1}\Sigma$ the induced integration measure. The vector n_μ normal to S points outward and $\epsilon := n_\mu n^\mu = 1$ (-1) if S is timelike (spacelike). If the surface is defined by $x^0 = $ cst, then

$$d^{D-1}\Sigma = \sqrt{g}\, d^{D-1}x, \qquad n_\mu = \delta_\mu^0.$$ (B.8)

We can write a generalization of (A.23) for a charge associated to a general surface S:

$$Q_S = \frac{1}{\lambda} \int_S d\Sigma_\mu\, J_a^\mu.$$ (B.9)

If the current J_a^μ is conserved, $\nabla_\mu J_a^\mu = 0$ (no source), Stokes' theorem (B.7) shows that the charge vanishes $Q_S = 0$ if S is a closed surface and that it is conserved $Q_{S_1} = -Q_{S_2}$ for two spacelike surfaces S_1 and S_2 extending to infinity (if J_a^μ vanishes at infinity) (see [19, chap. 3, 30, sec. 8.4] for more details).

B.2.1 Two Dimensions

Stokes' theorem (B.7) on flat space reads

$$\int d^2 x\, \partial_\mu v^\mu = \oint \epsilon_{\mu\nu}\, dx^\nu v^\mu = \oint (v^0 d\sigma - v^1 d\tau),$$ (B.10)

since $d\Sigma_\mu = \epsilon_{\mu\nu} dx^\nu$.

The integral of the curvature is a topological invariant

$$\chi_{g;b} := \frac{1}{4\pi} \int d^2\sigma \sqrt{g}\, R + \frac{1}{2\pi} \oint ds\, k$$
$$= 2 - 2g - b,$$

(B.11)

called the Euler characteristics and where g is the number of holes and b the number of boundaries.

B.3   Conformal Field Theory

In two dimensions, the energy–momentum tensor is defined by

$$T_{ab} = -\frac{4\pi}{\sqrt{g}} \frac{\delta S}{\delta g^{ab}}.$$

(B.12)

B.3.1   Complex Plane

Defining the real coordinates (x, y) from the complex coordinate on the complex plane

$$z = x + iy, \qquad \bar{z} = x - iy,$$

(B.13)

we have the formulas:

$$ds^2 = dx^2 + dy^2 = dz d\bar{z}, \qquad g_{z\bar{z}} = \frac{1}{2}, \qquad g_{zz} = g_{\bar{z}\bar{z}} = 0,$$

(B.14a)

$$\epsilon_{z\bar{z}} = \frac{i}{2}, \qquad \epsilon^{z\bar{z}} = -2i,$$

(B.14b)

$$\partial := \partial_z = \frac{1}{2}(\partial_x - i\partial_y), \qquad \bar{\partial} := \partial_{\bar{z}} = \frac{1}{2}(\partial_x + i\partial_y),$$

(B.14c)

$$V^z = V^x + iV^y, \qquad V^{\bar{z}} = V^x - iV^y,$$

(B.14d)

$$d^2x = dx dy = \frac{1}{2} d^2z, \qquad d^2z = dz d\bar{z},$$

(B.14e)

$$\delta(z) = \frac{1}{2}\delta^{(2)}(x), \qquad 1 = \int d^2z\, \delta^{(2)}(z) = \int d^2x\, \delta^{(2)}(x),$$

(B.14f)

$$\int_R d^2z \, (\partial_z v^z + \partial_{\bar{z}} v^{\bar{z}}) = -i \oint_{\partial R} (dz\, v^{\bar{z}} - d\bar{z}\, v^z) = -2i \oint_{\partial R} (v_z dz - v_{\bar{z}} d\bar{z}).$$

(B.14g)

B.3.2   General Properties

A primary holomorphic field $\phi(z)$ of weight h transforms as

$$f \circ \phi(z) = \left(\frac{\mathrm{d}f}{\mathrm{d}z}\right)^h \phi\big(f(z)\big) \tag{B.15}$$

for any local change of coordinates f. A quasi-primary operator transforms like this only for $f \in \mathrm{SL}(2, \mathbb{C})$. Its mode expansion reads

$$\phi(z) = \sum_n \frac{\phi_n}{z^{n+h}}, \qquad \phi_n = \oint_{C_0} \frac{\mathrm{d}z}{2\pi \mathrm{i}} \, z^{n+h-1} \phi(z), \tag{B.16}$$

where the integration is counter-clockwise around the origin.

The $\mathrm{SL}(2, \mathbb{C})$ vacuum $|0\rangle$ is defined by

$$\forall n \geq -h + 1 : \quad \phi_n |0\rangle = 0. \tag{B.17}$$

Its BPZ conjugate $\langle 0|$ satisfies

$$\forall n \leq h - 1 : \quad \langle 0|\phi_n = 0. \tag{B.18}$$

The state–operator correspondence associates a state $|\phi\rangle$ to each operator $\phi(z)$:

$$|\phi\rangle := \phi(0) |0\rangle = \phi_{-h} |0\rangle . \tag{B.19}$$

The operator corresponding to the vacuum is the identity 1.[1] The Hermitian and BPZ conjugated states are

$$\langle \phi^{\ddagger}| := \langle 0|I \circ \phi^{\dagger}(0) = \lim_{z \to \infty} z^{2h} \langle 0|\phi^{\dagger}(z),$$

$$\langle \phi| := \langle 0|I^{\pm} \circ \phi(0) = (\pm 1)^h \lim_{z \to \infty} z^{2h} \langle 0|\phi(z). \tag{B.20}$$

The energy–momentum tensor is a quasi-primary operator of weight $h = 2$

$$T(z) = \sum_n \frac{L_n}{z^{n+2}}. \tag{B.21}$$

The OPE between T and a primary operator h of weight h is

$$T(z)\phi(w) \sim \frac{h\,\phi(w)}{(z-w)^2} + \frac{\partial\phi(w)}{z-w}. \tag{B.22}$$

---

[1]Exceptionally, the state $|0\rangle$ and the operator 1 do not have the same symbol.

The OPE of T with itself defines the central charge c

$$T(z)T(w) \sim \frac{c/2}{(z-w)^4} + \frac{2T(w)}{(z-w)^2} + \frac{\partial T(w)}{z-w}.$$
(B.23)

B.3.3   Hermitian and BPZ Conjugations

Both conjugations do not change the ghost number of a state.

Hermitian
The Hermitian conjugate of a general state built from n operators A_i and a complex number λ is

$$(\lambda\, A_1 \cdots A_n\, |0\rangle)^\dagger = \lambda^*\, \langle 0| A_n^\dagger \cdots A_1^\dagger.$$
(B.24)

BPZ
The BPZ conjugate of modes is

$$\phi_n^t = (I^\pm \circ \phi)_n = (-1)^h (\pm 1)^n \phi_{-n},$$
(B.25)

where $I^\pm(z) = \pm 1/z$. The plus sign is usually used for the closed string, and the minus sign for the open string. Given a general state built from n operators and a complex number λ, the conjugation does not change the order of the operators and does not conjugate complex numbers:

$$(\lambda\, A_1 \cdots A_n\, |0\rangle)^t = \lambda\, \langle 0| (A_1)^t \cdots (A_n)^t.$$
(B.26)

However, it reverses radial ordering such that operators must be (anti-)commuted in radial ordered expressions.

The BPZ product satisfies

$$\langle A, B \rangle = (-1)^{|A||B|} \langle B, A \rangle.$$
(B.27)

Moreover, the inner product is non-degenerate, so

$$\forall A: \quad \langle A|B\rangle = 0 \quad \Longrightarrow \quad |B\rangle = 0.$$
(B.28)

Denoting by $\{|\phi_r\rangle\}$ a complete basis of states, then the conjugate basis $\{\langle\phi_r^c|\}$ is defined by the BPZ product as

$$\langle \phi_r^c | \phi_s \rangle = \delta_{rs}.$$
(B.29)

We have

$$\langle \phi_r | \phi_s^c \rangle = (-1)^{|\phi_r|} \delta_{rs}. \tag{B.30}$$

B.3.4   Scalar Field

The simplest matter CFT is a set of D scalar fields $X^\mu(z, \bar{z})$ such that $i\partial X^\mu$ and $i\bar{\partial} X^\mu$ are of weights $h = (1, 0)$ and $h = (0, 1)$

$$i\partial X^\mu = \sum_n \frac{\alpha_n^\mu}{z^{n+1}}, \qquad i\bar{\partial} X^\mu = \sum_n \frac{\bar{\alpha}_n^\mu}{\bar{z}^{n+1}}. \tag{B.31}$$

The commutation relations between the modes are

$$[\alpha_m^\mu, \alpha_n^\nu] = m\delta_{m+n,0}\eta^{\mu\nu}, \qquad [\bar{\alpha}_m^\mu, \bar{\alpha}_n^\nu] = m\delta_{m+n,0}\eta^{\mu\nu}, \qquad [\alpha_m^\mu, \bar{\alpha}_n^\nu] = 0. \tag{B.32}$$

The zero-modes of both operators are equal and correspond to the (centre-of-mass) momentum

$$\alpha_0^\mu = \bar{\alpha}_0^\mu = \sqrt{\frac{\alpha'}{2}} \, p^\mu. \tag{B.33}$$

The conjugate of p^μ is the centre-of-mass position x^μ:

$$[x^\mu, p^\nu] = \eta^{\mu\nu}. \tag{B.34}$$

Vertex operators are defined by

$$V_k(z, \bar{z}) = \, :e^{ik \cdot X(z, \bar{z})}:, \qquad h = \bar{h} = \frac{\alpha'^2 k^2}{4}. \tag{B.35}$$

The scalar vacuum $|k\rangle$ is annihilated by all positive-frequency oscillators, and it is characterized by its eigenvalue for the zero-mode operator

$$p^\mu |k\rangle = k^\mu |k\rangle, \qquad \forall n > 0: \quad \alpha_n^\mu |k\rangle = 0, \quad \bar{\alpha}_n^\mu |k\rangle = 0. \tag{B.36}$$

The vacuum is associated to the vertex operator V_k:

$$|k\rangle = V_k(0, 0) |0\rangle = e^{ik \cdot x} |0\rangle. \tag{B.37}$$

The conjugate vacuum is

$$\langle k| p^\mu = \langle k| k^\mu, \qquad \langle k| = |k\rangle^\dagger, \qquad \langle -k| = |k\rangle^t. \tag{B.38}$$

B.3.5   Reparametrization Ghosts

The reparametrization ghosts are described by an anti-commuting first-order system with the parameters (Chap. 7 and Table 7.1):

$$\epsilon = 1, \qquad \lambda = 2, \qquad c_{gh} = -26, \qquad q_{gh} = -3, \qquad a_{gh} = -1. \tag{B.39}$$

We focus on the holomorphic sector.
   The b and c ghosts have weights:

$$h(b) = 2, \qquad h(c) = -1, \tag{B.40}$$

such that the mode expansions are

$$b(z) = \sum_{n\in\mathbb{Z}} \frac{b_n}{z^{n+2}}, \qquad c(z) = \sum_{n\in\mathbb{Z}} \frac{c_n}{z^{n-1}}, \tag{B.41a}$$

$$b_n = \oint \frac{dz}{2\pi i} z^{n+1} b(z), \qquad c_n = \oint \frac{dz}{2\pi i} z^{n-2} c(z). \tag{B.41b}$$

The anti-commutators between the modes b_n and c_n read

$$\{b_m, c_n\} = \delta_{m+n,0}, \qquad \{b_m, b_n\} = 0, \qquad \{c_m, c_n\} = 0. \tag{B.42}$$

The energy–momentum tensor and the Virasoro modes are, respectively,

$$T = -2 : b\partial c : - : \partial b\, c :, \tag{B.43a}$$

$$L_m = \sum_n (n+m) : b_{m-n} c_n : = \sum_n (2m - n) : b_n c_{m-n} :. \tag{B.43b}$$

The expression of the zero-mode is

$$L_0 = -\sum_n n : b_n c_{-n} : = \sum_n n : b_{-n} c_n :. \tag{B.44}$$

The commutators between the L_n and the ghost modes are

$$[L_m, b_n] = (m - n) b_{m+n}, \qquad [L_m, c_n] = -(2m + n) c_{m+n}. \tag{B.45}$$

In particular, L_0 commutes with the zero-modes:

$$[L_0, b_0] = 0, \qquad [L_0, c_0] = 0. \tag{B.46}$$

The anomalous global U(1) symmetry for the ghost number N_{gh} is generated by the ghost current:

$$j = -:bc:, \qquad N_{\text{gh},L} = \oint \frac{dz}{2\pi i} j(z), \tag{B.47}$$

such that

$$N_{\text{gh}}(c) = 1, \qquad N_{\text{gh}}(b) = -1. \tag{B.48}$$

Remember that $N_{\text{gh}} = N_{\text{gh},L}$ in the left sector, such that we omit the index L. The modes of the ghost current are

$$j_m = -\sum_n :b_{m-n}c_n: = -\sum_n :b_n c_{m-n}:, \qquad N_{\text{gh},L} = j_0 = -\sum_n :b_{-n}c_n:. \tag{B.49}$$

The commutator of the current modes with itself and with the Virasoro modes are

$$[j_m, j_n] = m\,\delta_{m+n,0}, \qquad [L_m, j_n] = -n j_{m+n} - \frac{3}{2}m(m+1)\delta_{m+n,0}. \tag{B.50}$$

Finally, the commutators of the ghost number operator are

$$[N_{\text{gh}}, b(w)] = -b(w), \qquad [N_{\text{gh}}, c(w)] = c(w). \tag{B.51}$$

The level operators N^b and N^c and the number operators N_n^b and N_n^c are defined as

$$N^b = \sum_{n>0} n\, N_n^b, \qquad N^c = \sum_{n>0} n\, N_n^c, \tag{B.52a}$$

$$N_n^b = :b_{-n}c_n:, \qquad N_n^c = :c_{-n}b_n:. \tag{B.52b}$$

The commutator of the number operators with the modes is

$$[N_m^b, b_{-n}] = b_{-n}\delta_{m,n}, \qquad [N_m^c, c_{-n}] = c_{-n}\delta_{m,n}. \tag{B.53}$$

The OPEs between the ghosts and different currents are

$$c(z)b(w) \sim \frac{1}{z-w}, \qquad b(z)c(w) \sim \frac{1}{z-w}, \qquad b(z)b(w) \sim 0, \quad c(z)c(w) \sim 0,$$
$$\text{(B.54a)}$$

$$T(z)b(w) \sim \frac{2b(w)}{(z-w)^2} + \frac{\partial b(w)}{z-w}, \quad T(z)c(w) \sim \frac{-c(w)}{(z-w)^2} + \frac{\partial c(w)}{z-w}.$$
$$\text{(B.54b)}$$

$$j(z)b(w) \sim -\frac{b(w)}{z-w}, \quad j(z)c(w) \sim \frac{c(w)}{z-w}. \quad j(z)\mathcal{O}(w) \sim N_{\text{gh}}(\mathcal{O})\frac{\mathcal{O}(w)}{z-w},$$
$$\text{(B.54c)}$$

$$j(z)j(w) \sim \frac{1}{(z-w)^2}.$$
$$\text{(B.54d)}$$

$$T(z)j(w) \sim \frac{-3}{(z-w)^3} + \frac{j(w)}{(z-w)^2} + \frac{\partial j(w)}{z-w},$$
$$\text{(B.54e)}$$

any operator $\mathcal{O}(z)$ is defined by
The OPE (B.54e) implies that the ghost number is not conserved on a curved space:

$$N^c - N^b = 3 - 3g \qquad \text{(B.55)}$$

and leads to a shift between the ghost numbers on the plane and the cylinder:

$$N_{\text{gh},L} = N_{\text{gh},L}^{\text{cyl}} + \frac{3}{2}. \qquad \text{(B.56)}$$

The $SL(2, \mathbb{C})$ vacuum $|0\rangle$ is defined by

$$\forall n > -2: \quad b_n |0\rangle = 0, \qquad \forall n > 1: \quad c_n |0\rangle = 0. \qquad \text{(B.57)}$$

The mode c_1 does not annihilate the vacuum, and the two degenerate energy vacua are

$$|\downarrow\rangle := c_1 |0\rangle, \qquad |\uparrow\rangle := c_0 c_1 |0\rangle. \qquad \text{(B.58)}$$

The zero-point energy of these states is

$$L_0 |\downarrow\rangle = a_{\text{gh}} |\downarrow\rangle, \qquad L_0 |\uparrow\rangle = a_{\text{gh}} |\uparrow\rangle, \qquad a_{\text{gh}} = -1. \qquad \text{(B.59)}$$

The energy for the normal ordering of the different currents is

$$L_m = \sum_n \left(n - (1 - \lambda)m\right) {}^\star_\star b_{m-n} c_n {}^\star_\star + a_{\mathrm{gh}}\, \delta_{m,0}, \tag{B.60a}$$

$$j_m = \sum_n {}^\star_\star b_{m-n} c_n {}^\star_\star + \delta_{m,0}. \tag{B.60b}$$

The energy–momentum and ghost current zero-modes are explicitly

$$L_0 = \sum_n n\, {}^\star_\star b_{-n} c_n {}^\star_\star + a_{\mathrm{gh}} = \widehat{L}_0 - 1, \tag{B.61a}$$

$$N_{\mathrm{gh},L} = j_0 = \sum_n {}^\star_\star b_{-n} c_n {}^\star_\star + 1 = \widehat{N}_{\mathrm{gh},L} + \frac{1}{2}\left(N_0^c - N_0^b\right) - \frac{3}{2}, \tag{B.61b}$$

$$\widehat{L}_0 = N^b + N^c, \qquad \widehat{N}_{\mathrm{gh},L} := \sum_{n>0}\left(N_n^c - N_n^b\right). \tag{B.61c}$$

Then, one can straightforwardly compute the ghost number of the vacua:

$$N_{\mathrm{gh}}\,|0\rangle = 0, \qquad N_{\mathrm{gh}}\,|\downarrow\rangle = |\downarrow\rangle, \qquad N_{\mathrm{gh}}\,|\uparrow\rangle = 2|\uparrow\rangle. \tag{B.62}$$

Using (B.56) allows to write the ghost numbers on the cylinder:

$$N_{\mathrm{gh}}^{\mathrm{cyl}}\,|\downarrow\rangle = -\frac{1}{2}|\downarrow\rangle, \qquad N_{\mathrm{gh}}^{\mathrm{cyl}}\,|\uparrow\rangle = \frac{1}{2}|\uparrow\rangle. \tag{B.63}$$

The b_n and c_n are Hermitian:

$$b_n^\dagger = b_{-n}, \qquad c_n^\dagger = c_{-n}. \tag{B.64}$$

The BPZ conjugates of the modes are

$$b_n^t = (\pm 1)^n b_{-n}, \qquad c_n^t = -(\pm 1)^n c_{-n}, \tag{B.65}$$

using $I^\pm(z)$ with (6.111).

The conjugates of the vacuum read

$$|\downarrow\rangle^\ddagger = \langle 0|c_{-1}, \qquad |\uparrow\rangle^\ddagger = \langle 0|c_{-1} c_0. \tag{B.66}$$

The BPZ conjugates of the vacua are

$$\langle\downarrow| := |\downarrow\rangle^t = \mp\,\langle 0|c_{-1}, \qquad \langle\uparrow| := |\uparrow\rangle^t = \pm\,\langle 0|c_0 c_{-1}. \tag{B.67}$$

We have the following relations:

$$\langle\downarrow| = \mp|\downarrow\rangle^{\ddagger}, \qquad \langle\uparrow| = \mp|\uparrow\rangle^{\ddagger}. \tag{B.68}$$

The ghost are normalized with

$$\langle\uparrow|\downarrow\rangle = \langle\downarrow|c_0|\downarrow\rangle = \langle 0|c_{-1}c_0c_1|0\rangle = 1, \tag{B.69}$$

which selects the minus sign in the BPZ conjugation. The conjugate of the ghost vacuum is

$$\langle 0^c| = \langle 0|c_{-1}c_0c_1. \tag{B.70}$$

Considering both the holomorphic and anti-holomorphic sectors, we introduce the combinations:

$$b_n^{\pm} = b_n \pm \bar{b}_n, \qquad c_n^{\pm} = \frac{1}{2}(c_n \pm \bar{c}_n). \tag{B.71}$$

The normalization of $b_m^{\pm}$ is chosen to match the one of $L_m^{\pm}$ (B.75) and the one of $c_m^{\pm}$ such that

$$\{b_m^+, c_n^+\} = \delta_{m+n}, \qquad \{b_m^-, c_n^-\} = \delta_{m+n}. \tag{B.72}$$

We have the following useful identities:

$$b_n^- b_n^+ = 2b_n\bar{b}_n, \qquad c_n^- c_n^+ = \frac{1}{2}c_n\bar{c}_n. \tag{B.73}$$

B.4   Bosonic String

The BPZ conjugates of the scalar and ghost modes are

$$(\alpha_n)^t = -(\pm 1)^n\,\alpha_{-n}, \qquad (b_n)^t = (\pm 1)^n\,b_{-n}, \qquad (c_n)^t = -(\pm 1)^n\,c_{-n}. \tag{B.74}$$

Combinations of holomorphic and anti-holomorphic modes are

$$L_n^{\pm} = L_n \pm \bar{L}_n, \qquad b_n^{\pm} = b_n \pm \bar{b}_n, \qquad c_n^{\pm} = \frac{1}{2}(c_n \pm \bar{c}_n). \tag{B.75}$$

The closed string inner product is defined from the BPZ product by an additional insertion of c_0^-

$$\langle A, B \rangle = \langle A | c_0^- | B \rangle ,$$ (B.76)

while the open string inner product is equal to the BPZ product

$$\langle A, B \rangle = \langle A | B \rangle .$$ (B.77)

The vacuum for the matter and ghosts is

$$|k, 0\rangle := |k\rangle \otimes |0\rangle , \qquad |k, \downarrow\rangle := |k\rangle \otimes |\downarrow\rangle .$$ (B.78)

The vacuum is normalized as

open: $\langle k, \downarrow | c_0 | k, \downarrow \rangle = \langle k', 0 | c_{-1} c_0 c_1 | k, 0 \rangle = (2\pi)^D \delta^{(D)}(k + k'),$ (B.79a)

closed: $\langle k, \downarrow\downarrow | c_0 \bar{c}_0 | k, \downarrow\downarrow \rangle = \langle k', 0 | c_{-1} \bar{c}_{-1} c_0 \bar{c}_0 c_1 \bar{c}_1 | k, 0 \rangle = (2\pi)^D \delta^{(D)}(k + k').$ (B.79b)

Quantum Field Theory

C

In this appendix, we gather useful information on quantum field theories. The first section describes how to compute with path integral with non-trivial measures, generalizing techniques from finite-dimensional integrals. Then, we summarize the important concepts from the BRST and BV formalisms.

C.1   Path Integrals

In this section, we explain how analysis, algebra and differential geometry are generalized to infinite-dimensional vector spaces (fields).

C.1.1   Integration Measure

In order to construct a path integral for the field Φ, one needs to define a notion of distance on the space of fields. The distance between a field Φ and a neighbouring field $\Phi + \delta\Phi$ is

$$|\delta\Phi|^2 = G(\Phi)(\delta\Phi, \delta\Phi), \tag{C.1}$$

where G is the (field-dependent) metric on the field tangent space (the field dependence will be omitted when no confusion is possible). This induces a metric on the field space itself

$$|\Phi|^2 = G(\Phi)(\Phi, \Phi), \tag{C.2}$$

from which the integration measure over the field space can be defined as

$$d\Phi\sqrt{\det G(\Phi)}. \tag{C.3}$$

© Springer Nature Switzerland AG 2021
H. Erbin, *String Field Theory*, Lecture Notes in Physics 980,
https://doi.org/10.1007/978-3-030-65321-7

Moreover, the field metric also defines an inner product between two different elements of the tangent space or field space:

$$(\delta\Phi_1, \delta\Phi_2) = G(\Phi)(\delta\Phi_1, \delta\Phi_2), \qquad (\Phi_1, \Phi_2) = G(\Phi)(\Phi_1, \Phi_2). \qquad (C.4)$$

Remark C.1 (Metric in Component Form) If one has a set of spacetime fields $\Phi_a(x)$, then a local norm is defined by

$$|\delta\Phi_a|^2 = \int dx\, \rho(x)\gamma_{ab}\big(\Phi(x)\big)\delta\Phi_a(x)\delta\Phi_b(x), \qquad (C.5)$$

which means that the metric in component form is

$$G_{ab}(x, y)(\Phi) = \delta(x - y)\rho(x)\gamma_{ab}\big(\Phi(x)\big). \qquad (C.6)$$

Locality means that all fields are evaluated at the same point. On a curved space, it is natural to write γ only in terms of the metric g and set $\rho(x) = \sqrt{\det g(x)}$, such that the inner product is diffeomorphism invariant.

Since a Gaussian integral is proportional to the square root of the operator determinant, the integration measure can be determined by considering the Gaussian integral over the tangent space:

$$\int d\delta\Phi\, e^{-G(\Phi)(\delta\Phi, \delta\Phi)} = \frac{1}{\sqrt{\det G(\Phi)}}. \qquad (C.7)$$

Note that one needs to work on the tangent space because $G(\Phi)$ can depend on the field, which means that the integral

$$\int d\Phi\, e^{-G(\Phi)(\Phi, \Phi)} \qquad (C.8)$$

is not Gaussian.

Having constructed the Gaussian measure with respect to the metric $G(\Phi)$, it is now possible to consider the path integral of general functional F of the fields:

$$\int d\Phi\, \sqrt{\det G(\Phi)}\, F(\Phi). \qquad (C.9)$$

The (effective) action $S(\Phi)$ provides a natural metric on the field space by defining $\sqrt{\det G} = e^{-S}$, or

$$S = -\frac{1}{2} \operatorname{tr} \ln G(\Phi). \qquad (C.10)$$

However, it can be simpler to work with a Gaussian measure by considering only the quadratic terms in S and expanding the rest in a power series. In particular, the partition function is defined from the classical action S_{cl} by

$$Z = \int d\Phi \, e^{-S_{cl}(\Phi)}.$$

(C.11)

Given an operator D, its adjoint $D^\dagger$ is defined with respect to the metric as

$$G(\delta\Phi, D\delta\Phi) = G(D^\dagger\delta\Phi, \delta\Phi).$$

(C.12)

The *free-field measure* is such that the metric on the field space is independent from the field itself: $G(X) = G_0$. In particular, this implies that the metric is flat and its determinant can be absorbed in the measure, setting $\det G_0 = 1$. In this case, the measure is invariant under shift of the field:

$$\Phi \to \Phi + \varepsilon$$

(C.13)

such that

$$\int d\Phi \, e^{-\frac{1}{2}|\Phi+\varepsilon|^2} = \int d\Phi \, e^{-\frac{1}{2}|\Phi|^2}.$$

(C.14)

This property allows to complete squares and shift integration variables (for example, to generate a perturbative expansion and to derive the propagator).

Computation: Equation (C.14)

$$\int d\Phi \, e^{-\frac{1}{2}|\Phi+\varepsilon|^2} = \int d\widetilde{\Phi} \, \det\frac{\delta\Phi}{\delta\widetilde{\Phi}} \, e^{-\frac{1}{2}|\widetilde{\Phi}|^2} = \int d\widetilde{\Phi} \, e^{-\frac{1}{2}|\widetilde{\Phi}|^2}.$$

(C.15)

The first equality follows by setting $\widetilde{\Phi} = \Phi + \varepsilon$, and the result (C.14) follows by the redefinition $\widetilde{\Phi} = \Phi$.

C.1.2   Field Redefinitions

Under a field redefinition $\Phi \to \Phi'$, the norm and the measure are invariant:

$$d\Phi\sqrt{\det G(\Phi)} = d\widetilde{\Phi}\sqrt{\det \widetilde{G}(\widetilde{\Phi})}, \qquad G(\Phi)(\delta\Phi, \delta\Phi) = \widetilde{G}(\widetilde{\Phi})(\delta\widetilde{\Phi}, \delta\widetilde{\Phi}).$$

(C.16)

Conversely, one can find the Jacobian $J(\Phi, \widetilde{\Phi})$ between two coordinate systems by writing

$$d\Phi = J(\Phi, \widetilde{\Phi})d\widetilde{\Phi}, \qquad J(\Phi, \widetilde{\Phi}) = \left| \det \frac{\partial \Phi}{\partial \widetilde{\Phi}} \right| = \sqrt{\frac{\det \widetilde{G}(\widetilde{\Phi})}{\det G(\Phi)}}. \qquad (C.17)$$

If the measure of the initial field coordinate is normalized such that $\det G = 1$, or equivalently

$$\int d\delta\Phi\, e^{-|\delta\Phi|^2} = 1, \qquad (C.18)$$

one can determine the Jacobian by performing explicitly the integral

$$J(\widetilde{\Phi})^{-1} = \int d\delta\widetilde{\Phi}\, e^{-\widetilde{G}(\delta\widetilde{\Phi},\delta\widetilde{\Phi})}. \qquad (C.19)$$

Remark C.2 (Identity of the Jacobian for Φ and $\delta\Phi$) The Jacobian agrees on the space of fields and on its tangent space. This is most simply seen by using a finite-dimensional notation: considering the coordinates x^μ and a vector $v = v^\mu \partial_\mu$, the Jacobian for changing the coordinates to $\tilde{x}^\mu$ is equivalently

$$J = \det \frac{\partial \tilde{x}^\mu}{\partial x^\mu} = \det \frac{\partial \tilde{v}^\mu}{\partial v^\mu} \qquad (C.20)$$

since the vector transforms as

$$\tilde{v}^\mu = v^\nu \frac{\partial \tilde{x}^\mu}{\partial x^\nu}. \qquad (C.21)$$

C.1.3 Zero-Modes

A zero-mode Φ_0 of an operator D is a field such that

$$D\Phi_0 = 0. \qquad (C.22)$$

In the definition of the path integral over the space of fields Φ, the measure is defined over the complete space. However, this will lead, respectively, to a divergent or vanishing integral if the field is bosonic or fermionic, because the integration over the zero-modes can be factorized from the rest of the integral. Writing the field as

$$\Phi = \Phi_0 + \Phi', \qquad (\Phi_0, \Phi') = 0, \qquad (C.23)$$

where Φ' is orthogonal to the zero-mode Φ_0, a Gaussian integral of an operator D reads

$$Z[D] = \int d\Phi \sqrt{\det G}\, e^{-\frac{1}{2}(\Phi, D\Phi)} = \left(\int d\Phi_0 \right) \int d\Phi'\, e^{-\frac{1}{2}(\Phi', D\Phi')}. \qquad (C.24)$$

A first solution could be to simply strip the first factor (for example, by absorbing it in the normalization), but this is not satisfactory. In particular, the partition function with source

$$Z[D, J] = \int d\Phi \sqrt{\det G}\, e^{-\frac{1}{2}(\Phi, D\Phi) - (J, \Phi)} \qquad (C.25)$$

will depend on the zero-modes through the sources. But, since the zero-modes are still singled out, it is interesting to factorize the integration

$$Z[D, J] = \int d\Phi_0\, e^{-(J, \Phi_0)} \int d\Phi'\, e^{-\frac{1}{2}(\Phi', D\Phi') - (J, \Phi')} \qquad (C.26)$$

and understand what makes it finite. Ensuring that zero-modes are correctly inserted is an important consistency and leads to powerful arguments. Especially, this can help to guess an expression when it cannot be derived easily from first principles.

To exemplify the problem, consider the cases where there is a single constant zero-mode denoted as x (bosonic) or θ (fermionic). The integral over x is infinite:

$$\int dx = \infty. \qquad (C.27)$$

Oppositely, the integral of a Grassmann variable θ vanishes

$$\int d\theta = 0. \qquad (C.28)$$

A Grassmann integral also satisfies

$$\int d\theta\, \theta = \int d\theta\, \delta(\theta) = 1, \qquad (C.29)$$

such that an integral over a zero-mode does not vanish if there is one zero-mode in the integrand (due to the Grassmann nature of θ, the integrand can be at most linear). By analogy with the fermionic case, a possibility for getting a finite bosonic integral is to insert a delta function:

$$\int dx\, \delta(x) = 1. \qquad (C.30)$$

We will see that this is exactly what happens for the ghosts and super-ghosts in (super)string theories.

Since $\ker D$ is generally finite-dimensional, it is interesting to decompose the zero-mode on a basis and integrate over the coefficients in order to obtain a finite-dimensional integral. Writing the zero-mode as

$$\theta_0(x) = \theta_{0i}\psi_i(x), \qquad \ker D = \mathrm{Span}\{\psi_i\}, \tag{C.31}$$

where the coefficients θ_{0i} are constant Grassmann numbers, the change of variables $\theta \to (\theta_{0i}, \theta')$ implies

$$d\theta = \frac{1}{\sqrt{\det(\psi_i, \psi_j)}}\, d\theta' \prod_{i=1}^{n} d\theta_{0i}, \tag{C.32}$$

where $n = \dim \ker D$.

Next, according to the discussion above, one can ask if it is possible to rewrite an integration over $d\theta'$ in terms of an integration over $d\theta$ together with zero-mode insertions. This is indeed possible, and one finds

$$d\theta \prod_{i=1}^{n} \theta(x_i) = \frac{\det \psi_i(x_j)}{\sqrt{\det(\psi_i, \psi_j)}}\, d\theta'. \tag{C.33}$$

Computation: Equation (C.32)

$$1 = \int d\theta\, e^{-|\theta|^2} = \int d\theta' d\theta_0\, e^{-|\theta|^2 - |\theta_0|^2}$$

$$= J \int d\theta' \prod_i d\theta_{0i}\, e^{-|\theta'|^2 - |\theta_{0i}\psi_i|^2} = J\sqrt{\det(\psi_i, \psi_j)}.$$

Computation: Equation (C.33)
The simplest approach is to start with the LHS. This formula is motivated from the previous discussion: if the integration measure contains n zero-modes, it will vanish unless there are n zero-mode insertions. Moreover, one can replace each of them by the complete field since only the zero-mode part can contribute

$$\int d\theta_0 \prod_{j=1}^{n} \theta(x_j) = \int d\theta_0 \prod_{j=1}^{n} \theta_0(x_j) = \frac{1}{\sqrt{\det(\psi_i, \psi_j)}} \int d^n\theta_{0i} \prod_{j=1}^{n} \left[\theta_{0i}\psi_i(x_j)\right]$$

$$= \frac{\det \psi_i(x_j)}{\sqrt{\det(\psi_i, \psi_j)}} \int \prod_i d\theta_{0i}\, \theta_{0i} = \frac{\det \psi_i(x_j)}{\sqrt{\det(\psi_i, \psi_j)}}.$$

The third equality follows by developing the product and ordering the θ_{0i}: minus signs result from anti-commuting the θ_{0i} such that one gets the determinant of the basis elements.

C.2 BRST Quantization

Consider an action $S_m[\phi^i]$ that depends on some fields ϕ^i subject to a gauge symmetry:

$$\delta\phi^i = \epsilon^a \delta_a \phi^i = \epsilon^a R_a^i(\phi), \tag{C.34}$$

where ϵ^a are the (local) bosonic parameters, such that the action is invariant

$$\epsilon^a \delta_a S_m = 0. \tag{C.35}$$

The gauge transformations form a Lie algebra with structure coefficients f_{ab}^c

$$[\delta_a, \delta_b] = f_{ab}^c \delta_c. \tag{C.36}$$

It is important that (1) the algebra closes off-shell (without using the equations of motion), (2) the structure coefficients are field independent and (3) the gauge symmetry is irreducible (each gauge parameter is independent).

Remark C.3 (Interpretation of the R_a^i Matrices) If the ϕ^i transforms in a representation **R** of the gauge group, then the transformation is linear in the field

$$R_a^i(\phi) = (T_a^{\mathbf{R}})^i{}_j \phi^j, \tag{C.37}$$

with $T_a^{\mathbf{R}}$ the generators in the representation **R**. But, in full generality, this is not the case: for example, the gauge fields A_μ^a do not transform in the adjoint representation even if they carry an adjoint index (only the field strength does), and in this case,

$$R_{a\mu}^b = \delta_a^b \partial_\mu + f_{ac}^b A_\mu^c. \tag{C.38}$$

When the fields ϕ^i form a non-linear sigma models, the $R_a^i(\phi)$'s correspond to Killing vectors of the target manifold.

In order to fix the gauge symmetry in the path integral

$$Z = \Omega_{\text{gauge}}^{-1} \int d\phi^i \, e^{-S_m}, \tag{C.39}$$

gauge fixing conditions must be imposed:

$$F^A(\phi^i) = 0. \tag{C.40}$$

Indeed, without gauge fixing, the integration is performed over multiple identical configurations and the result diverges. The index A is different from the gauge index a because they can refer to different representations, but for the gauge fixing to be possible, they should run over as many values. s

Next, ghost fields c^a (fermionic) are introduced for every gauge parameter, anti-ghosts b_A (fermionic) and auxiliary (Nakanishi–Laudrup) fields B_A (bosonic) for every gauge condition. The gauge fixing and ghost actions are then defined by

$$S_{\text{gh}} = b_A c^a \, \delta_a F^A(\phi^i), \tag{C.41a}$$

$$S_{\text{gf}} = -\mathrm{i}\, B_A F^A(\phi^i) \tag{C.41b}$$

such that the original partition function is equivalent to

$$Z = \int \mathrm{d}\phi^i \, \mathrm{d}b_A \, \mathrm{d}c^a \, \mathrm{d}B_A \, e^{-S_{\text{tot}}}, \tag{C.42}$$

where

$$S_{\text{tot}} = S_m + S_{\text{gf}} + S_{\text{gh}}. \tag{C.43}$$

The total action is invariant

$$\delta_\epsilon S_{\text{tot}} = 0 \tag{C.44}$$

under the (global) BRST transformations

$$\delta_\epsilon \phi^i = \mathrm{i}\epsilon\, c^a \delta_a \phi^i, \qquad \delta_\epsilon c^a = -\frac{\mathrm{i}}{2} \epsilon\, f^a_{bc} c^b c^c, \qquad \delta_\epsilon b_A = \epsilon\, B_A, \qquad \delta_\epsilon B_A = 0, \tag{C.45}$$

where ϵ is an anti-commuting constant parameter. Note that the original action S_m is invariant by itself since the transformation acts like a gauge transformation with parameter ϵc^a. The transformation of c^a follows because it transforms in the adjoint representation of the gauge group. Direct computations show that this transformation is nilpotent

$$\delta_\epsilon \delta_{\epsilon'} = 0. \tag{C.46}$$

These transformations are generated by a (fermionic) charge Q_B called the BRST charge

$$\delta_\epsilon \phi^i = i[\epsilon Q_B, \phi^i] \tag{C.47}$$

and similarly for the other fields (stripping the ϵ outside the commutator turns it to an anti-commutator if the field is fermionic). Taking the ghosts to be Hermitian leads to a Hermitian charge.

An important consequence is that the two additional terms of the action can be rewritten as the BRST exact terms:

$$S_{\text{gf}} + S_{\text{gh}} = \{Q_B, b_A F^A\}. \tag{C.48}$$

A small change in the gauge fixing condition δF leads to a variation of the action

$$\delta S = \{Q_B, b_A \delta F^A\}. \tag{C.49}$$

The BRST charge should commute with the Hamiltonian in order to be conserved: this should hold in particular when changing the gauge fixing condition

$$[Q_B, \{Q_B, b_A \delta F^A\}] = 0 \quad \Longrightarrow \quad Q_B^2 = 0. \tag{C.50}$$

Some vocabulary is needed before proceeding further. A state $|\psi\rangle$ is said to be BRST *closed* if it is annihilated by the BRST charge

$$|\psi\rangle \text{ closed} \quad \Longleftrightarrow \quad |\psi\rangle \in \ker Q_B \quad \Longleftrightarrow \quad Q_B |\psi\rangle = 0. \tag{C.51}$$

States that are in the image of Q_B (i.e. they can be written as Q_B applied on some other states) are said to be *exact*

$$|\psi\rangle \text{ exact} \quad \Longleftrightarrow \quad |\psi\rangle \in \operatorname{Im} Q_B \quad \Longleftrightarrow \quad \exists |\chi\rangle : |\psi\rangle = Q_B |\chi\rangle. \tag{C.52}$$

The *cohomology* $\mathcal{H}(Q_B)$ of Q_B is the set of closed states that are not exact

$$|\psi\rangle \in \mathcal{H}(Q_B) \quad \Longleftrightarrow \quad |\psi\rangle \in \ker Q_B, \quad \nexists |\chi\rangle : |\psi\rangle = Q_B |\chi\rangle. \tag{C.53}$$

Hence, the cohomology corresponds to

$$\mathcal{H}(Q_B) = \frac{\ker Q_B}{\operatorname{Im} Q_B}. \tag{C.54}$$

Two elements of the cohomology differing by an exact state are in the same equivalence class

$$|\psi\rangle \simeq |\psi\rangle + Q_B |\chi\rangle. \tag{C.55}$$

Considering the S-matrix $\langle \psi_f | \psi_i \rangle$ between a set of physical initial states ψ_i and final states ψ_f, a small change in the gauge fixing condition leads to

$$\delta_F \langle \psi_f | \psi_i \rangle = \langle \psi_f | \{ Q_B, b_A \delta F^A \} | \psi_i \rangle \qquad (C.56)$$

after expanding the exponential to first order. Since the S-matrix should not depend on the gauge, this implies that a physical state ψ must be BRST closed (i.e. invariant)

$$Q_B | \psi \rangle = 0. \qquad (C.57)$$

Conversely, this implies that any state of the form $Q_B | \chi \rangle$ cannot be physical because it is orthogonal to every physical state $| \psi \rangle$

$$\langle \psi | Q_B | \chi \rangle = 0. \qquad (C.58)$$

This implies, in particular, that the amplitudes involving $| \psi \rangle$ and $| \psi \rangle + Q_B | \chi \rangle$ are identical and any amplitude for which an external state is exact vanishes. As a conclusion, physical states are in the BRST cohomology

$$| \psi \rangle \text{ physical} \quad \Longleftrightarrow \quad | \psi \rangle \in \mathcal{H}(Q_B). \qquad (C.59)$$

If there is a gauge where the ghosts decouple from the matter field, then the invariance of the action and the S-matrix under changes of the gauge fixing ensures that this statement holds in any gauge (but, one still needs to check that the gauge preserves the other symmetries). If such a gauge does not exist, then one needs to employ other methods to show the desired result.

Note that B_A can be integrated out by using its equations of motion

$$\frac{\delta F^A}{\delta \phi^i} B_A = -\frac{\delta S_m}{\delta \phi^i}, \qquad (C.60)$$

and this modifies the BRST transformation of the anti-ghost to

$$\delta_\epsilon b_A = -\epsilon \left(\frac{\delta F^A}{\delta \phi^i} \right)^{-1} \frac{\delta S_m}{\delta \phi^i}. \qquad (C.61)$$

It is also possible to introduce a term

$$\{ Q_B, b_A B_B M^{AB} \} = i B_A M^{AB} B_B \qquad (C.62)$$

for any constant matrix M^{AB}. Since this is also a BRST exact term, the amplitudes are not affected. Integrating over B_A produces a Gaussian average instead of a delta function to fix the gauge.

In the previous discussion, the BRST symmetry was assumed to originate from the Faddeev–Popov gauge fixing. But, in fact, it is possible to start directly with an action of the form

$$S[\phi, b, c, B] = S_0[\phi] + Q_B \Psi[\phi, b, c, B], \tag{C.63}$$

where Ψ has the ghost number of -1. It can be proven that this is the most general action invariant under the BRST transformations (C.45). This can describe gauge fixed action that cannot be described by the Faddeev–Popov procedure: in particular, the latter yields actions that are quadratic in the ghost fields (by definition of the Gaussian integral representation of the determinant), but this does not exhaust all the possibilities. For example, the background field method applied to Yang–Mills theory requires using an action quartic in the ghosts.

In this section, several hypotheses have been implicit (off-shell closure, irreducibility and constant structure coefficients). If one of them breaks, then it is necessary to employ the more general BV formalism.

C.3   BV Formalism

The Batalin–Vilkovisky (BV or also field–antifield) formalism is the most general framework to quantize theories with a gauge symmetry. While the BRST formalism (Appendix C.2) is sufficient to describe simple systems, it breaks down when the structure of the gauge symmetry is more complicated, for example, in systems implying gravity. The BV formalism is required in the following three cases (which can occur simultaneously):

1. the gauge algebra is open (on-shell closure);
2. the structure coefficients depend on the fields;
3. the gauge symmetry is reducible (not all transformations are independent).

The BV formalism is also useful for standard gauge symmetries to demonstrate renormalizability and deal with anomalies.

As explained in the previous section, the ghosts and the BRST symmetry are crucial to ensure the consistency of the gauge theory. The idea of the BV formalism is to put on an equal footing the physical fields and all the required auxiliary and ghost fields (before gauge fixing). The introduction of antifields—one for each of the fields—and the description of the full quantum dynamics in terms of a quantum action (constrained by the quantum master equation) ensure the consistency of the system. Additional benefits are the presence of a (generalized) BRST symmetry, the existence of a Poisson structure (which allows to bring concepts from the Hamiltonian formalism), the covariance of the formalism and the simple interpretation of counter-terms as corrections to the classical action.

For giving a short intuition, the BV formalism can be interpreted as providing a (anti)canonical structure in the Lagrangian formalism, the role of the Hamiltonian being played by the action.

C.3.1   Properties of Gauge Algebra

Before explaining the BV formalism, we review the situations listed above. The classical action for the physical fields ϕ^i is denoted by $S_0[\phi]$ and the associated equations of motion by

$$\mathcal{F}_i(\phi) = \frac{\partial S_0}{\partial \phi^i}. \tag{C.64}$$

Then, a gauge algebra is open and has field-dependent structure coefficients $F_{ab}^c(\phi)$ if

$$[T_a, T_b] = F_{ab}^c(\phi) T_c + \lambda_{ab}^i \mathcal{F}_i(\phi). \tag{C.65}$$

On-shell, $\mathcal{F}_i = 0$, and the second term is absent, such that the algebra closes. The fields themselves are constants from the point of view of the gauge algebra, but their presences in the structure coefficients complicate the analysis of the theory. Moreover, the path integral is off-shell, and for this reason, one needs to take into account the last term.

Finally, the gauge algebra can be reducible: in brief, it means that there are gauge invariances associated to gauge parameters—and correspondingly ghosts for ghosts—, and this pattern can repeat indifinitely. Since there is one independent ghost for each generator, there are too many ghosts if the generators are not all independent, and there is a remnant gauge symmetry for the ghost fields (in the standard Faddeev–Popov formalism, the ghosts are not subject to any gauge invariance). This originates from relations between the generators R_a^i: denoting by m_0 the number of level-0 gauge transformations, the number of independent generators is rank R_a^i. Then, the

$$m_1 = m_0 - \text{rank } R_a^i \tag{C.66}$$

relations between the generators translate into a level-1 gauge invariance of the ghosts. This symmetry can be gauge fixed by performing second time the Faddeev–Popov procedure, yielding commuting ghosts. This symmetry can also be reducible, and the procedure can continue without end. If one finds that the gauge invariance at level $n = \ell$ is irreducible, one says that the gauge invariance is ℓ-reducible. If this does not happen, one defines $\ell = \infty$. The number of generators at level n is denoted by m_n.

A p-form gauge theory is written in terms of a gauge field A_p with a gauge invariance

$$\delta A_p = d\lambda_{p-1}. \tag{C.67}$$

But, due to the nilpotency of the derivative, deformations of the gauge parameter satisfying

$$\delta \lambda_{p-1} = d\lambda_{p-2} \tag{C.68}$$

do not translate into a gauge invariance of A_p. Similarly from this should be excluded the transformation

$$\delta \lambda_{p-2} = d\lambda_{p-3}, \tag{C.69}$$

and so on until one reaches the case $p = 0$. Hence, a p-form field has a p-reducible gauge invariance. ◄

C.3.2   Classical BV

Denoting the fields collectively as

$$\psi^r = \{\phi^i, B_A, b_A, c^a\}, \tag{C.70}$$

the simplest BV action reads

$$S[\psi^r, \psi_r^*] = S_0[\phi] + Q_B \psi^r \, \psi_r^* \tag{C.71}$$

with the antifields

$$\psi_r^* = \{\phi_i^*, B^{A*}, b^{A*}, c_a^*\}. \tag{C.72}$$

The action (C.63) is recovered by writing

$$\psi_r^* = \frac{\partial \Psi}{\partial \psi^r}. \tag{C.73}$$

This indicates that the general BRST formalism could be rephrased in the BV language. But, in the same way that the BRST formalism generalizes the Faddeev–Popov formalism, it is in turn generalized by the BV formalism. Indeed, the above action is linear in the antifields: this constraint is not required, and one can write more general actions. In the rest of this section, we explain how this works at the level of the action (classical level) and how the sets of fields and antifields are defined.

Consider a set of physical fields ϕ^i with the gauge invariance

$$\delta\phi^i = \epsilon_0^{a_0} R_{a_0}^i(\phi^i). \tag{C.74}$$

Then, associate a ghost field c^{a_0} to each of the gauge parameters ϵ^{a_0}. If the gauge symmetry is reducible, the new gauge invariance is associated to the ghosts

$$\delta c_0^{a_0} = \epsilon_1^{a_1} R_{a_1}^{a_0}(\phi^i, c^{a_0}). \tag{C.75}$$

This structure is recurring, the ghosts of the level-n gauge invariance are denoted by c^{a_n} and they satisfy

$$\delta c_n^{a_n} = \epsilon_{n+1}^{a_{n+1}} R_{a_{n+1}}^{a_n}(\phi^i, c_0^{a_0}, \ldots, c_n^{a_n}). \tag{C.76}$$

Thus, the set of fields is

$$\psi^r = \{c_n^{a_n}\}_{n=-1,\ldots,\ell}, \qquad c_{-1} := \phi. \tag{C.77}$$

A ghost number is introduced

$$N_{\text{gh}}(\phi^i) = 0, \qquad N_{\text{gh}}(c_n^{a_n}) = n + 1, \tag{C.78}$$

and the Grassmann parity of the ghosts is defined to be opposite (resp., identical) of the parity of the associated gauge parameter for even (resp., odd) n

$$|c_n| = |\epsilon_n^{a_n}| + n + 1. \tag{C.79}$$

To each of these fields is associated an antifield ψ_r^* of opposite parity as ψ^r and such that their ghost numbers sum to -1

$$N_{\text{gh}}(\psi_r^*) = -1 - N_{\text{gh}}(\psi^r), \qquad |\psi_r^*| = -|\psi^r|. \tag{C.80}$$

The fields and antifields together are taken to define a graded symplectic structure

$$\omega = \sum_r d\psi^r \wedge d\psi_r^* \tag{C.81}$$

with respect to which they are conjugated to each other

$$(\psi^r, \psi_s^*) = \delta_{rs}, \qquad (\psi^r, \psi^s) = 0, \qquad (\psi_r^*, \psi_s^*) = 0. \tag{C.82}$$

The antibracket (graded Poisson bracket) $(\cdot, \cdot)$ reads

$$(A, B) = \frac{\partial_R A}{\partial \psi^r} \frac{\partial_L B}{\partial \psi_r^*} - \frac{\partial_R A}{\partial \psi_r^*} \frac{\partial_L B}{\partial \psi^r}, \tag{C.83}$$

where the L and R indices indicate left and right derivatives. It is graded symmetric, which means that

$$(A, B) = -(-1)^{(|A|+1)(|B|+1)}(B, A). \tag{C.84}$$

It also satisfies a graded Jacobi identity and the property

$$N_{\mathrm{gh}}((A, B)) = N_{\mathrm{gh}}(A) + N_{\mathrm{gh}}(B) + 1, \qquad |(A, B)| = |A| + |B| + 1 \mod 2. \tag{C.85}$$

Moreover, the antibracket acts as a derivative

$$(A, BC) = (A, B)C + (-1)^{|B||C|}(A, C)B. \tag{C.86}$$

The dynamics of the theory is described by the (classical) master action $S[\psi^r, \psi_r^*]$, which satisfies

$$N_{\mathrm{gh}}(S) = 0, \qquad |S| = 0. \tag{C.87}$$

In order to reproduce correctly the dynamics of the classical system without ghosts, this action is required to satisfy the boundary condition

$$S[\psi^r, \psi_r^* = 0] = S_0[\phi^i], \qquad \left.\frac{\partial_L \partial_R S}{\partial c_{n-1,a_{n-1}}^* \partial c_n^{a_n}}\right|_{\psi^*=0} = R_{a_n}^{a_{n-1}}. \tag{C.88}$$

Indeed, if the antifields are set to zero, the ghost fields cannot appear because they all have positive ghost numbers, and it is not possible to build terms with vanishing ghost numbers from them.

In analogy with the Hamiltonian formalism, the master action can be used as the generator of a global fermionic symmetry, and inspection will show that it corresponds to a generalization of the BRST symmetry. Writing the generalized and classical BRST operator as s, the transformations of the fields and antifields read

$$\delta_\theta \psi^r = \theta \, s\psi^r = -\theta \, (S, \psi_r) = \theta \, \frac{\partial_R S}{\partial \psi_r^*}, \tag{C.89a}$$

$$\delta_\theta \psi_r^* = \theta \, s\psi_r^* = -\theta \, (S, \psi_r^*) = -\theta \, \frac{\partial_R S}{\partial \psi^r}, \tag{C.89b}$$

where θ is a constant Grassmann parameter. The variation of a generic functional $F[\psi^r, \psi_r^*]$ is

$$\delta_\theta F = \theta \, s F = -\theta \, (S, F). \tag{C.90}$$

For the BRST transformation to be a symmetry of the action, the action must satisfy the classical master equation

$$(S, S) = 0. \tag{C.91}$$

This equation can easily be solved by expanding S in the ghosts: the various terms can be interpreted in terms of properties of the gauge algebra. Then, the Jacobi identity used with two S and an arbitrary functional gives

$$(S, (S, F)) = 0, \tag{C.92}$$

and this implies that the transformation is nilpotent

$$s^2 = 0. \tag{C.93}$$

A classical observable $\mathcal{O}$ satisfies

$$s\mathcal{O} = 0. \tag{C.94}$$

Due to the BRST symmetry, the action is not uniquely defined, and the action

$$S' = S + (S, \delta F) \tag{C.95}$$

also satisfies the master equation, where δF is arbitrary up to the condition $N_{\text{gh}}(\delta F) = -1$. This can be interpreted as the action S in a new coordinate system $(\psi'^r, \psi_r'^*)$ with

$$\psi' = \psi - \frac{\delta F}{\delta \psi^*}, \qquad \psi'^* = \psi^* + \frac{\delta F}{\delta \psi} \tag{C.96}$$

such that

$$S'[\psi, \psi^*] = S\left[\psi - \frac{\delta F}{\delta \psi^*}, \psi^* + \frac{\delta F}{\delta \psi}\right]. \tag{C.97}$$

Indeed, for $F = F[\psi, \psi^*]$, one has

$$S'[\psi, \psi^*] = S[\psi, \psi^*] + (S, \psi)\frac{\delta F}{\delta \psi} + (S, \psi^*)\frac{\delta F}{\delta \psi^*} = S[\psi, \psi^*] - \frac{\partial_R S}{\partial \psi^*}\frac{\delta F}{\delta \psi} + \frac{\partial_R S}{\partial \psi}\frac{\delta F}{\delta \psi^*}. \tag{C.98}$$

It can be shown that this transformation preserves the antibracket and the master equation

$$(\psi'^r, \psi'^*_s) = \delta_{rs}, \qquad (S', S') = 0. \tag{C.99}$$

More generally, any transformation preserving the antibracket is called an (anti)canonical transformation. One can also consider generating functions depending on both the old and new coordinates, as is standard in the Hamiltonian formalism. Under a transformation, any object depending on the coordinates changes as

$$G' = G + (\delta F, G). \tag{C.100}$$

One can consider finite transformation without problems.

In order to perform the gauge fixing, one needs to eliminate the antifields. A convenient condition is

$$S_\Psi[\psi^r] = S\left[\psi^r, \frac{\partial \Psi}{\partial \psi^r}\right], \qquad \psi^*_r = \frac{\partial \Psi}{\partial \psi^r}, \tag{C.101}$$

where $\Psi[\psi^r]$ is called the gauge fixing fermion and satisfies

$$N_{\mathrm{gh}}(\Psi) = -1, \qquad |\Psi| = 1. \tag{C.102}$$

From the discussion on coordinate transformations, this amounts to work in new coordinates where $\psi'^*_r = 0$. But such a function Ψ cannot be built from the fields because they all have positive ghost numbers. One needs to introduce *trivial pairs* of fields.

A trivial pair $(B, \bar{c})$ is defined by the properties

$$|B| = -|\bar{c}|, \qquad N_{\mathrm{gh}}(B) = N_{\mathrm{gh}}(\bar{c}) + 1, \tag{C.103a}$$

$$s\bar{c} = B, \qquad sB = 0, \tag{C.103b}$$

and the new action reads

$$\bar{S} = S[\psi^r, \psi^*_r] - B\bar{c}^* \tag{C.104}$$

(the position dependence is kept implicit). In this context, ψ^r and ψ^*_r are sometimes called minimal variables. From this, one learns that

$$(\bar{S}, \bar{S}) = (S, S) = 0. \tag{C.105}$$

At level 0, one introduces the pair

$$(B_{0a_0}, \bar{c}_{0a_0}) := (B^0_{0a_0}, \bar{c}^0_{0a_0}) \tag{C.106}$$

and the associated antifields. The field $\bar{c}_0 := b$ is the Faddeev–Popov anti-ghost associated to c_0, and the trivial pair satisfies

$$|B_0| = |\epsilon_0|, \qquad |\bar{c}_0| = -|\epsilon_0|, \qquad N_{\mathrm{gh}}(B_0) = 0, \qquad N_{\mathrm{gh}}(c_0) = -1. \tag{C.107}$$

For level 1, two additional pairs

$$(B^0_{1a_1}, \bar{c}^0_{1a_1}), \qquad (\bar{B}^{1a_1}_1, c^{1a_1}_1), \tag{C.108}$$

and the corresponding antifields are introduced. The motivation for adding an additional pair is that the level-0 pair only fixes $m_0 - m_1$ of the generators: the additional m_1 extra ghosts $c^{1a_1}_1$ can be fixed by the residual level-0 symmetry. The first level-1 pair fixes the level-1 symmetry.

Then, the gauge fixed action enjoys a BRST symmetry acting only on the fields

$$\delta_\theta \psi^r = \theta \, s\psi^r = \theta \left. \frac{\partial_R S}{\partial \psi^*_r} \right|_{\psi^*_r = \partial_r \Psi}. \tag{C.109}$$

Note that this BRST operator is generically nilpotent only on-shell

$$s^2 \propto \mathrm{eom.} \tag{C.110}$$

C.3.3   Quantum BV

At the quantum level, one considers the path integral

$$Z = \int \mathrm{d}\psi^r \, \mathrm{d}\psi^*_r \, e^{-W[\psi^r, \psi^*_r]/\hbar}, \tag{C.111}$$

where W is called the quantum master action. The reason for distinguishing it from the classical master action S is that the measure is not necessarily invariant by itself under the generalized BRST transformation—this translates into a non-gauge invariance of the measure of the physical fields, i.e. a gauge anomaly.

Quantum BRST transformation is generated by the quantum BRST operator σ

$$\delta_\theta F = \theta \, \sigma F = (W, F) - \hbar \, \Delta F, \tag{C.112}$$

where

$$\Delta = \frac{\partial_R}{\partial \psi_r^*} \frac{\partial_L}{\partial \psi^r}. \tag{C.113}$$

Then, the path integral is invariant if W satisfies the quantum master equation

$$(W, W) - 2\hbar \Delta W = 0, \tag{C.114}$$

which can also be written as

$$\Delta e^{-W/\hbar} = 0. \tag{C.115}$$

This can be interpreted as the invariance of Z under changes of coordinates; indeed, one finds that

$$\delta W = \frac{1}{2}(W, W), \tag{C.116}$$

and the integration measure picks a Jacobian

$$\text{sdet } J \sim 1 + \Delta W. \tag{C.117}$$

In the limit $\hbar \to 0$, one recovers the classical master equation. More generally, the action can be expanded in powers of $\hbar$

$$W = S + \sum_{p \geq 1} \hbar^p W_p. \tag{C.118}$$

Observables are given by operators $\mathcal{O}[\psi, \psi^*]$ invariant under σ:

$$\sigma \mathcal{O} = 0, \tag{C.119}$$

which ensures that the expectation value is invariant under changes of Ψ

$$\delta \langle \mathcal{O} \rangle = 0. \tag{C.120}$$

Note that if $\mathcal{O}$ depends just on ψ, the condition reduces to $s\mathcal{O} = 0$, but generically there are no such operators (except constants) satisfying this condition for open algebra.

Consider the gauge fixed integral

$$Z = \int d\psi^r \, e^{W_\Psi[\psi^r]}, \qquad W_\Psi[\psi^r] = W\left[\psi^r, \frac{\partial \Psi}{\partial \psi^r}\right]. \tag{C.121}$$

Varying the gauge fixing fermion by $\delta\Psi$ gives

$$Z = \int \mathrm{d}\psi^r \, \mathrm{e}^{W_\Psi[\psi^r]} \left(\frac{\partial_R S}{\partial\psi_r^*}\right)_{\psi^*=\partial_\psi\Psi} \frac{\partial(\delta\Psi)}{\partial\psi^r}. \tag{C.122}$$

Integrating by part gives the quantum master equation.

C.4   Suggested Readings

- Manipulations of functional integral are given in [11, sec. 15.1, 22.1, 17, chap. 14, 7, 20].
- Zero-modes are discussed in [3].
- A general summary of path integrals for bosonic and fermionic fields can be found in [21, app. A].
- BRST formalism: most QFT books contain an introduction; more complete references are [27, chap. 15, 12, 26];
- BV formalism [27, chap. 15, 6, 9, 10, 12, 26] (several explicit examples are given in [10, sec. 3], and see [1, 24, 29] for more specific details).

References

1. I.A. Batalin, G.A. Vilkovisky, Quantization of gauge theories with linearly dependent generators. Phys. Rev. D **28**(10), 2567–2582 (1983). https://doi.org/10.1103/PhysRevD.28.2567
2. K. Becker, M. Becker, J.H. Schwarz, *String Theory and M-Theory: A Modern Introduction*, 1st edn. (Cambridge University Press, Cambridge, 2006)
3. S.K. Blau, M. Visser, A. Wipf, Determinants, Dirac operators, and one-loop physics. Int. J. Mod. Phys. A **4**(6), 1467–1484 (1989) https://doi.org/10.1142/S0217751X89000625
4. R. Blumenhagen, D. Lüst, S. Theisen, *Basic Concepts of String Theory*. English. 2013 edn. (Springer, Berlin, 2014)
5. R. Blumenhagen, E. Plauschinn, *Introduction to Conformal Field Theory: With Applications to String Theory*. en. Lecture Notes in Physics (Springer, Berlin, 2009). https://www.springer.com/de/book/9783642004490
6. B. DeWitt, *The Global Approach to Quantum Field Theory*. English. 1st edn. (Oxford University Press, Oxford, 2014)
7. E. D'Hoker, D.H. Phong, The geometry of string perturbation theory. Rev. Mod. Phys. **60**(4), 917–1065 (1988) https://doi.org/10.1103/RevModPhys.60.917
8. P. Di Francesco, P. Mathieu, D. Senechal, *Conformal Field Theory*, 2nd edn. (Springer, Berlin, 1999)
9. A. Fuster, M. Henneaux, A. Maas, BRST-antifield quantization: a short review. Int. J. Geom. Methods Mod. Phys. **2**, 939–964 (2005) https://doi.org/10.1142/S0219887805000892. arXiv: hep-th/0506098
10. J. Gomis, J. Paris, S. Samuel, Antibracket, antifields and gauge-theory quantization. Phys. Rep. **259**(1–2), 1–145 (1995). https://doi.org/10.1016/0370-1573(94)00112-G. arXiv: hep-th/9412228
11. B. Hatfield, *Quantum Field Theory of Point Particles and Strings*. English. (Addison Wesley, Reading, 1998)

12. M. Henneaux, C. Teitelboim, *Quantization of Gauge Systems*. English (Princeton University Press, Princeton, 1994)
13. C.V. Johnson, *D-Branes*. English (Cambridge University Press, Cambridge, 2006)
14. M. Kaku, *Introduction to Superstrings and M-Theory* (Springer, Berlin, 1999)
15. S.V. Ketov, *Conformal Field Theory*. English (World Scientific, Singapore, 1995)
16. E. Kiritsis, *String Theory in a Nutshell*. English (Princeton University Press, Princeton, 2007)
17. M. Nakahara, *Geometry, Topology and Physics*, 2nd edn (Institute of Physics Publishing, Bristol, 2003)
18. K. Ohmori, A review on tachyon condensation in open string field theories (2001). arXiv: hep-th/0102085
19. E. Poisson, *A Relativist's Toolkit: The Mathematics of Black-Hole Mechanics*. English. 1st edn. (Cambridge University Press, Cambridge, 2007)
20. J. Polchinski, Evaluation of the one loop string path integral. Commun. Math. Phys. **104**(1), 37–47 (1986). https://doi.org/10.1007/BF01210791
21. J. Polchinski, *String Theory: Volume 1, An Introduction to the Bosonic String* (Cambridge University Press, Cambridge, 2005)
22. V. Schomerus, *A Primer on String Theory*. English. 1st edn. (Cambridge University Press, Cambridge, 2017)
23. A. Sen, *String Theory 1* (2011). http://www.hri.res.in/~sen/combind.pdf
24. A. Sen, B. Zwiebach, A note on gauge transformations in Batalin-Vilkovisky theory. Phys. Lett. B **320**(1–2), 29–35 (1994). https://doi.org/10.1016/0370-2693(94)90819-2. arXiv: hep-th/9309027
25. D. Tong, Lectures on string theory (2009). arXiv: 0908.0333
26. J.W. van Holten, Aspects of BRST quantization (2002). arXiv: hep-th/0201124
27. S. Weinberg, *The Quantum Theory of Fields, Volume 2: Modern Applications*. English (Cambridge University Press, Cambridge, 2005)
28. P. West, *Introduction to Strings and Branes*. English. 1st edn. (Cambridge University Press, Cambridge, 2012)
29. E. Witten, A note on the antibracket formalism. Mod. Phys. Lett. A **5**(7), 487–494 (1990). https://doi.org/10.1142/S0217732390000561
30. B. Zwiebach, *A First Course in String Theory*. English, 2nd edn. (Cambridge University Press, Cambridge, 2009)

Index

© Springer Nature Switzerland AG 2021
H. Erbin, *String Field Theory*, Lecture Notes in Physics 980,
https://doi.org/10.1007/978-3-030-65321-7

Printed in the United States
by Baker & Taylor Publisher Services